Assessment of Potential Migration of Radionuclides and Trace Elements from the White Mesa Uranium Mill to the Ute Mountain Ute Reservation and Surrounding Areas, Southeastern Utah

By David L. Naftz, Anthony J. Ranalli, Ryan C. Rowland, and Thomas M. Marston

Prepared in cooperation with the Ute Mountain Ute Tribe and
U.S. Environmental Protection Agency (Region 8)

Series Name 2011–5231

U.S. Department of the Interior
U.S. Geological Survey

U.S. Department of the Interior
KEN SALAZAR, Secretary

U.S. Geological Survey
Marcia K. McNutt, Director

U.S. Geological Survey, Reston, Virginia: 2012

For more information on the USGS—the Federal source for science about the Earth, its natural and living resources, natural hazards, and the environment, visit http://www.usgs.gov or call 1–888–ASK–USGS.

For an overview of USGS information products, including maps, imagery, and publications, visit http://www.usgs.gov/pubprod

To order this and other USGS information products, visit http://store.usgs.gov

Suggested citation:
Naftz, D.L., Ranalli, A.J., Rowland, R.C., and Marston, T.M., 2011, Assessment of potential migration of radionuclides and trace elements from the White Mesa uranium mill to the Ute Mountain Ute Reservation and surrounding areas, Southeastern Utah: U.S. Geological Survey Science Investigations Report 2011–5231, 146 p.

Acknowledgements

The project benefited substantially from project oversight and guidance by J. Sam Vance, Tribal 106 Coordinator for EPA Region 8. Field sampling assistance during the study by Colin Larrick and Scott Clow, Josh Maloney, Jeremiah Cuthair, and Tomoe Natori of the Ute Mountain Ute Tribe Environmental Department is gratefully acknowledged. Assistance by White Mesa mill and Denison Mines personnel (Harold Roberts, David Frydenlund, David Turk, and Ryan Palmer) for sampling access and guidance on mill operations is gratefully acknowledged. Funding for the study was provided by the EPA, Ute Mountain Ute Tribe, and USGS. Dissolved gas analyses by Lawrence Livermore Laboratory are gratefully acknowledged.

Contents

Figures

Tables

Conversion Factors and Abbreviations

SI to Inch/Pound

Multiply	By	To obtain
Length		
centimeter (cm)	0.3937	inch (in)
millimeter (mm)	0.03937	inch (in)
meter (m)	3.281	foot (ft)
kilometer (km)	0.6214	mile (mi)
meter (m)	1.094	yard (yd)
Area		
hectare (ha)	2.471	acre
hectare (ha)	0.003861	square mile (mi^2)
Volume		
liter (L)	33.82	ounce, fluid (fl. oz)
liter (L)	2.113	pint (pt)
liter (L)	1.057	quart (qt)
liter (L)	0.2642	gallon (gal)
cubic centimeter (cm^3)	0.06102	cubic inch (in^3)
liter (L)	61.02	cubic inch (in^3)
Flow rate		
meter per second (m/s)	3.281	foot per second (ft/s)
liter per second (L/s)	15.85	gallon per minute (gal/min)
Mass		
gram (g)	0.03527	ounce, avoirdupois (oz)
kilogram (kg)	2.205	pound avoirdupois (lb)
megagram (Mg)	1.102	ton, short (2,000 lb)
megagram (Mg)	0.9842	ton, long (2,240 lb)
megagram per year (Mg/yr)	1.102	ton per year (ton/yr)
metric ton per year	1.102	ton per year (ton/yr)
Pressure		
kilopascal (kPa)	0.009869	atmosphere, standard (atm)
kilopascal (kPa)	0.01	bar

Temperature in degrees Celsius (°C) may be converted to degrees Fahrenheit (°F) as follows:

°F=(1.8×°C)+32

Temperature in degrees Fahrenheit (°F) may be converted to degrees Celsius (°C) as follows:

°C=(°F-32)/1.8

Vertical coordinate information is referenced to the North American Vertical Datum of 1988 (NAVD 88)

Horizontal coordinate information is referenced to the North American Datum of 1983 (NAD 83)

Specific conductance is given in microsiemens per centimeter at 25 degrees Celsius (µS/cm at 25°C).

Concentrations of chemical constituents in water are given either in milligrams per liter (mg/L) or micrograms per liter (µg/L).

Acronyms

AR	activity ratio
BLM	Bureau of Land Management
CF-IRMS	continuous flow isotope-ratio mass spectrometer
CMERSC	Central Mineral and Environmental Resources Science Center
DI-IRMS	dual inlet isotope-ratio mass spectrometer
DOI	Department of the Interior
HGAAS	hydride generation atomic absorption spectrometry
ICP-AES	inductively coupled plasma-atomic emission spectrometer
ICP-MS	inductively coupled plasma-mass spectrometer
LT-MDL	long term method-detection limit
LRL	lower reporting limit
lsd	land surface datum
MCL	maximum contaminant level
MCLG	maximum contaminant level goals
NFM	National Field Manual
NIST	National Institute of Standards and Technology
NURE	National Uranium Resource Evaluation
NWIS	National Water Information System
NWQL	National Water Quality Laboratory
PCA	principal component analysis
PVC	polyvinyl chloride
pzc	point of zero change
TU	tritium units
UCL	upper confidence limit
UMTRA	Uranium Mill Tailing Remediation Action
U of U	University of Utah
US EPA	US Environmental Protection Agency
USGS	US Geological Survey
VCDT	Vienna Canyon Diablo Troilite
VSMOW	Vienna Standard Mean Ocean Water
XRD	X-ray Diffraction

List of Elements and Symbols

Ag	Silver	H	Hydrogen	Na	Sodium	Ti	Titanium
Al	Aluminum	He	Helium	Ne	Neon	Tl	Thallium
Ar	Argon	In	Indium	Ni	Nickel	U	Uranium
As	Arsenic	K	Potassium	Pb	Lead	V	Vanadium
Ca	Calcium	Kr	Krypton	S	Sulfur	W	Tungstun
Cr	Chromium	Mg	Magnesium	Sb	Antimony	Xe	Xenon
Cs	Cesium	Mn	Manganese	Se	Selenium	Y	Yttrium
Cu	Copper	Mo	Molybdenum	Ta	Tantalum	Zn	Zinc
Fe	Iron	N	Nitrogen	Te	Tellurium		

Assessment of Potential Migration of Radionuclides and Trace Elements from the White Mesa Uranium Mill to the Ute Mountain Reservation and Surrounding Areas, Southeastern Utah

By David L. Naftz, Anthony J. Ranalli, Ryan C. Rowland, and Thomas M. Marston

Abstract

In 2007, the Ute Mountain Ute Tribe requested that the U.S. Environmental Protection Agency and U.S. Geological Survey conduct an independent evaluation of potential offsite migration of radionuclides and selected trace elements associated with the ore storage and milling process at an active uranium mill site near White Mesa, Utah. Specific objectives of this study were (1) to determine recharge sources and residence times of groundwater surrounding the mill site, (2) to determine the current concentrations of uranium and associated trace elements in groundwater surrounding the mill site, (3) to differentiate natural and anthropogenic contaminant sources to groundwater resources surrounding the mill site, (4) to assess the solubility and potential for offsite transport of uranium-bearing minerals in groundwater surrounding the mill site, and (5) to use stream sediment and plant material samples from areas surrounding the mill site to identify potential areas of offsite contamination and likely contaminant sources.

The results of age-dating methods and an evaluation of groundwater recharge temperatures using dissolved-gas samples indicate that groundwater sampled in wells in the surficial aquifer in the vicinity of the mill is recharged locally by precipitation. Tritium/helium age dating methods found a "modern day" apparent age in water samples collected from springs in the study area surrounding the mill. This apparent age indicates localized recharge sources that potentially include artificial recharge of seepage from constructed wildlife refuge ponds near the mill. The stable oxygen isotope-ratio, delta oxygen-18, or $\delta(^{18}O/^{16}O)$, known as $\delta^{18}O$, and hydrogen isotope-ratio, delta deuterium, or $\delta(^{2}H/^{1}H)$, known as δD, data indicate that water discharging from Entrance Spring is isotopically enriched by evaporation and has a similar isotopic fingerprint as water from Recapture Reservoir, which is used as facilities water on the mill site. Water from Recapture Reservoir also is used to irrigate fields surrounding the town of Blanding and infiltration of this irrigated water also could contribute to the enriched isotopic fingerprint observed for Entrance Spring. Similarities in the delta sulfur-34$_{sulfate}$ values in water samples from the wildlife ponds and tailings cells indicate a potential contaminant linkage between the tailings cells and the refuge ponds that could be related to wind carried (eolian) transport of aerosols from the tailings cells. To date (2010), neither the delta sulfur-34$_{sulfate}$ nor the delta oxygen-18$_{sulfate}$ values measured in the wells and springs surrounding the uranium mill site have an isotopic signature characteristic of water from the tailings cells.

Except for Entrance Spring and Mill Spring, all groundwater samples collected at down-gradient sample sites during this study had dissolved-uranium concentrations in the range expected for naturally-occurring uranium. The uranium-isotope data indicate that the mill is not a source of uranium in the groundwater in the unconfined-aquifer at any site monitored during the study, with the possible exception of Entrance Spring. The uranium-234 to uranium-238 activity ratios measured in water samples collected at Entrance Spring, and the decrease in this ratio associated with an increase in the concentration of dissolved uranium indicate potential mixing of uranium ore with groundwater at the spring through eolian transport of small particles from ore-storage pads and uncovered ore trucks, with subsequent deposition in the Entrance Spring drainage, followed by dissolution in the unconfined groundwater. The isotopic values of uranium found in other water samples collected during the study do not appear to be related to uranium ore deposits.

Water samples collected from Entrance Spring contained the highest median uranium concentrations relative to water samples collected from the other wells and springs monitored during the study. Water samples collected from Entrance Spring also contained elevated concentrations of selenium and vanadium. Sediment samples collected from three ephemeral drainages east of the uranium mill site (including Entrance Spring) contained uranium concentrations exceeding background values downwind of the predominant wind directions at the site. Sediment samples collected from ephemeral drainages on the south and west boundaries of the uranium mill site generally did not exceed background-uranium concentrations. Elevated concentrations of uranium and vanadium, indicating offsite transport, were found in plant tissue samples collected north-northeast, east, and south of the mill site, downwind of the predominant wind directions at the site. The uranium and vanadium concentrations in plant tissue samples collected west of the uranium mill site were low.

On the basis of the study results, consideration should be given to future monitoring programs in areas surrounding the uranium mill site to address current and future environmental concerns. These potential monitoring programs should consider (1) quarterly monitoring of major- and trace-element concentrations in selected springs and wells; (2) annual monitoring of Entrance Spring for uranium isotopes, delta sulfur-34$_{sulfate}$, delta oxygen-18, and delta deuterium; (3) annual monitoring of background water quality at selected spring and monitoring well sites; (4) periodic sampling and chemical analyses of sagebrush in areas east of the uranium mill site coupled with off-site fugitive dust monitoring; (5) installation of a new monitoring well upgradient from the East and West wells; (6) the addition of non-routine chemical constituents to ongoing monitoring programs within the uranium mill site that could provide additional insight(s) into potential contaminant sources and processes; and (7) archiving future monitoring data into a maintained database that is easily accessible to all project stakeholders.

Introduction

Legacy uranium (U) mining and milling operations have resulted in soil and water contamination at many sites throughout the western United States. In 1978, Congress passed the Uranium Mill Tailings Radiation Control Act (UMTRCA) that directed government agencies to stabilize, dispose of, and control materials contaminated by uranium milling operations (Peterson and others, 2008). There are a total of 23 former uranium mill sites in the western United States that have required active remediation in the Department of Energy's Uranium Mill Tailings Remediation Action (UMTRA) program (Jordan and others, 2008). Liquid wastes associated with these legacy mill sites typically contain radionuclides, heavy metals, ammonia, nitrate, and sulfates that have seeped into the vadose zone and sometimes reached underlying aquifers. A few examples of soil and water contamination from milling operations include (1) groundwater from a uranium-mill tailings repository near Durango, Colorado, which is contaminated with As, Mn, Mo, Se, U, V and Zn (Morrison and others, 2002); (2) groundwater from the Bear Creek mill site in northeastern Wyoming, which is contaminated with uranium (U) and sulfate (SO_4^{2-}), and has an unnaturally low pH (Zhu and Burden, 2001); and (3) U and vanadium (V) contaminated soil and groundwater from a uranium mill site near Naturita, Colorado (Davis and others, 2006). While UMTRCA has addressed the remediation of legacy uranium milling sites, there are over 4,000 mines with a history of uranium production in the western United States that also can pose environmental risks (Peterson and others, 2008).

The White Mesa uranium mill is an active facility that is operated by Denison Mines. This facility is a fully licensed, conventional processing mill with a V co-product recovery circuit (Denison Mines, 2010). The mill site is located in San Juan County, Utah, about 10 kilometers (km) south of the city of Blanding and 6 km north of the Ute Mountain Ute Reservation (fig. 1). Ore material processed at the mill is obtained from Denison mine properties in the Colorado Plateau, the Henry Mountains Complex, and the Arizona Strip. The mill site is currently (2010) the only conventional uranium mill operating in the United States (Denison Mines, 2010).

Construction of the mill began in 1979 and the first U/V ore was processed during May 1980 (Denison Mines, 2010). The mill uses sulfuric acid (H_2SO_4) leaching and a solvent extraction recovery process to extract and recover U and V from the ore material. The mill is currently licensed to process an average of 2,000 tons of ore per day and produce 3.6 million kilograms (kg) of triuranium octoxide (U_3O_8) per year (Denison Mines, 2010). The mill is also licensed to process alternate feed materials, which include U-bearing materials derived from U conversion, tantalum (Ta) and other metal processing facilities or material from U.S. government cleanup projects. In 2007, the mill produced approximately 115,300 kg of U_3O_8 from alternate feed materials (Denison Mines, 2010).

An evaluation of the concentration of major ions and metals measured in the groundwater up- and down-gradient of the mill reveals complex spatial variations in (1) the concentration of U and other metals in bedrock, soils, and groundwater; (2) the geochemical conditions favorable for either U solubility or precipitation in groundwater; and (23) geologic conditions that can influence groundwater-residence times in White Mesa. This spatial variability makes it extremely difficult to assess the environmental effects of the mill by using trace-element concentration data alone.

A groundwater study by independent scientists to characterize groundwater flow, chemical composition, noble gas composition, and apparent age was conducted because of increasing and elevated trace-metal concentrations in monitoring wells within the White Mesa mill site. On the basis of apparent recharge dates from chlorofluorocarbons (CFCs) and tritium (^3H) concentrations, most groundwater beneath the mill was estimated to be more than 50 years in age. An exception to this trend, measurable levels of tritium found in some monitoring wells in the northeast part of the site, likely resulted from leakage of constructed wildlife ponds on mill property (fig. 1). Hurst and Solomon (2008) concluded that active vertical and horizontal groundwater flow is clearly evident beneath the mill; however, trace-metal concentrations, age-dating methods, and stable-isotope fingerprinting did not detect leakage from the tailing cells. Because of active groundwater flow, continued monitoring of the groundwater to evaluate the future performance of the tailing cells within the mill was strongly recommended.

Although personnel and contractors for the White Mesa mill have been collecting groundwater- and air-quality data since 1980, the Ute Mountain Ute Tribe requested the U.S.

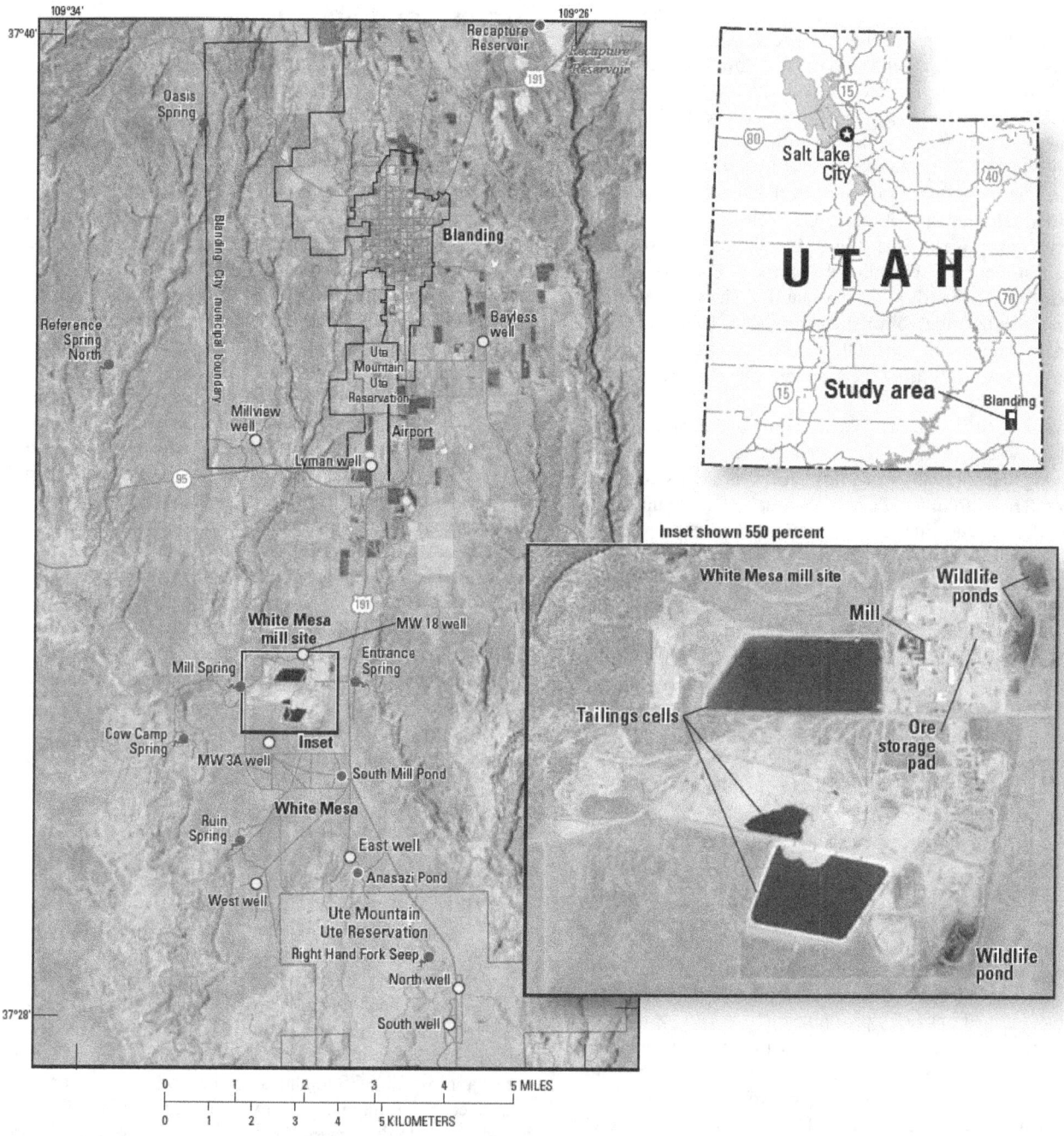

Figure 1. Location of White Mesa mill site relative to the town of Blanding and the Ute Mountain Ute Reservation, San Juan County, Utah, and tailings cells, ore-storage pad, and wildlife ponds on the mill property.

Environmental Protection Agency (EPA) and the U.S. Geological Survey (USGS) to perform an independent evaluation of the potential offsite migration of radionuclides and trace elements associated with the ore storage and milling process. Potential air- and water-exposure pathways of U and other trace elements to tribal members include (1) airborne dust from uncovered ore storage pads; (2) airborne emissions from drying ovens at the mill; (3) dissolution of airborne dust deposited on soil and plant surfaces; (4) transport of material from the ore storage pads into ephemeral channels draining the mill site during rain and snowmelt events; and (5) leakage from the tailings ponds to shallow aquifers beneath the mill, resulting in offsite migration toward the reservation.

Inspections of quarterly reports produced by the White Mesa mill of groundwater and air monitoring data led the Ute Mountain Ute Tribe to request this independent evaluation. Large spatial variability in the concentration and composition of major ions and ranges in the concentrations of U from 5 to 10 micrograms per liter (μg/L) in many wells to exceeding the EPA maxium contaminant level (MCL) of 30 μg/L, consistently in a few wells both up- and down-gradient of the mill, prompted the tribe to question if the concentrations of U measured in the wells are background concentrations or evidence of contamination by the mill. A review by the USGS of reports describing the geology and hydrology of White Mesa, quarterly reports produced by the mill, data collected by the Ute Mountain Tribe, and data collected by the USGS at Fry Canyon west of the White Mesa indicated existing data were insufficient to determine the source of U in the Dakota Sandstone and the Burro Canyon aquifer. Therefore, an evaluation of the potential for offsite migration of U and other metals from the mill toward the reservation along the potential exposure pathways using the available data is difficult for several reasons.

The use of U concentration data only to determine if the mill is a source of the U in the groundwater is ambiguous. Although Hem (1989) stated that U concentrations in groundwater derived from natural sources usually fall within 1 to 10 μg/L, the range in the concentrations of U measured in monitoring wells by the mill reflects concentrations measured in groundwater in Fry Canyon (up to 40 μg/L) near White Mesa, which have been determined to be derived from natural sources (Wilkowske and others, 2002). Adding to this ambiguity is the fact that concentrations of U above the EPA MCL of 30 μg/L have been measured in wells up- and down-gradient of the mill. Thus, it is difficult to determine the source of U, given the spatial variation in concentrations of U in the Dakota Sandstone/Burro Canyon Formation aquifer.

Evaluation of the potential for offsite migration of U and other metals in groundwater from the mill toward the reservation is difficult because the available data are not sufficient to determine the mobility of U entering the Dakota Sandstone/ Burro Canyon aquifer. For example, if leakage from a tailings cell were to occur, would U remain in solution? If ore material was blown off the ore-storage pad and deposited on White Mesa, would U dissolve in the groundwater? Or, would U be removed from solution through adsorption to minerals in the soil and/or bedrock or by precipitation?

Finally, another important consideration for the effect of mill operations on groundwater quality in the Ute Mountain Ute Reservation is the length of time would it take for U released by the mill entering the Dakota Sandstone/Burro Canyon Formation aquifer to migrate to the reservation. The mill estimated a travel time of 3,000 years from one of the tailing cells to the reservation boundary using Darcy's Law. There are limitations to this calculation, however, because the permeability tests were performed in wells only on mill property north of the reservation and would not have measured permeability in the Dakota Sandstone and Burro Canyon Formation south of the mill property. use of Darcy's Law to estimate groundwater velocity assumes a homogeneous medium. Given that the sediments that compose the Dakota Sandstone/Burro Canyon Formation aquifer are stream deposits, it is possible that there are preferential flow channels and that groundwater velocities in the aquifer vary.

Purpose and Scope

Although monitoring the concentration of U in groundwater up- and down-gradient of the mill is a scientifically valid technique, it is the opinion of the USGS and EPA that the monitoring of groundwater using concentration data only is not sufficient to determine either the source of U in the Dakota Sandstone/Burro Canyon Formation aquifer or to fully evaluate the potential of offsite migration of U and other metals from the mill toward the reservation along the potential exposure pathways. The overall objective of this report is to better understand and document past, present, and possible future transport of U and associated trace-element emissions from the White Mesa uranium mill to the surrounding tribal and non-tribal lands. Specific study objectives are to (1) use tritium activity, noble gas concentrations, and stable isotopes of oxygen and hydrogen to better understand recharge sources and residence times of groundwater surrounding the mill site; (2) determine the current concentrations of U and associated trace elements in groundwater surrounding the mill site; (3) use isotopes of U and sulfur to differentiate natural and anthropogenic contaminant sources to groundwater resources surrounding the mill site; (4) use geochemical modeling methods to assess the solubility of U-bearing minerals in groundwater surrounding the mill site and potential for offsite transport; and (5) use major- and trace-element concentration data in stream sediments and plant materials from areas surrounding the mill site to identify potential contaminant sources.

Methodology

Water Sample Collection

Water samples were collected using techniques described in the USGS National Field Manual (NFM; U.S. Geological Survey, variously dated). Samples were collected for analysis of major ions, trace metals, nutrients (nitrate + nitrite and orthophosphate), U isotopes, hydrogen and oxygen isotopes of water, sulfur and oxygen isotopes of dissolved sulfate, dissolved gases, and tritium. The quality assurance/quality control plan for the White Mesa uranium project includes the use of approved USGS methods for the collection and analysis of surface and groundwater samples, the collection of field blanks and field duplicates, the addition of matrix spikes to the metal samples, and adherence to stringent chain-of-custody procedures (U.S. Geological Survey, 2010b and 2010c). An overview of water sampling procedures at springs and stock ponds, monitoring wells, and domestic and public supply wells is provided. Techniques used to collect dissolved gas and tritium samples are discussed separately.

Springs, Stock Ponds, and Reservoir

Water-quality samples were collected from springs during seven quarterly sampling events. Samples also were collected from stock ponds near the mill during one quarterly sampling event.

Springs are located at geologic contacts along diffuse seepage zones (fig. 1). It was not possible to collect water samples from springs that were not in contact with the atmosphere. Clean-sampling procedures described in the USGS NFM, chapter A4 (2006) were adapted to the conditions at each spring. Samples were collected from small (7.6–15 centimeters [cm] wide, less than 2.5 cm deep) drainage channels at Cow Camp and Mill Springs, from small pools (0.9–3 meters [m] in diameter, up to 15 cm deep) that had formed naturally at the base of Oasis and Entrance Springs, and from an acrylonitrile-butadiene-styrene (ABS) plastic pipe draining an approximately 0.9-m diameter galvanized tub placed beneath dripping water from Ruin Spring. For Cow Camp, Mill, Oasis, and Entrance Springs, grab samples were collected by filling a 250-milliliter (mL) pre-rinsed and field-rinsed plain poly-ethylene bottle with sample water and transferring the water to a pre-rinsed and field rinsed 3.8-liter (L) plain polyethylene bottle. The process was repeated until the 3.8-L bottle was filled. Samples were collected at Ruin Spring by simply filling a pre-rinsed and field-rinsed 3.8-L polyethylene bottle at the ABS plastic pipe.

Physical and chemical field parameters (pH, specific conductance, water temperature, and dissolved oxygen) were measured after each sample was collected with a calibrated In-Situ Troll 9000TM multiparameter water-quality sonde equipped with a magnetic stir bar. For Oasis and Entrance Springs, the sonde was placed in the small pond at the base of the springs. Field parameters at Cow Camp and Ruin Springs were measured by placing the sonde in a clean 1,000-mL graduated cylinder oriented to capture flow. Because of low-flow conditions at Mill Spring, field parameters were measured by filling the calibration cup for the sonde with water from the spring. Field parameters were recorded when five consecutive readings were within USGS stability criteria (Wilde, 2008). When there was adequate flow, volumetric flow measurements were completed at Mill, Cow Camp, and Ruin Springs. Flow at Entrance Spring was measured with a 7.6-cm modified

Parshall flume about 6-m downstream from its source. Flow at Oasis Spring was too diffuse to quantify. Samples were placed in a cooler for transportation to the mobile laboratory trailer where they were processed for shipment to the laboratory.

Three ponded water samples also were collected during the study period. A point sample was collected at each site about 0.9 m from shore in water about 0.6-m deep using a 3.8-L pre-rinsed and field-rinsed plain polyethylene bottle. Field parameters were collected by placing the calibrated sonde in the water at mid-sample depth after the sample was collected, and parameters were recorded once USGS stability criteria were achieved. Samples were placed in a cooler for transportation to the mobile laboratory trailer where they were processed for laboratory shipment.

Groundwater Monitoring Wells

Table 1 summarizes physical characteristics, including well depth and screened intervals, of wells sampled in this study. Water-quality samples were collected from two low-yield wells during seven quarterly sampling events. Because of their low yield, a low-flow sampling technique was used to sample these wells. A Grundfos Redi-Flo2TM stainless-steel submersible pump with a 1.3-cm inner-diameter reinforced, clear polyvinyl chloride (PVC) discharge line was used to sample the wells. Prior to sampling each well, the pump and discharge line were cleaned using procedures for stainless-steel submersible pumps described in chapter A3 of the USGS NFM (Wilde, 2004). Purge rates ranged from 150 to 300 milliliters per minute (mL/min) in order to avoid pumping the wells dry. Field parameters were measured with a calibrated multiparameter water-quality sonde equipped with an air-tight flow chamber. Water level, purge rate, purge volume, and field parameters were recorded every 5 minutes during the purging procedure. Samples were collected when three to five consecutive field-parameter readings were within USGS stability criteria.

Table 1.   Physical characteristics of wells sampled near the White Mesa uranium mill, San Juan County, Utah 2007–09.

[**Abbreviations**: ft, foot; —, not available; *, approximate]

| Station number | Field name | Station name | Aquifer | Primary use of water | Altitude of land surface, (ft) | Depth of well, (ft) | Well open interval | |
							Depth to top of openings, (ft)	Depth to bottom of openings, (ft)
372954109293601	East well	(D–38–22)10bcc–1 WM East monitoring well	Surficial	Monitoring	5,440	110	70	90
372930109310701	West well	(D–38–22) 8dcd–1 WM West monitoring well	Surficial and Morrison formation	Monitoring	5,450	110	89	109
373442109291501	Lyman well	(D–37–22) 10cdc–1 LY well	Surficial	Domestic	5,790	120	—	—
373612109273201	Bayless well	(D–37–22) 2aad–1 BAY well	Surficial	Domestic	5,860	—	—	—
372817109275701	North well	(D–38–22) 23acb–1 WM North well	Navajo aquifer	Public supply	5,280	1,515	927	1,135
372756109280901	South well	(D–38–22) WM South well	Navajo aquifer	Public supply	5,300	1,739	1,277	1,739
373501109310801	Millview well	(D–37–22) 8dba– 1 Millview well [1]	—	Livestock	5,830	300*	—	—
373116109305601	MW3A	(D–37–22) 32ddc–1 MW3A	Surficial	Monitoring	5,550	95	75	95
373233109301001	MW18	(D–37–22) 28acc–1 MW18	Surficial	Monitoring	5,650	148	—	134

[1] This well was sampled once in September 2007. The casing collapsed, or an object was lodged in the casing, sometime between September 2007 and March 2008.

Domestic and Public Supply Wells

Two domestic wells (Lyman and Bayless) were sampled during one quarterly sampling event and two public supply wells (North and South wells; fig. 1) were sampled during three quarterly sampling events. Standard procedures described in the USGS NFM, chapter A4 (USGS, 2006), were used to collect water samples and field parameters from domestic and public-supply wells.

Dissolved Gas and Tritium Water Samples

Water samples were collected for analysis of dissolved gases (N_2, ^{40}Ar, ^{84}Kr, ^{20}Ne, ^{129}Xe, ^3He, and ^4He) and tritium (^3H) from springs and wells during September 2007 to determine recharge temperatures and apparent groundwater-age dates. Four dissolved-gas samples were collected with passive-diffusion samplers using methods described by Sheldon (2002). The diffusion sampler is constructed of 0.3-cm inner-diameter copper tubing and a semipermeable gas-diffusion sampling membrane. The sampler was placed directly into the well or spring and allowed to equilibrate for about 24 hours. After equilibration, the sampler was removed and immediately sealed (ends of the copper tubing were sealed using a crimping device). Six water samples were collected in copper tubes from wells using standard techniques. Samples were collected by connecting the sample vessel (8-millimeter (mm) inner-diameter copper tubing, 250-mm long) to the wellhead of pumping wells with clear Tygon tubing at full wellhead pressure. Water flowed for several minutes to purge air bubbles. The copper tubing was tapped lightly to dislodge bubbles and a visual inspection for bubbles was made. Steel clamps pinched the copper tubing flat in two locations to secure the sample. Tritium samples, including one sample collected in October 2009 that is not associated with noble gas data, were collected in either 1-L glass or 1-L polythethylene bottles, and sealed with a polyseal cap, leaving no air space in the bottle. A calibrated multiparameter water-quality probe was used to measure physical and chemical field parameters, including total dissolved-gas pressure.

Field Processing of Water Samples

Water samples were processed in the field using standard techniques (U.S. Geological Survey, variously dated). Samples were processed in a dust-free processing chamber using "clean hands" procedures. Samples analyzed for dissolved constituents were filtered with 0.45-micrometer (μm) pore-size disposable capsule filters. Trace-element samples were preserved with 7.7 normal (N), ultrapure nitric acid. Table 2 summarizes the bottle type, preservation method, and storage environment used for each category of analytes measured in water samples.

Alkalinity titrations were completed in the field using filtered-water samples within 2 hours of sample collection. A HachTM digital-titration kit using 0.16 N or 1.6 N sulfuric acid titration cartridges, calibrated Radiometer pH meter, and magnetic stirrer were used for alkalinity titrations. Dissolved iron and dissolved sulfide were measured in filtered samples from groundwater wells with a Chemometric portable photometer immediately after bottles to be analyzed for dissolved constituents were filled.

Ephemeral Stream Sediment and Consolidated Rock Samples

Sediments from 31 sites in dry-ephemeral streams near the mill were collected in June 2008 to evaluate potential geochemical anomalies. Three ephemeral-stream sites located 6 km north of the White Mesa uranium mill also were sampled to quantify current geochemistry in ephemeral stream sediments on White Mesa. At all the sites, samples were composited from 3-m transects to a depth of 0.5 cm. Sampling equipment (plastic spoon and tub) were cleaned between sample sites with deionized water and lint-free paper towels. Technicians wore a new pair of powder-free latex gloves at each sample site. Samples were double bagged and stored at room temperature for shipment to the laboratory. Two standard reference samples, to assess analytical quality control, were submitted to the laboratory with the environmental samples. The USGS Central Mineral and Environmental Resources Science Center (CMERSC) analyzed the samples for 43 elements using techniques described in the "Analytical Methods" section of this report. Chain of custody protocols for sediment and vegetation samples sent to the CMERSC were used (Murphy and others, 1997). Samples of consolidated rock from the Burro Canyon and Brushy Basin Formations were collected from several sites in June 2008 for mineralogic analysis. A rock hammer was used to remove weathered material. Freshly exposed samples were stored in plastic bags for shipment to the laboratory.

Vegetation

Samples of big sagebrush (*Artemisia tridentata*) were collected from 64 sites during September 1–3, 2009, to identify potential geochemical anomalies in plant tissue. A sampling grid covering areas adjacent to the mill was used to guide the sampling effort. Several grid cells included multiple sample sites to help evaluate geochemical variability at various geographic scales. Each sample was a composite of young stems and leaves, generally the terminal 10–20 cm of the branches, representing growth less than 1-year old (Gough and Erdman, 1980). Samples were clipped from up to six plants within a 15-m radius. Approximately 150 grams (g) of vegetation was collected for each sample. Stainless-steel pruning shears were used to clip the samples. Samples were placed in cloth sample bags and stored at room temperature in a ventilated box. Sampling personnel wore powder-free latex gloves while sampling, and the stainless steel pruning shears were wiped down between sample sites. Quality-control samples consisted of six split replicates and four standard reference samples. The CMERSC analyzed the samples for 43 elements using techniques described in the "Analytical Methods" section of this report.

Table 2.   Summary of water sample bottle type, preservative, storage environment, and laboratory used for analysis of water samples collected near the White Mesa uranium mill, San Juan County, Utah, 2007–09.

[**Abbreviations**: ICP-MS, inductivley coupled plasma mass spectrometry; mL, milliliter; N, acid normalilty; NAU, Northern Arizona University; NWQL, National Water Quality Laboratory; U of U, University of Utah; USGS, U S Geological Survey]

Analyte	Bottle type	Filtered	Preservative	Storage	Laboratory
Major anions, dissolved	500-mL plain polyethylene	Yes	None	Room temperature	NWQL
Trace metals and major cations, dissolved	250-mL polyethylene, acid rinsed	Yes	7.7 N ultra-pure nitric acid	Room temperature	NWQL
Trace metals, total	250-mL polyethylene, acid rinsed	No	7.7 N ultra-pure nitric acid	Room temperature	NWQL
Nutrients (nitrate+nitrite and orthophosphate), dissolved	125-mL polyethylene, opaque	Yes	None	4 degrees Celsius	NWQL
Oxygen/deuterium stable isotopes in water	60-mL glass	No	None	Room temperature	USGS Reston Stable Isotope Lab
Sulfur-34/Sulfur-32 and oxygen stabe isotopes in dissolved sulfate	1,000-mL plain polyethylene	No	None	Room temperature	USGS Reston Stable Isotope Lab
Uranium-234, 235, 236, and 238 isotopes	1,000-mL plain polyethylene	Yes	7.7 N ultra-pure nitric acid	Room temperature	NAU ICP-MS lab
Tritium	1,000-mL plain polyethylene with polylseal cap	No	None	Room temperature	U of U Dissolved Gas Lab and NWQL
Tritium	1,000-mL glass with polyseal cap	No	None	Room temperature	Lawrence Livermore Lab
Dissolved Gases	Passive-diffusion sampler	No	None	Room temperature	U of U Dissolved Gas Lab
Dissolved Gases	Copper tube	No	None	Room temperature	Lawrence Livermore Laboratory

Cores were collected from live Cottonwood trees (*Populus*, species not identified) near several springs in November 2008 to evaluate potential correlation between U concentrations in springs and core tissue. A three-thread increment borer (0.5-cm diameter core) was used to extract the cores. The outer 1.9 cm of selected cores, representing relatively younger growth, was submitted to the USGS National Water Quality Laboratory (NWQL) for analysis of U in core tissue. Dendrochronology was established on selected tree cores by Dr. Tom Yanosky, USGS (retired), to determine whether or not the cored trees were alive prior to mill operations.

Analytical Methods

USGS National Water Quality Laboratory

Analyses of major and minor ions, trace elements, and nutrients in water samples were completed by the USGS NWQL in Lakewood, Colorado, using standard analytical techniques described by Fishman and Friedman (1989). One water sample was submitted to NWQL for analysis of tritium by electrolytic enrichment and gas counting. Selected tree cores also were submitted to NWQL and analyzed for U by inductively coupled plasma-mass spectrometry (ICP-MS) after drying and microwave assisted acid digestion (US Environmental Protection Agency, 1996). All data are stored in the USGS National Water Information System (NWIS) database and are available on the internet at http://waterdata.usgs.gov/ut/nwis/qw.

USGS Central Mineral and Environmental Resources Science Center

Sagebrush Analytical Methods

Sagebrush samples were submitted to the USGS CMERSC in Denver, Colorado. At CMERSC, unwashed sagebrush samples were dried at room temperature for 24 to 48 hours and then milled. The milled samples were converted to ash in a drying oven held at 500 degrees Celsius (°C) for 13 hours. Detailed methods for plant material ashing are provided by Peacock and Crock (2002). Ashed samples were decomposed using a mixture of hydrochloric, nitric, perchloric, and hydrofluoric acids at low temperature prior to analysis. Aliquots of the digested plant material were aspirated into both an inductively coupled plasma-mass spectrometer (ICP-MS) and an inductively coupled plasma-atomic emission spectrometer (ICP-AES). Forty-two major-, minor-, and trace-element concentrations were determined. Calibration of the ICP-MS is done with aqueous standards and internal standards that are used to compensate for matrix affects and internal drift. The ICP-AES is calibrated by standardizing with digested rock reference materials and a series of multi-element solution standards. Arsenic (As) and selenium (Se) concentrations in sagebrush were measured by hydride generation atomic absorption spectrometry (HGAAS) after drying (no ashing) and acid digestion (Hageman and others, 2002).

Stream Sediment Analytical Methods

USGS CMERSC separated the fine fraction of the sediment (passing a 200-mesh sieve) using methods described by Peacock and others (2002). The fine fraction of each sample was decomposed using a mixture of hydrochloric, nitric, perchloric, and hydrofluoric acids at low temperature, and analyzed for forty-two elements using ICP-MS and ICP-AES, as described above. Selenium also was measured in fine sediment samples by the contract laboratory using HGAAS (Hageman and others, 2002) after total digestion with the same acids used for the ICP-MS and ICP-AES sample preparation.

USGS Reston Stable Isotope Laboratory

Stable oxygen and hydrogen isotope ratios in water molecules and stable sulfur and oxygen isotope ratios in dissolved sulfate were measured by the USGS Stable Isotope Laboratory in Reston, Virginia. The isotope ratios are reported as delta (δ) values, which are equivalent to parts per thousand, in units permil. The δ value for an isotope ratio, R, is computed using the following equation:

$$\delta R = [(R_{sample}/R_{standard}) - 1] \cdot 1{,}000 \qquad (1)$$

where

δR	is the δ value for a specific isotope in the sample,
R_{sample}	is the ratio of the rare isotope to the common isotope for a specific element in the sample, and
$R_{standard}$	is the ratio of the rare isotope to the common isotope for the same element in the standard reference material.

A brief summary of analytical methods used to measure these stable isotope ratios follows.

The hydrogen isotope-ratio, delta deuterium, or $\delta(^{2}H, {}^{1}H)$, known as δD, of water was measured by equilibrating the sample with gaseous hydrogen using a platinum catalyst. To do this, the water and platinum catalyst were placed in glass tubes on a manifold; air from each sample vessel was exhausted, and the vessels were filled with gaseous hydrogen, and the equilibrated hydrogen from each sample vessel was expanded into a dual inlet isotope-ratio mass spectrometer (DI-IRMS), which determines stable hydrogen isotopic composition (Révész and Coplen, 2008a). δD values are relative to the Vienna Standard Mean Ocean Water (VSMOW) standard.

Water samples analyzed for the stable oxygen isotope-ratio $\delta(^{18}O/^{16}O)$, or $\delta^{18}O$, were loaded into glass sample containers on a vacuum manifold to allow for equilibration with carbon dioxide (CO_2) at 25°C. When isotopic equilibra-

tion was obtained, an aliquot of CO_2 was extracted from each sample container, separated from water vapor using a dry ice trap, and injected into a DI-IRMS, which measures the $\delta^{18}O$ value (Révész and Coplen, 2008b). $\delta^{18}O$ values are relative to the VSMOW standard.

Dissolved sulfate (SO_4^{2-}) in water samples was precipitated as barium sulfate ($BaSO_4$) using barium chloride ($BaCl_2$) at pH 3–4 in the laboratory. Any dissolved organic sulfur (S) in the sample was oxidized to SO_2 and degassed from the sample prior to precipitation of $BaSO_4$. Filtered $BaSO_4$ was injected into an elemental analyzer to convert sulfur in $BaSO_4$ into SO_2 gas. SO_2 gas was then injected into a continuous flow isotope-ratio mass spectrometer (CF-IRMS) to determine $\delta^{34}S$ (Révész and Qi, 2006). $\delta^{34}S$ values are relative to the Vienna Canyon Diablo Troilite (VCDT) standard.

For determination of $\delta^{18}O$ values in sulfate, continuous flow isotope ratio analysis was completed after sample preparation of $BaSO_4$ by conversion to carbon monoxide with a thermal combustion/elemental analyzer system. $\delta^{18}O$ values are relative to the VSMOW standard.

Northern Arizona University Inductively Coupled Plasma Mass Spectrometry Laboratory

Uranium isotope ratios were measured by Dr. Michael Ketterer at Northern Arizona University's ICP-MS laboratory. Water samples collected from September 2007 to November 2008 were analyzed by sector-field ICP-MS. Samples collected in April and September 2009 were analyzed by quadrapole ICP-MS. For sector-field ICP-MS, high purity ^{233}U was used as an internal standard. A mass bias-correction factor determined from a known standard was used to correct raw U isotope ratios. Further details for sector-field ICP-MS measurements of U isotopes can be found in Ketterer and others (2000, 2003). For quadrupole ICP-MS, $^{238}U/^{235}U$ ratios were measured in unspiked sample aliquots. A control of known, naturally-occurring U was used to measure $^{238}U/^{235}U$ and to correct the ratios in the samples for mass bias effects. Appropriate blank subtractions were performed (M. Ketterer, written commun., 2009). A separate aliquot was taken for analysis of $^{234}U/^{235}U$ and $^{236}U/^{235}U$ ratios and was spiked with high purity ^{233}U. Isotopic ratios were corrected for minor interference of $^{232}Th^{1}H^{+}$ on ^{233}U, and appropriate blank subtractions were performed (M. Ketterer, written commun., 2009).

The activity ratio (AR) of ^{234}U to ^{238}U can be computed by dividing the measured atom ratio of ^{234}U to ^{238}U by 0.00005472 or by dividing the measured atom ratio of ^{234}U to ^{235}U by 0.0075448. The divisor 0.00005472 is derived from the relationship between the amount of a radionuclide and its activity, as shown in the following equation:

$$A = \lambda \cdot N \qquad (2)$$

where

A	is the activity (disintegrations per unit time) of the radionuclide,
λ	is its decay constant, and
N	is the number of atoms of the radionuclide.

When secular equilibrium is achieved, each daughter radionuclide has the same activity as the head of the decay chain, which is the case $A1 = A2$, where $A1$ and $A2$ are activities for radionuclides in a decay chain (Kraemer and Genereux, 1998). For example, with ^{234}U and ^{238}U, the divisor 0.0075448, used to compute the AR of ^{234}U to ^{238}U from the measured atom ratio of ^{234}U to ^{235}U, is the ^{234}U to ^{235}U atom ratio that develops in a closed system left to equilibrate for more than 10^6 years.

Lawrence Livermore Laboratory

Dissolved concentrations of ^4He, Ar, Kr, Ne, and Xe were measured in water samples by the Lawrence Livermore Laboratory. Reactive gases were removed with multiple reactive metal getters. Known quantities of isotopically enriched ^{22}Ne, ^{86}Kr, and ^{136}Xe were added to provide internal standards. The isotope dilution protocol used for measuring noble gas concentrations is insensitive to potential isotopic composition variation in dissolved gases (especially Ne) from diffusive gas exchange. Noble gases were separated from one another using cryogenic adsorption. Helium was analyzed using a VG-5400 noble gas mass spectrometer. Other noble gas isotopic compositions were measured using a quadrupole mass spectrometer. The argon (Ar) abundance was determined by measuring the total noble gas sample pressure using a high-sensitivity capacitive manometer. The procedure was calibrated using water samples equilibrated with the atmosphere at a known temperature and pressure. Tritium (^3H) concentrations were determined on 500-g subsamples by the ^3He in-growth method (approximately 15-day accumulation time). Analytical uncertainties are approximately 1 percent for $^3He/^4He$; 2 percent for He, Ne, and Ar; and 3 percent for Kr and Xe.

University of Utah Dissolved Gas Service Center

Dissolved concentrations of N_2, ^{40}Ar, ^{84}Kr, ^{20}Ne, and ^{129}Xe were analyzed by the University of Utah's (U of U) Dissolved Gas Service Center using both quadrupole and sector-field mass spectrometers. The mass spectrometer analysis provides the relative mole fractions of these dissolved gases. The sector-field mass spectrometer is used to precisely measure abundances of ^3He and ^4He. An electron multiplier is used to measure low-abundance ions, and a Faraday cup measures more abundant ions. The dissolved-gas concentrations of the water sample are then calculated on the basis of Henry's Law by using field measurements of total dissolved-gas pressure and water temperature. Calibrations are made using dry atmosphere and air equilibrated water samples, collected at different temperatures. A rigorous daily calibration procedure is followed. Four standards are usually analyzed for every six environmental samples.

Tritium samples were analyzed with the tritium in-growth method (Clarke and others, 1976) at the University of Utah's Dissolved Gas Service Center. Tritium is analyzed by measuring the ratio of the heavier and less-abundant isotope to the lighter and more-abundant isotope. Tritium concentrations are reported in tritium units (TU), where one TU equals one molecule of $^3H^1HO$ in 10^{18} molecules of ^1H_2O.

USGS X-Ray Diffraction Laboratory

Rock samples were analyzed by the U.S. Geological Survey Geologic Division X-Ray Diffraction Laboratory (XRD) in Denver, Colorado. Each sample was evaluated for zones of inhomogeneity, and if present, subsamples were taken from these zones. Samples were lightly crushed and passed through a riffle splitter. One-hundred grams of material from the splitter was milled in a ball mill for approximately 8 minutes so that particles would pass a 100-mesh screen. Two grams of material that passed the 100-mesh screen were placed in a McCrone Micronizing mill with 10 mL of 2-propanol for 4 minutes, which reduced the particle size to near 1 micron. The slurry was dried overnight. A 2-gram aliquot of the dried sample was passed through a 60-mesh sieve and then side packed into a sample holder for analysis. Samples were analyzed with a PANalytical Xpert Pro-MPD X-ray Diffractometer. Identification of mineral phases was done with Material Data Inc. Jade 9.1 software using ICDD's 2009-PDF-4 and National Institute of Standards and Technology FIZ/NIST Inorganic ICSD databases (W. Benzel, written commun., 2010).

Pattern-Recognition Modeling

The software package Pirouette (version 4.0, revision 1.0; Infometrix, 2010) was used for pattern-recognition modeling of the stream-sediment multivariate data set. Values below the lower reporting limit (LRL) were assigned a value of 0.75 times the LRL. Histograms and probability plots were used to evaluate data normality of the raw and log-transformed data sets. Log transformation of the data sets resulted in near-normal distributions for most constituents. The data were mean-centered prior to pattern-recognition modeling.

Quality Assurance and Quality Control

The quality-assurance/quality-control plan for this study included the use of approved USGS methods for the collection and analysis of surface and groundwater samples (U.S. Geological Survey, variously dated), USGS chain of custody protocol for the shipping of samples to the laboratory and tracking samples in the laboratory (U.S. Geological Survey, 2010c and 2010b), the computation of a cation/anion balance for each sample, a comparison of the dissolved to total metal ratios, the collection of field blanks and field duplicates, and the addition of matrix spikes to the metal samples.

Cation/Anion Balances

The accuracy of the analysis of major dissolved ions was evaluated by calculating a cation/anion balance for each sample. A fundamental principle of solution chemistry is that a condition of electroneutrality exists for the major ions dissolved in water, which means that when measured in milliequivalents per liter (meq/L), the sum of the positive charges equals the sum of the negative charges. The equation used to calculate a cation/anion balance is as follows:

$$\text{cation/anion balance} = \frac{(\text{sum of the cations} - \text{sum of the anions})}{(\text{sum of the cations} + \text{sum of the anions})} \cdot 100 \quad (3)$$

Ideally, the result of this calculation should equal zero, but in practice some deviation from zero is acceptable. If significant deviation from zero occurs, there must be either errors in the analytical measurement or the presence of an ionic specie or species at significant concentrations that were not included in the analysis. The criteria used for the determination of an acceptable cation/anion balance in this study are based on the results of analyses from the USGS NWQL and are shown in table 3. This table shows the error that is acceptable as a function of the total cations and anions. Since the total cations and anions in all samples analyzed for this study was greater than 1.71 meq/L, a cation/anion balance within 5 percent was considered acceptable. The range of values for the cation/anion balance calculation for the 52 samples collected in this study was 2.26 percent below to 7.52 percent above balance, with only 3 samples greater than 5 percent (6.10, 6.52, and 7.52 percent). Given the range of values for the samples analyzed in this study, we consider all samples to have an acceptable cation/anion balance; therefore, any analytical errors present are small enough not to affect the interpretation of the data, and all major dissolved ionic species were included in the analyses. An analysis of the cation/anion balance cannot be used as the only means of detecting measurement error because an acceptable cation/anion balance could occur in situations where large errors in the individual ion analyses balance one another. The cation/anion balance also does not evaluate the quality of the analysis for dissolved and total metals. Therefore, the results of the analysis of dissolved/total metals ratios, field blanks, field replicates, and matrix spikes also will be discussed.

Total and Dissolved Metals

Analytical results for the total concentration of a metal were compared to the dissolved concentration when both fractions were analyzed. Ideally, the total concentration of a metal should be greater than or equal to the dissolved fraction of the metal; however, as a result of variability that can occur as a result of sample collection, processing, transport, and analysis, the dissolved fraction can sometimes be greater than the total fraction. This situation commonly occurs at concentrations that approach the analytical method detection limit. For concentrations less than 1 µg/L, analytical results for the total and dissolved fraction of a given metal were considered acceptable if the results were within twice the long-term method detection limit (LT-MDL) of the least precise method (the least precise method is usually associated with analysis of the total concentration of a metal). For example, if the LT-MDL for dissolved copper (Cu) is 0.5 µg/L, and the LT-MDL for total Cu is 0.6 µg/L, dissolved Cu could exceed total Cu by two times 0.6 µg/L, or 0.12 µg/L. For concentrations equal to or greater than 1.0 µg/L, analytical results for the total and dissolved fraction of a given metal were considered acceptable if the results were within 10 percent. If analytical results for a given sample failed the criteria described above, reanalysis of the total and dissolved fraction was requested of the NWQL. There were a few instances where analytical results for total and dissolved concentrations of metals did not meet the criteria described above, even after re-runs were performed. Because these instances involved concentrations near method detection limits, they were accepted and should be viewed with caution.

Field Blanks and Field Duplicates

Field blanks and field duplicates were collected during this study to quantify the errors involved in collecting, processing, transporting, and analyzing samples. Every measurement has an error associated with it that cannot be eliminated, but the error can be quantified so that appropriate interpretations of the environmental data can be made. Bias and variability are two components of error associated with any water-quality measurement. Bias is the systematic error inherent in a method or measurement system and can be either positive (contamination) or negative (loss). Variability is the random error in independent measurements that results from repeated application of the measurement process under specified conditions.

In a water-quality study, two types of samples are needed: environmental samples and quality-control samples. Environmental samples fulfill the scientific objective(s) of the study. Quality-control samples provide estimates of the bias and variability of the environmental data. Field blanks are samples that are intended to be free of the analyte(s) of interest and are analyzed to test for bias from the introduction of contamination into environmental samples in any stage of the sample-collection and analysis processes. Field replicates are a group of samples that are collected in a manner such that the samples are thought to be essentially identical in composition and are used to estimate the variability of the sample-collection and analysis process. Field blanks and field replicates are collected in the same manner as the environmental samples.

Once a data set is established with an estimated amount of bias and variability, it is necessary to determine how the bias and variability affect the interpretation of the environmental

Table 3. Acceptance criteria for cation/anion balances, White Mesa mill study area, Utah.

[**Abbreviations**: meq/L, milliequivalents per liter; >, greater than; ±, plus or minus; %, percent]

Ionic strength (meq/L)	Acceptable cation/ anion balance
0–0.2809	± 28%
0.281–0.561	± 22%
0.561–0.8309	± 15%
0.831–1.109	± 10%
1.11–1.409	± 8%
1.41–1.709	± 6%
>1.71	± 5%

data. Thus, the analysis of quality-control sample data supports the interpretations of the environmental data by establishing, with a known level of confidence, the amount (if any) of sample contamination that has occurred during the study and by establishing the range of variability in the quality-control sample data relative to the range of variability in the environmental data.

Analysis of Field Blanks

Under ideal conditions any contamination present in field blanks would be so small that concentrations would be less than the detection limit. In practice, although concentrations measured in many field blanks are less than the detection limit, some blanks contain concentrations greater than the detection limit. Therefore, as stated in Mueller and Titus (2005),

"The objective in analyzing data from blanks is to determine the amount of contamination that is not likely to be exceeded in a large percentage of the water samples represented by the blanks. This objective can be achieved by constructing an upper confidence limit (UCL) for a high percentile of contamination in the population of water samples that includes environmental samples and blanks. This UCL is the maximum contamination expected in the specified percentage of water samples. For example, the 95-percent UCL for the 90th percentile of concentrations in blanks is the maximum contamination expected in 90 percent of all water samples. The 95-percent confidence level indicates there is only a 5-percent chance that this contamination has been underestimated. Another way to express this is that we are 95-percent confident that this amount of contamination would be exceeded in no more than 10 percent of all samples (including environmental samples) that were collected, processed, and analyzed in the same manner as the blanks."

In calculating the UCL for the blank data, all estimated values and values that were detected but were within the range of two or more detection limits were censored to the highest detection limit. A review of the field blank data in tables 4 and 5 shows that all the blanks analyzed for dissolved beryllium, boron, cadmium, chloride, cobalt, fluoride, iron, lithium, nitrate + nitrite, selenium, silver, sodium, sulfate, thallium, and U and total selenium were reported as less than the detection limit. Thus, contamination by each of these analytes is estimated with about 92-percent confidence to be no greater than the detection limit in at least 70 percent of all samples. The 92-percent confidence level indicates that there is only an 8-percent chance that this contamination has been underestimated. For those analytes that had measurable concentrations in the blanks, we are 92-percent confident that the amount of contamination listed in table 4 would be exceeded in no more than 30 percent of all samples. For example, for dissolved U there is 92-percent confidence that contamination is no greater than the detection limit of 0.02 µg/L in at least 70 percent of all samples. For total U there is 92-percent confidence that contamination is no greater than 0.024 µg/L in at

least 70 percent of all samples. Another way to express this is that contamination by total U is estimated, with 92-percent confidence, to exceed 0.024 µg/L in no more than 30 percent of all samples.

This amount of contamination can then be compared to environmentally important concentrations of each analyte to determine the likelihood that contamination has affected interpretation of the environmental data. Mueller and Titus (2005) state that "in general, if potential contamination is less than 10 percent of a measured value, the effect of contamination bias on that measured value can be ignored." The detection limit for all of the analytes that were never measured above the detection is at least 10 times less than the environmental concentrations measured in this study or EPA drinking water MCLs. For example, the detection limit of dissolved U (0.02 µg/L) is 1,500 times less than the EPA drinking water MCL of 30 µg/L. Therefore, even if contamination were equal to or greater than 0.02 µg/L in 30 percent of all samples, the contamination would have to be two orders of magnitude greater than this value for potential bias to affect the interpretation of the U environmental data. We draw similar conclusions for all the other analytes that were never measured above the detection limit because the environmental concentrations of these analytes are greater than 10 times their respective detection limits. The same conclusions can be drawn for those analytes that had measurable concentrations in the field blanks, except for total Al, Cr, Cu, Fe, V, and Zn collected from the wells.

Typically, field-blank and field-replicate samples collected from the springs and the wells would be analyzed separately because different equipment is used to collect samples from these sites. As the analysis in this section demonstrated, however, except for total Al, Cr, Cu, Fe, V, and Zn in field blanks collected from the wells, there is no evidence of contamination affecting the interpretation of the environmental data. Therefore, the environmental data collected from all sites for all of the other analytes can be considered comparable, and the field-blank and field-replicate data can be pooled to determine the magnitude of bias and variability in the data. The concentrations of total Al, Cr, Cu, Fe, V, and Zn in field blanks collected from the wells are high enough that contamination of the environmental samples limits the utility of this data. Therefore, in this report, the total Al, Cr, Cu, Fe, V, and Zn data collected from the wells is interpreted with caution.

Analysis of Field Replicates

The field replicate data were analyzed to assess the amount of variability present in the environmental data by calculating a 95-percent confidence interval for a single sample and by determining the minimum significant difference that can be detected between any two individual measurements using the equations given in Mueller and Titus (2005). These calculations involved calculating a standard deviation for each field replicate pair and examining graphs of the standard deviation of each replicate pair as a function of the average concentration of each field replicate pair to determine if the standard

Table 4.   Upper 92–percent confidence limits for contamination by trace elements and nutrients in the 70th percentile of all samples on the basis of data from field blanks prepared at spring and groundwater sampling sites, White Mesa mill study area, Utah.

[**Abbreviations**: mg/L, milligrams per liter; μg/L, micrograms per liter; —, not available; <, less than; *, confidence interval for orthophosphate is 67 2 percent because of smaller sample size]

Analyte	Number of blanks	Most common detection limit filtered, (unfiltered)	Concentration units	Upper 92-percent confidence limit (filtered)	Upper 92-percent confidence limit (unfiltered)
Aluminum	7	4 (4)	μg/L	<4 0	248
Antimony	7	0 14	μg/L	0 09	—
Arsenic	7	0 06 (0 6)	μg/L	0 07	<0 60
Barium	7	0 4	μg/L	<0 4	—
Beryllium	7	0 01	μg/L	<0 02	—
Boron	7	6	μg/L	<6	—
Cadmium	7	0 04	μg/L	<0 04	—
Chromium	7	0 12 (0 40)	μg/L	0 25	7 5
Cobalt	7	0 02	μg/L	<0 02	—
Copper	7	1 0 (1 2)	μg/L	<1 0	<4 0
Iron	7	8 (6)	μg/L	<8	269
Lead	7	0 08 (0 06)	μg/L	<0 08	0 3
Lithium	7	1	μg/L	<1 0	—
Manganese	7	0 2 (0 4)	μg/L	0 9	7 8
Molybdenum	7	0 2 (0 1)	μg/L	<0 2	1 3
Nickel	7	0 2 (0 12)	μg/L	0 31	6 2
Selenium	7	0 04 (0 08)	μg/L	<0 06	<0 12
Silver	7	0 1	μg/L	<0 1	—
Strontium	7	0 8	μg/L	4 08	—
Thallium	7	0 04	μg/L	<0 04	—
Uranium	7	0 02 (0 02)	μg/L	<0 02	0 024
Vanadium	7	0 16 (1 6)	μg/L	0 2	0 61
Zinc	7	1 8 (2 0)	μg/L	<2 0	3 8
Nitrate + nitrite	7	0 04	mg/L	<0 04	—
Orthophosphate*	5	0 008	mg/L	<0 008	—

Table 5. Upper 92–percent confidence limits for contamination by major ions in the 70th percentile of all samples on the basis of data from field blanks prepared at spring and groundwater sampling sites, White Mesa mill study area, Utah.

[**Abbreviations**: mg/L, milligrams per liter; <, less than]

Analyte	Number of blanks	Most common detection limit	Concentration units	Upper 92–percent confidence limit
Calcium	7	0.04	mg/L	0.41
Chloride	7	0.12	mg/L	<0.12
Fluoride	7	0.12	mg/L	<0.12
Magnesium	7	0.02	mg/L	0.075
Potassium	7	0.06	mg/L	0.06
Silica	7	0.02	mg/L	0.03
Sodium	7	0.12	mg/L	<0.12
Sulfate	7	0.18	mg/L	<0.18

deviation is constant over the range of concentrations measured. Typically, the higher the constituent concentration, the greater the standard deviation; however, the relation between standard deviation for each replicate pair was constant over the range in concentration measured for each constituent, or only a weak relation with concentration existed. This consistency most likely is a result of relatively little variation in the environmental concentrations for all constituents (that is, concentrations were similar to each other, and, for most of the trace metals, concentrations were generally quite low).

Therefore, the average standard deviation of the replicate pairs for each constituent was substituted into the following equation to calculate a 95-percent confidence interval for a single sample:

$$C_{interval} = C_{sample} \pm Z_{0.95} \cdot SD \tag{4}$$

where

$C_{interval}$ is the confidence interval for a single measurement = $100(1-\alpha)$,

C_{sample} is the concentration of a single sample,

SD is the average standard deviation of the replicate pairs, and

$Z_{0.95}$ is the statistic for the 95-percentage point of the standard normal curve = 1.96.

When one of the replicate pairs was below the reporting limit but the other had measurable amounts of a constituent reported, the sample with a value of less than the reporting limit was assigned a value of one-half the reporting limit to perform the calculation. The 95-percent confidence interval data for a single sample are presented in tables 6 to 8 and can be interpreted in the following manner: there is 95-percent confidence that the true value of any individual measurement for any constituent listed in tables 6 to 8 will fall within the range in those tables.

To determine the minimum significant difference that can be detected between any two individual measurements, the following formula was used:

$$\Delta C \text{ (difference in concentration between two samples)} \geq 1.96 \cdot \sqrt{2} \cdot SD \tag{5}$$

If the difference in concentration between any two samples is equal to or greater than the values listed in tables 6, 7, and 8, there is a 95-percent probability that the difference is significant.

Matrix Spikes

An aliquot from one unfiltered sample collected during the September 2008, November 2008, April 2009, and September 2009 water-quality sampling events was spiked with trace metals at the USGS NWQL in order to evaluate whether or not the sample matrix (the overall chemical composition of the sample) affected the quality of the metal analyses .Trace metals that were spiked included Fe, Al, Pb, Mo, U, As, Cr, Cu,

Table 6. Estimates of variability of filtered trace elements and nutrients, White Mesa mill study area, Utah.

[**Abbreviations**: µg/L, micrograms per liter; —, not available]

Chemical constituent	Number of replicate sets	Concentration units	95-percent confidence interval for a single sample	Minimum significance difference between any two individual measurements
Aluminum	6	µg/L	0.2	0.3
Antimony	5	µg/L	0.01	0.02
Arsenic	6	µg/L	0.04	0.06
Barium	6	µg/L	2	2
Beryllium	6	µg/L	—	—
Boron	5	µg/L	5	7
Cadmium	6	µg/L	0.02	0.02
Chromium	6	µg/L	0.06	0.09
Cobalt	6	µg/L	0.01	0.02
Copper	6	µg/L	—	—
Iron	6	µg/L	1	2
Lead	6	µg/L	0.002	0.003
Lithium	6	µg/L	2.7	3.8
Manganese	6	µg/L	1.6	2.3
Molybdenum	6	µg/L	0.4	0.6
Nickel	5	µg/L	0.05	0.07
Selenium	6	µg/L	0.1	0.1
Silver	6	µg/L	—	—
Strontium	5	µg/L	31	44
Thallium	6	µg/L	—	—
Uranium	5	µg/L	0.35	0.5
Vanadium	6	µg/L	0.06	0.08
Zinc	6	µg/L	0.1	0.1
Nitrate + nitrite	5	µg/L	0.01	0.02
Orthophosphate	5	µg/L	0.002	0.002

Table 7.   Estimates of variability of unfiltered trace elements, White Mesa mill study area, Utah.

[**Abbreviations**: µg/L, micrograms per liter]

Chemical constituent	Number of replicate sets	Concentration units	95-percent confidence interval for a single sample	Minimum significance difference between any two individual measurements
Aluminum	6	µg/L	59.4	84
Arsenic	6	µg/L	0.1	0.2
Chromium	6	µg/L	0.1	0.2
Copper	6	µg/L	0.3	0.4
Iron	6	µg/L	59.1	83.6
Lead	6	µg/L	0.2	0.2
Manganese	6	µg/L	1.6	2.3
Molybdenum	6	µg/L	0.2	0.3
Nickel	6	µg/L	0.2	0.3
Selenium	6	µg/L	0.1	0.2
Uranium	6	µg/L	0.4	0.5
Vanadium	6	µg/L	0.4	0.6
Zinc	6	µg/L	0.3	0.5

Table 8.   Estimates of variability of major ions, White Mesa mill study area, Utah.

[**Abbreviations**: mg/L, milligram per liter]

Chemical constituent	Number of replicate sets	Concentration units	95-percent confidence interval for a single sample	Minimum significance difference between any two individual measurements
Bicarbonate	3	mg/L	8	12
Calcium	6	mg/L	2.5	3.6
Chloride	5	mg/L	0.7	0.9
Fluoride	5	mg/L	0.13	0.18
Magnesium	6	mg/L	0.4	0.6
Potassium	5	mg/L	0.1	0.14
Silica	6	mg/L	0.2	0.2
Sodium	6	mg/L	1.7	2.4
Sulfate	5	mg/L	8	11

Ni, Se, V, and Zn. With the exception of Fe that was analyzed by ICP-AES, spiked elements in the samples were analyzed by ICP-MS. Table 9 summarizes the spike amount for each trace metal, analytical results for both spiked and unspiked samples, and the percent recoveries associated with each analysis. Percent recoveries computed for all samples and elements ranged from 63 to 131, with an average of 98. The percent recovery for zinc (63) in the April 2009 sample and molybdenum (131) in the September 2009 sample both fall outside the US EPA percent recovery allowable limits for laboratory-spiked environmental samples analyzed by ICP-MS (US Environmental Protection Agency, 1994a) or ICP-AES (US Environmental Protection Agency, 1994b). On the basis of this observation, analytical results for total zinc (Zn) in water samples from the West well and total molybdenum (Mo) in water samples from Entrance Spring (fig. 1) could be compromised because of matrix effects and should be viewed with caution.

Quality Assurance and Quality Control Summary

The results of all of these quality-control calculations allow for a number of statements about the overall data quality. First, the amount of bias, as measured by field blanks, indicates that contamination of samples did not occur, except for the few total metals discussed. The amount of random error, as measured by the field replicates, is small enough that the comparison of samples to a water-quality standard, or the comparison of samples collected from different sites or from the same site at different times, is not compromised. For the major ions, this finding supports the interpretation of the cation/anion balance calculations that analytical errors are minimal and that all major dissolved ionic species are included in the analysis. For dissolved and total metals, the concentration of the dissolved metals consistently falling below the concentration of the total metals and the concentrations of spiked samples falling within acceptable percent recovery ranges in most samples indicate that the sample matrix did not significantly affect the analytical measurement of the metals. High total metal concentrations of Al, Cr, Cu, Fe, V, and Zn measured in field blanks for water-quality samples collected from the wells were the only parameters that indicated error could affect the interpretation of the environmental data. As a result, we conclude that any error resulting from the collection, processing, transporting, and analysis of the water-quality samples for major ions and dissolved and total metals does not affect the overall interpretation of the environmental data.

Table 9. Analytical results for spiked and unspiked samples, and comparison of precent recoveries to EPA percent recovery allowable limits for analytical methods 200.7 and 200.8. For unspiked results that are less than the analytical detection limit, one half the detection limit was used to compute percent recoveries. Unfiltered samples were spiked.

[**Abbreviations**: mm/dd/yyyy, month/day/year; µg, micrograms; µg/L, micrograms per liter; US EPA, United States Environmental Protection Agency; <, less than]

Local identifier	Field identifier	Station number	Sample date (mm/dd/yyyy)	Parameter	Spike amount, (µg)	Analytical result, unspiked sample, (µg/L)	Analytical result, spiked sample, (µg/L)	Percent recovery	US EPA percent recovery allowable limits
(D–37–22)31dcb–S1	Cow Camp Spring	373122109321501	09/17/2008	Iron	100	323	431	108	70–130
				Aluminum	50	499	552	106	70–130
				Lead	50	0.5	52.8	105	70–130
				Molybdenum	50	1.6	55	107	70–130
				Uranium	50	9.18	65.7	113	70–130
				Arsenic	50	2	50.4	97	70–130
				Chromium	50	0.5	51	101	70–130
				Copper	50	0.574	46.2	91	70–130
				Nickel	50	0.43	47.4	94	70–130
				Selenium	50	1.6	44	85	70–130
				Vanadium	50	1.56	54.4	106	70–130
				Zinc	50	1.04	40.8	80	70–130
(D–37–22)27ccc–S1	Entrance Spring	373202109293401	11/11/2008	Iron	100	43.5	150	107	70–130
				Aluminum	50	66.1	116	99	70–130
				Lead	50	0.114	53.0	106	70–130
				Molybdenum	50	3.87	58.3	109	70–130
				Uranium	50	25.7	80.0	109	70–130
				Arsenic	50	2.32	53.0	101	70–130
				Chromium	50	<0.4	49.9	99	70–130
				Copper	50	<4	45.2	86	70–130
				Nickel	50	0.41	46.3	92	70–130
				Selenium	50	8.95	55.9	94	70–130
				Vanadium	50	5.35	56.3	102	70–130
				Zinc	50	<2	43.1	84	70–130
(D–38–22)8dcd–1	West well	372930109310701	04/21/2009	Iron	100	219	313.0	94	70–130
				Aluminum	50	<18	54.9	92	70–130
				Lead	50	4.74	57.3	105	70–130
				Molybdenum	50	43.5	85.6	84	70–130
				Uranium	50	18	73	110	70–130
				Arsenic	50	2.5	47.8	91	70–130
				Chromium	50	4.1	55.5	103	70–130
				Copper	50	13	57.9	90	70–130
				Nickel	50	10	53.4	87	70–130
				Selenium	50	0.52	42.3	84	70–130
				Vanadium	50	<4.8	53	101	70–130
				Zinc	50	24.8	56.4	63	70–130
(D–37–22)27ccc–S1	Entrance Spring	373202109293401	09/23/2009	Iron	50	46.2	90.6	89	70–130
				Aluminum	50	51.6	107.4	112	70–130
				Lead	50	0.28	47.5	94	70–130
				Molybdenum	50	3.87	69.2	131	70–130
				Uranium	50	20.2	80.4	120	70–130
				Arsenic	50	2.31	53	101	70–130
				Chromium	50	<0.4	52.7	105	70–130
				Copper	50	<4	46.6	89	70–130
				Nickel	50	0.64	47.4	94	70–130
				Selenium	50	8.1	56.8	97	70–130
				Vanadium	50	5.24	58.1	106	70–130
				Zinc	50	2.07	47.4	91	70–130

Description of Study Area

The White Mesa uranium mill is located on White Mesa in San Juan County, Utah (fig. 1). White Mesa is composed of Quaternary eolian deposits that overlie a sequence of Mesozoic rocks (fig. 2). Two aquifers are used by Ute Mountain Ute tribal members in the vicinity of the White Mesa uranium mill. A shallow, unconfined aquifer exists in the Dakota Sandstone and Burro Canyon Formation, which extends to a depth of about 23 m. The water in this aquifer is the source of numerous springs located on the reservation south of the mill. The water in these springs is used by tribal members for drinking and watering cattle, and by wildlife hunted by tribal members. Below the Burro Canyon Formation are about 366 m of low-permeability rocks (Morrison Formation) overlying the Entrada and Navajo Sandstones, which support the aquifer supplying drinking water to tribal members in the town of White Mesa.

To evaluate the potential for dissolution of airborne material (potential sources include ore-storage piles, alternative feed-storage area, and drying stacks) deposited on soil and leakage from the tailings ponds to contaminate the groundwater of White Mesa, it is necessary to understand (1) the direction of groundwater flow, (2) the residence time of groundwater, and (3) whether the geochemistry of the groundwater enhances or retards transport of U. To understand these factors, knowledge of the mineralogy (chemical composition) and hydrologic properties of the rocks composing White Mesa is essential. In the next section, a summary of the lithology of the rocks in the White Mesa, described by Witkind (1964) and Johnson and Thordarson (1966), is given, beginning at the bottom of the stratigraphic column with the Navajo Sandstone and progressing up through the stratigraphic column to end with the Eolian Sand. A summary of the hydrologic properties of these rocks, described by Whitfield and others (1983) and Freethey and Cordy (1991), is given in the "Hydrology" section.

Lithology of the Rocks Composing White Mesa

Navajo Sandstone

The Navajo Sandstone is Late Triassic/Early Jurassic in age, eolian in origin, and is about 125 m thick near the Abajo Mountains. It is a very pale orange, massive crossbedded, fine- to medium-grained, quartz sandstone that is generally well sorted and is characterized by long, sweeping tangential sets of cross strata. The Navajo Sandstone is composed primarily of subround to round, frosted quartz grains ranging in diameter from 0.05 to 0.36 mm, with most grains having a diameter of about 0.15 mm. All of the quartz grains are covered by a thin film of iron oxide. The Navajo Sandstone is poorly cemented and friable, with silica acting as the principal cement, but calcite and iron oxide also act as cement. Near the top of the Navajo Sandstone, several limy sandstone beds, most of them about 4 feet thick and of limited lateral extent, occur. Calcite is the dominant cement in these deposits.

Entrada Sandstone

The Entrada Sandstone is Late Jurassic in age, eolian in origin, and 91 to 122 m thick in southeast Utah. It is a very pale orange massive friable crossbedded, very fine to medium-grained sandstone. The Entrada Sandstone is composed

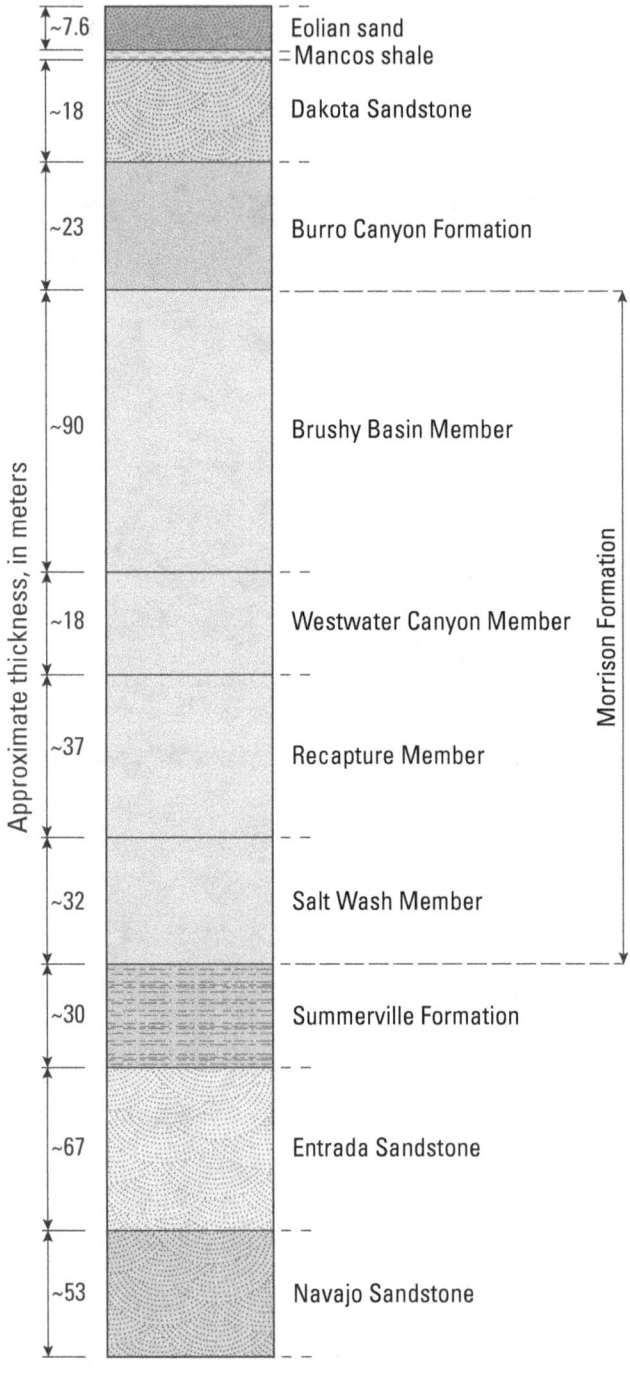

Figure 2. Stratigraphic column for White Mesa, San Juan County, Utah (Titan Environmental Corporation, 1994).

primarily of angular to well-rounded quartz grains ranging in diameter from about 0.05 to 0.3 mm, with most of the grains having a diameter of about 0.15 mm. The sandstone beds are weakly cemented by calcite, silica, and iron oxide.

Summerville Formation

The Summerville Formation is Late Jurassic in age and ranges in thickness from 20 to 38 m but in most places is 26 m thick. The sediments composing the Summerville Formation were deposited in a marine and marginal marine environment and consist of alternating beds of pale reddish-brown to moderate reddish-brown shaly siltstone and very fine to fine-grained sandstone.

Morrison Formation

The Morrison Formation of Late Jurassic age overlies the Summerville Formation and has been divided into the Salt Wash Member, the Recapture Member, the Westwater Canyon Member, and the Brushy Basin Member. All of the sediments composing the Morrison Formation were deposited by streams whose source was a highland area midway along the state line between Arizona and Utah.

The Salt Wash Member of the Morrison Formation is composed of lenticular sandstone beds that alternate at irregular intervals with beds of silty claystone, mudstone, and siltstone, and it averages about 300 feet in thickness in southeastern Utah. The sandstone beds of the Salt Wash Member range in color from moderate grayish yellow to light gray. All of the sandstone beds are crossbedded and moderately friable and range in thickness from 0.3 to 12 m but can be as much as 61 m thick in the few places where the intervening claystone beds become sandy and form a continuous sandstone sequence. The sandstone is composed primarily of fine (0.20 mm) to coarse (0.65 mm), angular to round, frosted grains of quartz with small amounts of microcline and chert that are moderately to well cemented by calcite, silica, and iron oxide. Stringers of conglomerate, claystone, and carbonaceous material are scattered unevenly throughout the sandstone. The claystone, mudstone, and siltstone beds are chiefly pale reddish brown but locally are altered to yellowish gray. Pale reddish-brown thin-bedded, very fine, fine-, and medium-grained sandstone beds that laterally grade into the claystone-siltstone sequence are interbedded through the claystone and siltstone beds of the basal part of the Salt Wash.

The Recapture Member of the Morrison Formation is composed of interbedded grayish-red, silty and sandy claystone and thin lenses of light brown fine- to medium-grained sandstone; it ranges in thickness from 0 to 61 m in southeast Utah. The Recapture Member intertongues with and grades into the Salt Wash near Blanding, Utah, becoming unrecognizable as a separate formation. Several facies, including a conglomeratic sandstone facies, an intermediate sandstone facies, and an outer claystone and sandstone facies, have been identified in the Recapture Member. In southeast Utah, the Recapture is predominantly claystone containing a few isolated lenses of sandstone or conglomerate.

The Westwater Canyon Member of the Morrison is composed of interbedded yellowish-brown fine- to coarse-grained sandstone and minor amounts of greenish-gray to reddish-brown silty and sandy claystone, and it is as much as 76 m thick in southeastern Utah. The Westwater Canyon Member intertongues with, and grades into the lower part of the Brushy Basin Member between Blanding and Monticello, Utah.

The Brushy Basin Member of the Morrison Formation is composed of beds of impure structureless, variegated claystone, mudstone, and siltstone that range in thickness from 84 to 107 m. It has an average thickness of about 91 m in the area surrounding the Abajo Mountains. These beds are described as a moderate greenish yellow, streaked irregularly by pale red, light red, and light brownish gray. In general, the claystone matrix consists of minute (0.01 mm and smaller) angular grains of quartz cemented by calcite and silica. Angular to subround quartz grains that range from 0.05 to about 0.21 mm in diameter, with most being about 0.1 mm, are scattered irregularly through the matrix. Much bentonitic clay of volcanic origin is also present. Johnson and Thordarson (1966) state that, locally, the Brushy Basin Member contains thin beds of limestone and beds of grayish-red to greenish-black siltstone that were probably deposited in small fresh-water lakes.

Burro Canyon Formation and Dakota Sandstone

Witkind (1964) discusses these two formations as a single unit because of the poor exposures and the indiscernible contact between them in the Abajo Mountains area. The Burro Canyon Formation is of late Cretaceous age, and the Dakota Sandstone is of early Cretaceous age. In the vicinity of the Abajo Mountains, the Burro Canyon Formation consists of alternating beds of conglomerate, conglomeratic sandstone, and sandstone. The sandstone beds are light gray and pale grayish-orange, friable, massive in places, but locally, thin to thick bedded, crossbedded, and channeled. The dominant mineral is quartz, with small amounts of microcline and chert present. The shape of the grains range from angular to well rounded, and they have diameters ranging from 0.02 to 0.5 mm, with most being about 0.1 mm in diameter. Calcite is the dominant cement, with silica and iron oxide also functioning as cement. The conglomerate and conglomeratic sandstone are normally at the base of the Burro Canyon Formation, and the rocks become less coarse near the top.

The Dakota Sandstone is described as a pale grayish-orange to yellowish brown, massive, intricately crossbedded, friable fine- to coarse-grained sandstone. Scattered irregularly through the Dakota Sandstone are lenses of conglomerate, dark-gray claystone seams, and lenticular carbonaceous seams. The sandstone consists chiefly of quartz grains that are cemented by silica and calcite. The grains are of two sizes; most common are angular grains about 0.06 mm in diameter that surround large numbers of well-rounded quartz grains about 0.40 mm in diameter.

Eolian Sand

The surface of the eastern portion of the White Mesa, including the mill site area, has been mapped as an eolian deposit by Haynes and others (1972). Witkind (1964) describes this deposit in the area north of Blanding, Utah, as "unconsolidated pale reddish brown dune sand composed of angular to well-rounded quartz grains that range from 0.02 to 0.20 mm in diameter. All the grains are covered by a film of iron oxide that gives them a distinctive reddish brown appearance. The iron oxide also acts as a weak cement, and many of the feebly held grains form aggregates as much as 0.4 mm in diameter."

Uranium Deposits

Johnson and Thordarson (1966) state that U deposits in southeast Utah occur as tabular deposits nearly parallel to the bedding in fluvial sandstones that range in thickness from a few centimeters to 6 m or more, and in width from 0.6 to more than 305 m.

The source of the U and other metals in Colorado Plateau U deposits is not known; however, lead-uranium ratios indicate that these ores are about 65 million years old, and the enclosing rocks are much older: the Morrison Formation is 130 million years old, so the ore minerals had to have been epigenetically introduced or redistributed (Johnson and Thordarson, 1966). Thus, the metals apparently were deposited from solutions that mostly traveled laterally through the rocks until confinement caused precipitation of the ore minerals in a favorable host rock. The distribution of the ore-bearing solutions over large areas on the Colorado Plateau is indicated by the widespread occurrence of uranium, vanadium, and copper deposits.

Of the formations present in the White Mesa, Johnson and Thordarson (1966) state that U deposits in amounts suitable for economic recovery occur only in the Salt Wash Member of the Morrison Formation. Johnson and Thordarson (1966) state that "significant ore deposits, however, are not evenly distributed through the Salt Wash but rather are clustered in eastward-trending belts of relatively favorable ground thought to represent the traces of ancient stream channels or channel systems on the Salt Wash fan." Favorable ground is defined as areas within the Salt Wash Formation that contain a greater percentage of sandstone and have sandstone lenses that are thicker than average, which can indicate the position of rather persistent trunk channel systems.

Among other formations found on the White Mesa, the Navajo and Entrada Formations, the Recapture and Westwater Canyon Members of the Morrison Formation, and the Summerville Formation are thought to contain no appreciable potential U reserves. The Brushy Basin could contain appreciable potential reserves of low-grade ore and sub ore-grade uranium-bearing rock because of the presence of uranium deposits 1,000 to 10,000 tons or more in size in the vicinity of the study area. The Burro Canyon Formation and Dakota Sandstone are not known to contain significant uranium deposits in the report area, even though the sandstone beds of the Burro Canyon and Dakota Formations are similar in many respects to the ore-bearing rocks in the Salt Wash Member of the Morrison Formation. Most likely this was caused by the blanket like sandstone beds of the Burro Canyon and Dakota dispersing, rather than concentrating, uranium-bearing solutions.

Hydrology

The area surrounding the Ute Mountain Ute community of White Mesa experiences a climate characterized by meager and undependable rainfall, with large annual ranges in temperature and a season of severe cold. Average yearly precipitation measured at Blanding, Utah, from 1904 to 2005, was 34 cm (Western Regional Climate Center, 2010).

A conceptual model of the hydrologic cycle on the White Mesa is shown in figure 3. This model was developed using information presented in Whitfield and others (1983), Freethey and Cordy (1991), Kirby (2008), quarterly monitoring reports produced by the mill for the State of Utah, and our observations. Groundwater in the White Mesa occurs within each formation shown in figure 2; however, not all of these formations function as aquifers. According to Freethey and Cordy (1991), the Dakota Sandstone and Burro Canyon Formation support an unconfined aquifer. The Westwater Canyon, Recapture, and Salt Wash members of the Morrison Formation house a confined aquifer that is not used by tribal members. The Navajo Sandstone contains a confined aquifer that provides drinking water to the towns of White Mesa and Blanding. The Brushy Basin Member of the Morrison Formation and the Summerville Formation act as aquitards that prevent the mixing of groundwater with the formations above and below them. The conceptual model and the rest of the discussion in this section focus on the unconfined aquifer in the Dakota Sandstone and Burro Canyon Formation because it is the potential for mill contamination of this aquifer that concerns the Ute Mountain Ute Tribe. Groundwater in this aquifer flows south/southeast from the mill to the Ute Mountain Ute Reservation (Kirby, 2008), and the springs, emanating primarily from the Burro Canyon Formation, are used by tribal members.

Precipitation falling on the White Mesa is a major source of recharge to the unconfined aquifer. There are no permanent streams on the White Mesa within these formations, and these formations do not extend to the Abajo Mountains; thus, they are isolated from the Abajo Mountains and cannot be recharged from precipitation falling on the Abajo Mountains. The infiltration of precipitation on the White Mesa is facilitated by the presence of eolian sand, which increases recharge potential because it is easily infiltrated and prevents rapid evaporation or runoff (Witfield and others, 1983). Groundwater recharge to this aquifer probably varies seasonally because of greater precipitation in winter on White Mesa, 19.8 cm in winter compared to 12.1 cm in summer, (National Oceanic and Atmospheric Administration (NOAA);

http://www.ncdc noaa.gov/oa/ncdc html) in combination with the lower winter air temperatures, likely results in most of the groundwater recharge occurring in the winter months.

Another source of groundwater recharge to the unconfined aquifer east and northeast of the mill is Recapture Reservoir. Two wildlife ponds constructed by the mill to attract birds away from the tailing cells are filled with water from Recapture Reservoir. Water from these wildlife ponds leaks downward into the unconfined aquifer and flows east toward Entrance Spring. Northeast of the mill and north of Entrance Spring, water from Recapture Reservoir is used to irrigate agricultural fields, which percolates down to the unconfined aquifer. Evidence for both of these sources of groundwater recharge is discussed in the "Hydrology" subsection of the "Results and Discussion."

Precipitation is probably the only source of recharge to the two aquifers beneath the unconfined aquifer. A major source of recharge for the Morrison Formation is most likely the Abajo Mountains because the Morrison Formation is exposed from White Mesa north, so precipitation falling on the Abajo Mountains could recharge this aquifer. One potential recharge area for the Entrada and Navajo Sandstones is precipitation falling on Comb Ridge to the west of the study area (Freethey and Cordy, 1991).

Groundwater discharge from the unconfined aquifer occurs primarily by evapotranspiration and discharge from the numerous springs around White Mesa that occur at the contact of the Burro Canyon Formation and the Brushy Basin Member of the Morrison Formation. Although most groundwater recharge probably occurs in winter, and the eolian sand facilitates groundwater recharge, most precipitation probably never reaches the water table. Whitfield and others (1983) state that in recharge areas in southeast Utah, an estimated 2 percent of average annual precipitation reaches the zone of saturation.

Freethey and Cordy (1993) state that in southeast Utah only about 1 percent of precipitation recharges aquifers exposed at the surface, which receive 20 to 25 cm of winter precipitation. Seepage to the underlying Brushy Basin Member of the Morrison Formation is thought to be negligible because the Brushy Basin Member is considered a confining unit, as described above and shown by the number of springs in the area.

One key concern with respect to the fate of any contaminant potentially released from the mill to groundwater is the speed at which it would migrate in groundwater and discharge to the springs around White Mesa that are used by tribal members. A consultant hired by the mill (Titan, 1994) estimated travel times between 8,900 and 13,400 years for groundwater to travel distances of 8,000 to 12,000 feet. These estimates were calculated with Darcy's Law using hydraulic conductivity data obtained from 12 single, well-pumping/recovery tests and from 30 packer tests.

The calculation of groundwater travel times with Darcy's Law is a valid method. The permeability tests, however, were performed in wells only on mill property north of the reservation and would not have measured permeability in the Dakota Sandstone and Burro Canyon Formation south of mill property. As a result of the heterogeneous composition of the stream sediments that compose the Dakota Sandstone and Burro Canyon Formation, it is possible that permeability on mill property is not representative of permeability south of the mill. In these formations, it is entirely possible that highly permeable pathways, such as joints, fractures, or paleo-stream channels, exist between the mill and the reservation, which could result in faster groundwater travel times than those

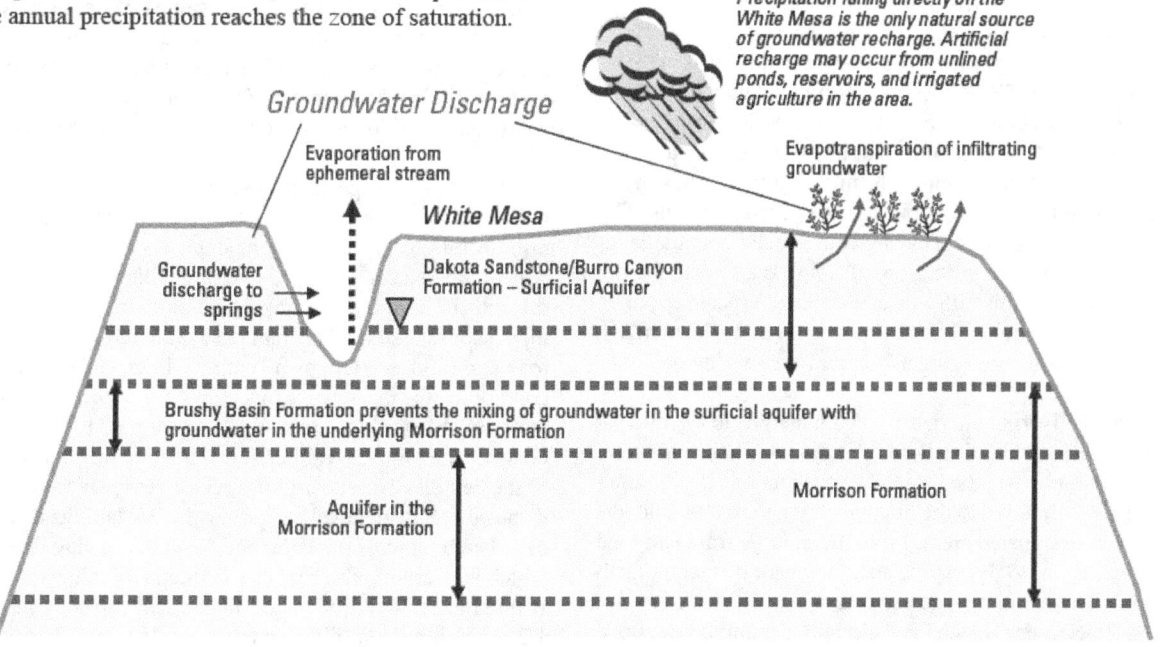

Figure 3. Conceptual model of the near-surface principal aquifers and occurrence of discharge and recharge on White Mesa, San Juan County, Utah.

calculated using permeability data measured only on mill property. Therefore, the calculated groundwater travel times potentially are not an accurate measurement of the time it takes groundwater to travel from the mill to the reservation. In this study, a different approach was taken to estimate groundwater travel times that used measurements of concentrations of noble gases and the isotopes, tritium and helium-3. These measurements are used to calculate the time elapsed since water infiltrated the aquifer and arrived at the sampling location, either a well or a spring. As a result, this method accounts for differences in permeability along the groundwater flow path. The results of this sampling are discussed in the "Noble Gases and Tritium/Helium-3" subsection of the "Results and Discussion."

Mill Operations

Production Circuit

The White Mesa uranium mill was originally designed for a capacity of 1,500 dry tons per day, but the capacity was boosted to the present rated design of 1,980 dry tons per day prior to commissioning (U.S. Department of Energy, 2005). Mill operations are periodic, and the periods of mill operation have been as follows:

May 6, 1980–February 4, 1983: 1,511,544 tons of ore and other materials were processed

October 1, 1985–December 7, 1987: 1,023,393 tons were processed

July 1988–November 1990: 1,015,032 tons were processed

August 1995–January 1996: 203,317 tons were processed

May 1996–September 1996: Processed 3,868 tons of calcium fluoride material

Since early 1997: The mill has processed over 100,000 tons from several additional feed stocks.

From November 1999 to April 2002 the mill was in standby status (INTERA, 2006). During this time, the mill received and stockpiled alternate feed materials. From April 2002 to May 2003, 266,690 tons of alternate feed materials were processed. Subsequently, the mill returned to standby mode but continued to stockpile alternate feed materials. The mill is currently operating, having commenced operations in March 2005, with the processing of Cameco alternate feed materials. During this mill run, additional alternate feed materials currently in stockpile will also be processed. The mill began processing conventionally mined ores during the first quarter of 2008.

Trucks delivering alternative feed materials to the mill arrive at the Blanding Ore Buying Station and drive up on large scales to be weighed (U.S. Department of Energy, 2005). The trucks then move to the buying station yard and unload their ore in designated areas. From there, large front end loaders move the ore to the buying station, where it is temporarily stockpiled on an ore storage pad that covers an area of approximately 8 hectares. The pad is underlain by compacted, mostly fine-grained material. Crushed limestone was reported to have been incorporated into the pad at the time of construction. The surface of the pad is sloped to promote drainage and prevent offsite movement of drainage. The alternative feed materials are temporarily staged until a sufficient quantity is received to run the mill. The period that materials are stockpiled varies but is typically about 2 years. Feeds currently stored on the site in piles typically cover an area of approximately 0.04 to 0.61 hectares and often merge. Pile thicknesses vary but can exceed 9 m.

Leaching U from crushed ore requires treating the ore with heat, a strong acid (sulfuric acid), and an oxidant (sodium chlorate). The resulting solution is referred to as "pregnant liquor." To extract U dissolved in the pregnant liquor, kerosene is added, which concentrates U in the organic phase. The organic and aqueous phases of this mixture separate (U laden kerosene floats to the top of the solution), after which the U is extracted from the kerosene by the addition of acidified brine. The U is precipitated from the acidified brine solution using ammonia, air, and heat. To complete the U extraction process, the precipitated U is dried at approximately 650°C, which dewaters the U oxide and burns off any additional impurities as well (International Uranium Corp., 2010).

Tailings Circuit

The Dakota Sandstone is the uppermost strata in which the tailings disposal cells are sited (Titan, 1994). The tailings facilities at White Mesa mill consist of four cells. Cell 1 is constructed with a 3.0-cm thick PVC earthen-covered liner and is used to store the process solution. Cell 2 is constructed with a 3.0-cm thick PVC earthen-covered liner and is used to store the barren tailings sands. Cell 3 is constructed with a 3.0-cm thick PVC earthen-covered liner and is used to store the barren tailings sands and solutions. Seams in the liner for Cell 4A were compromised as a result of thermal stress from years of exposure to full sunlight. Because of sunlight damage to the liner material in Cell 4A that started in the 1990s, relining of Cell 4A began in 2007, which now provides an additional 2 million tons of tailings capacity (Denison Mines, 2010).

Wet tailings disposal cells store slurried tailings, and dry tailings disposal cells store low-moisture-content tailings from mill operations (Titan Environmental Corporation, 1994). An engineered cap is placed over the tailings in the wet and dry cells to limit infiltration of precipitation. The wet tailings disposal cell has a 15-cm base/drainage layer of crushed rock and sand overlain by a synthetic liner. Under operational conditions, the tailings are placed within the cell as slurry; therefore, the tailings are completely saturated. The maximum depth of the tailings within the cell is three feet below the top of the cell dike (freeboard limit). The cap for the wet tailings disposal cell is identical to that for the dry tailings disposal cell. The bottom of the latter cell has a 0.3-m clay layer base, which is overlain with a synthetic liner. Dry tailings are placed within the cell over the liner. The dry cell cap consists of a 1.2-m thick random-fill base layer overlain by 0.3 m of clay, 0.3 m of filter material (capillary break), 1.1 m of random fill (protective layer), and a rock cover.

As a zero permitted discharge facility, the White Mesa mill must evaporate all of the liquids used during processing (Titan Environmental Corporation, 1994). This evaporation takes place in two areas: Cell 1, which is used for solutions only, and Cell 3, in which tailings and solutions exist. The original engineering design indicated that a net water gain to the cells would occur during mill operations. In addition to natural evaporation, spray systems occasionally have been used to enhance evaporation rates and control dust. To minimize net water gain, solutions are recycled from the active tailings cells to the maximum extent possible. Solutions from Cells 1 and 3 are brought back to the counter current decantation circuit, where additional extraction can be realized. Recycling to other parts of the mill circuit is not feasible because of the acid content of the solution.

Ongoing tailings reclamation occurs through the following processes. As each tailings cell is filled with tailings, solutions are separated from tailings solids and pumped to the evaporation pond. Tailings solids are allowed to dry in place. As each cell reaches final capacity, reclamation will begin with the placement of interim cover over the tailings. As additional cells are excavated, the overburden is used to reclaim previous cells. This sequential reclamation process is intended to reduce total reclamation time as well as reduce potential for adverse effects to human health and the environment.

An overview of mill operations has lead to the identification of a few potential exposure pathways of heavy metals from the mill to tribal members (fig. 4). These air and groundwater exposure pathways of U and other metals to tribal members include (1) airborne dust from ore storage pads and trucks delivering ore to the mill, as well as emissions from the mill's drying ovens; (2) dissolution of airborne dust deposited on the soil; and (3) leakage from the tailings ponds to the groundwater aquifer, which flows from the mill toward the reservation.

Results and Discussion

This section presents the data collected during the investigation and the interpretive results. The first topic presented and discussed is hydrology, which includes age dating and observed changes in water levels during the study period. Next, water-rock interaction is discussed, which describes the primary geochemical processes controlling groundwater quality in the study area. The two sections following that describe trace-element concentrations and distributions, and uranium mobility in the aquifer systems within the study area. After that is a discussion of the isotope geochemistry of uranium, oxygen, hydrogen, and sulfur and how these isotopes can help to identify contaminant and recharge sources to the groundwater. The final two topics are the concentration of trace elements in sediment samples associated with ephemeral drainages and vegetation samples adjacent to the mill site.

Hydrology

Noble Gases and Tritium/Helium-3

Dissolved-gas samples were collected and analyzed to evaluate the groundwater recharge temperature. Most noble gases that are dissolved in groundwater originate in the atmosphere. As water recharges the aquifer, it becomes isolated from the atmosphere, and the dissolved-gas concentrations are "fixed" on the basis of solubility relative to temperature, pressure, and salinity at the water table (Aeschbach-Hertig and

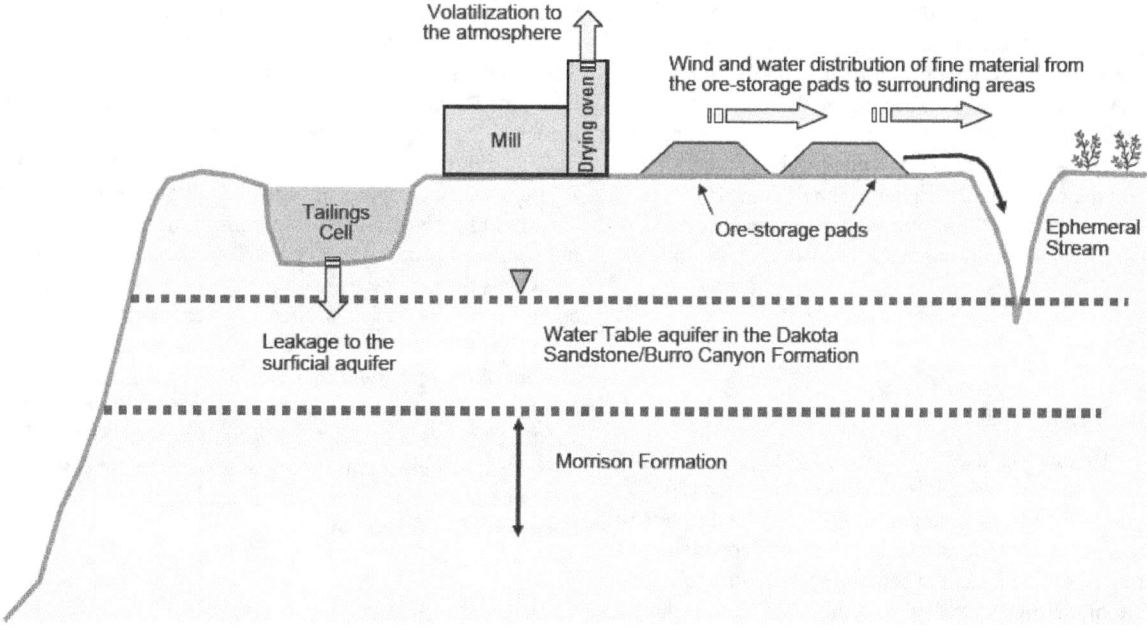

Figure 4. Potential sources of contamination from the mill site to surrounding areas.

others, 1999; Ballentine and Hall, 1999; Stute and Schlosser, 2001). Because these gases are generally nonreactive along flow paths in the subsurface, their dissolved concentrations measured in groundwater at points of discharge (wells and springs) provide a record of physical conditions (temperature and pressure) that reflect the altitude of the ground-water recharge location. For this study, dissolved concentrations of N_2, ^{40}Ar, ^{84}Kr, ^{20}Ne, and ^{129}Xe were used in the closed system equilibration model (Aeschbach-Hertig and others, 2000; Kipfer and others, 2002) to calculate estimated groundwater recharge temperature, pressure, excess air, and a fractionation factor (related to the partial dissolution of trapped air bubbles). Because there are five known parameters (the individual dissolved-gas concentrations) and four unknowns, this is an over-determined problem that can be solved (optimized) with a system of linear equations.

The dissolved-gas concentrations of N_2, ^{40}Ar, ^{84}Kr, ^{20}Ne, ^{4}He, and ^{129}Xe in groundwater are listed in table 10. Estimated most-probable recharge temperatures for wells completed in (East, West, and Millview wells; table 1) and springs (Entrance, Oasis, Ruin, and Cow Camp Springs) emanating from the shallow Dakota Sandstone and Burro Canyon aquifers range from 14 to 20°C. This temperature range is coincident with identified sources of local recharge through infiltration of precipitation or artificial recharge by wildlife refuge ponds adjacent to the mill site on White Mesa. Estimated most-probable recharge temperatures for two wells completed in the Navajo Sandstone aquifer (north and south wells) range from 8 to 9°C, and likely indicate water originating from higher elevations, such as known recharge areas near the Abajo Mountains northwest of the study area, not at altitudes common to the White Mesa area.

Tritium (^{3}H) is a radioactive isotope of hydrogen that decays to tritiogenic helium-3 ($^{3}He_{trit}$) and has a half-life of 12.3 years. Tritium is produced in the upper atmosphere and occurs naturally in precipitation at concentrations of less than about 8 tritium units (TU) in northern Utah (International Atomic Energy Agency, 2010). Testing of above-ground thermonuclear weapons in the 1950s and 1960s was the source for ^{3}H concentrations in precipitation, which peaked at more than 1,000 TU in the northern hemisphere. The ratio of ^{3}H to $^{3}He_{trit}$ yields the apparent age (time since recharge occurred) of a groundwater sample according to the following equation:

$$t = \lambda^{-1} \ln(({}^{3}He_{trit}/{}^{3}H)+1) \qquad (6)$$

where

  t is the apparent age in years, and
  λ is the ^{3}H decay constant of 0.0563 per year.

The $^{3}H/^{3}He$ method, used to date water younger than about 50 years, is explained in detail by Solomon and Cook (2000).

The age derived from equation (6) reflects mixed waters of different ages and, for that reason, is called the "apparent age" of a sample. Note that a sample containing a mixture of modern and pre-modern water (where "modern" refers to recharge that occurred during or after the period of above-ground

nuclear testing and "pre-modern" refers to recharge occurring before that time), however, always will appear to have the age of the modern fraction because dilution with pre-modern water does not change the ratio of ^{3}H to $^{3}He_{trit}$. The amount of mixing between modern and pre-modern recharge water can be determined with mixing curves using historic concentrations of tritium in rainfall.

$^{3}H/^{3}He$ age data for water sampled in the White Mesa area range from recent, or "modern," to very old, as indicated by the presence of elevated amounts of $^{4}He_{terr}$ derived from the decay of uranium to thorium over long periods (table 10). Samples from wells finished in the shallow Dakota Sandstone/Burro Canyon aquifer had apparent ages greater than 50 years (East, West, and Millview wells). Analysis of samples from wells finished in the Navajo Sandstone aquifer yielded ages greater than 50 years (North and South wells), and had elevated levels of terrigenic ^{4}He compared to other sites. Water from Cow Camp Spring had an apparent age of 12 to 19 years. Both Oasis and Entrance Springs had water with recent apparent ages (fig. 5). Cow Camp Spring was the only site that yielded $^{3}He_{trit}$, or dissolved helium derived from tritium decay, which allowed for calculation of apparent age by using the ratio of $^{3}He_{trit}$ to ^{3}H in the water. Sites categorized as "recent" have detectable amounts of ^{3}H but no $^{3}He_{trit}$, which results in a calculated apparent age equal to zero.

The apparent age and probable recharge temperatures of water derived from wells completed in the Dakota/Burro Canyon aquifer suggest that the aquifer is locally recharged by precipitation and that lateral water movement in the aquifer is low, given the isolated geographic conditions present on White Mesa. The apparent age of Entrance Spring could indicate a localized and possibly induced flow path from artificial recharge. A potential source for this artificial recharge includes infiltrating water from the unlined wildlife refuge ponds located to the northeast of the mill site and irrigated agriculture surrounding Blanding, Utah. This possibility is justified further by data presented in Hurst and Solomon (2008), who found measurable levels of ^{3}H in monitoring wells surrounding the wildlife refuge ponds within the mill site, indicating infiltration from the wildlife ponds. Other shallow wells located on White Mesa have apparent ages that are greater than 50 years and are indicative of areas where infiltration by precipitation is the dominant source of recharge. Two sites, Cow Camp Spring and Oasis Spring, have apparent ages of very recent (1990s) and modern, respectively. These sites are both located farther from the mill site and wildlife ponds than Entrance Spring and are likely recharged by water derived from precipitation and localized stream flow paths associated with the ephemeral stream channels in which they occur. Entrance Spring discharges in an ephemeral stream channel also but is within the area where water-levels are influenced by the wildlife ponds (Denison Mines, 2008; fig. 8).

Table 10. Dissolved-gas, recharge temperature, and tritium/helium–3 data for groundwater and spring water near White Mesa, Utah.

[Abbreviations mm/dd/yyyy, month/day/year; cm³STP/g, cubic centimeters per gram at standard temperature and pressure; °C, temperature in degress Celsius; TU, tritium units; R/Ra, Measured ^{3}He ^{4}He isotopic ratio relative to the helium isotopic ratio of air; ^{4}He$_{terr}$, terrigenic helium; ^{3}He$_{trit}$, tritiogenic helium; TU, tritium units; ND, no data; NC, not calculated; <, less than; >, greater than]

Site name	Sampling date (mm/dd/yyyy)	Sampling altitude (meters)	Dissolved gases						Excess air (cm³STP/g)	Most probable recharge temperature (°C)
			Nitrogen (cm³STP/g)	Argon-40 (cm³STP/g)	Krypton-84 (cm³STP/g)	Xenon-129 (cm³STP/g)	Neon-20 (cm³STP/g)	Helium-4 (cm³STP/g)		
Cow Camp Spring[1]	09/19/2007	1,511	1.25E–02	3.40E–04	4.30E–08	3.22E–09	1.61E–07	4.27E–08	0.103	9
Entrance Spring[1]	09/20/2007	1,691	9.62E–03	2.53E–04	2.98E–08	2.26E–09	1.35E–07	3.68E–08	0.100	20
Millview well[1]	09/18/2007	1,772	1.14E–02	2.84E–04	3.73E–08	2.65E–09	1.46E–07	3.90E–08	0.101	14
Oasis Seep[1]	09/19/2007	1,905	9.75E–03	2.45E–04	3.20E–08	2.18E–09	1.29E–07	3.28E–08	0.100	19
South well[2]	09/11/2007	1,615	ND	3.92E–04	8.36E–08	1.21E–08	2.69E–07	1.38E–07	0.006	9
North well[2]	09/11/2007	1,612	ND	3.88E–04	8.45E–08	1.23E–08	2.54E–07	1.24E–07	0.005	8
East well[2]	09/11/2007	1,662	ND	2.76E–04	6.07E–08	8.61E–09	1.77E–07	3.90E–08	0.001	17
West well[2]	09/11/2007	1,664	ND	2.56E–04	5.96E–08	8.20E–09	1.49E–07	3.37E–08	0.001	18
Ruin Spring[2]	09/11/2007	1,644	ND	2.82E–04	6.33E–08	8.21E–09	1.87E–07	4.18E–08	0.001	19
Cow Camp Spring[2]	09/19/2007	1,511	ND	2.88E–04	6.47E–08	9.46E–09	1.57E–07	3.57E–08	0.001	14

Site name	Sampling date (mm/dd/yyyy)	^{3}H (TU)	R/Ra	^{3}He/^{4}He	^{4}He$_{terr}$ (cm³STP/g)	^{3}He$_{trit}$ (TU)	Calculated Initial ^{3}H (TU)	Apparent ^{3}H/^{3}He age (years)	Apparent recharge year
Cow Camp Spring[1]	09/19/2007	5.3	1.369	1.89E–06	2.06E–09	9.9	15.2	18–19	1990
Entrance Spring[1]	09/20/2007	4.2	0.957	1.32E–06	5.07E–10	0.0	4.2	modern	Recent
Millview well[1]	09/18/2007	<0.3	0.927	1.28E–06	1.33E–09	0.0	<0.3	>50	Pre–1950s
Oasis Seep[1]	09/19/2007	3.6	0.992	1.37E–06	–2.62E–09	0.0	3.6	modern	Recent
South well[2]	09/11/2007	<0.3	0.492	6.80E–07	7.06E–08	ND	NC	>50	Pre–1950s
North well[2]	09/11/2007	<0.3	0.519	7.19E–07	6.07E–08	ND	NC	>50	Pre–1950s
East well[2]	09/11/2007	<0.3	0.963	1.33E–06	ND	ND	NC	>50	Pre–1950s
West well[2]	09/11/2007	0.5	0.978	1.35E–06	ND	ND	NC	>50	Pre–1950s
Ruin Spring[2]	09/11/2007	<0.3	0.971	1.34E–06	ND	ND	NC	>50	Pre–1950s
Cow Camp Spring[2]	09/19/2007	5.6	1.260	1.74E–06	ND	ND	11.0	[3]12	1995

[1] Sample analysis performed at the University of Utah Noble Gas Laboratory
[2] Sample analysis performed at Lawrence Livermore National Laboratory
[3] Lawrence Livermore National Laboratory reported value

Water Levels

Non-vented and vented pressure transducers were installed at the West and East wells, respectively (fig. 5), and data were logged hourly from late December 2007 to late April 2009 and late December 2007 to late September 2009, at the respective wells, except when equipment malfunctioned, resulting in periods of missing data. Also, barometric pressure was logged hourly at the West well, and subtracted from the water pressure transducer data logged at the site. Water pressure readings were transformed to water level, in feet below land surface datum (lsd), and verified with measurements of water level made in the field with an electronic tape. Water levels measured at each well varied by less than 0.37 m during the monitoring period and ranged from 25.60 to 25.93 m below lsd at the West well and from 16.67 to 17.03 m below lsd at the East well (Figs. 6 and 7). From late December 2007 to late June 2008 there was a slight trend toward increased water levels, on the order of 0.06 m, at both wells. On the basis of data from the West well, this trend appeared to level off by August 2008, and, on the basis of data from the East well, it was not repeated the following year.

Water-level fluctuations measured at the West and East wells are strongly correlated (fig. 8), indicating that the wells are screened in the same aquifer (Dakota aquifer). The relatively minor water-level fluctuations observed at both wells support the interpretation by consultants and others that the Dakota aquifer in the vicinity of the mill is perched and isolated from significant recharge from high-elevation precipitation and perennial streams (Titan Environmental Corp., 1994; Intera, Inc., 2006; Denison Mines Inc., 2008). If water levels in the wells were influenced by high-elevation precipitation and perennial streams, one would expect to see clear trends, such as increased water levels in response to precipitation events or seasonal factors, such as infiltration of snowmelt. The minor increase in water levels from December 2007 to June 2008 seen in figures 6 and 7 could be related to greater than normal precipitation that was measured on White Mesa at Blanding, Utah, from December 2007 to February 2008 (Fig. 9; National Oceanic and Atmospheric Administration, 2010). During this period, 24.1 cm of precipitation was measured, which is nearly 160 percent greater than normal.

Despite the use of a vented transducer at the East well, and subtracting barometric pressure from pressure values logged by the non-vented transducer in the West well, there is a strong correlation between water levels measured at both wells and barometric pressure measured at the West well (fig. 8). This could indicate relatively high barometric efficiency, suggesting that the Dakota aquifer is semi-confined in the vicinity of the mill (Fitts, 2002). This is consistent with the low porosity and hydraulic conductivity values associated with the Dakota aquifer (Freethey and Cordy, 1991; Denison Mines Inc., 2008).

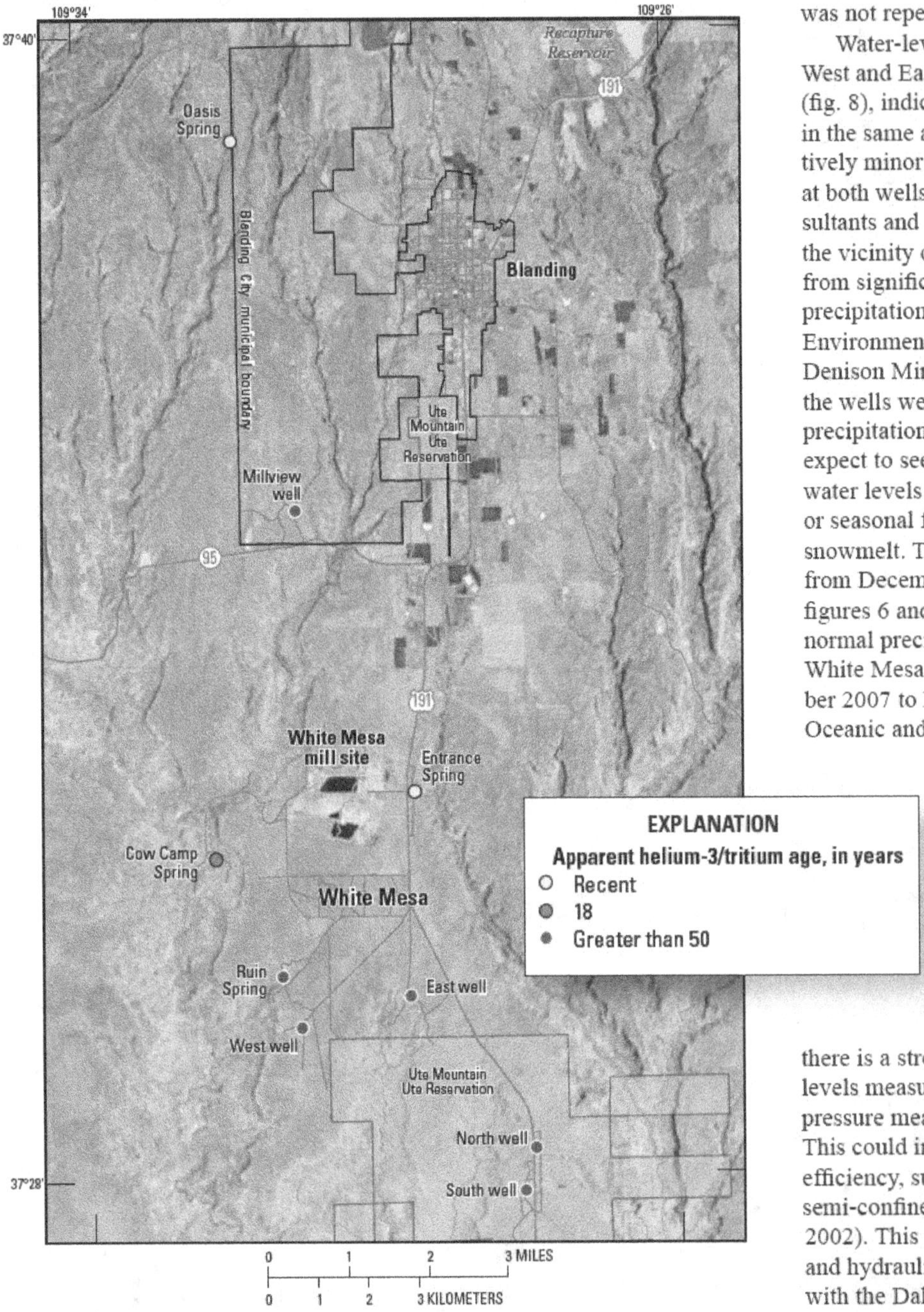

Figure 5. Apparent ages of water samples collected from wells and springs surrounding the White Mesa mill site in southeastern Utah.

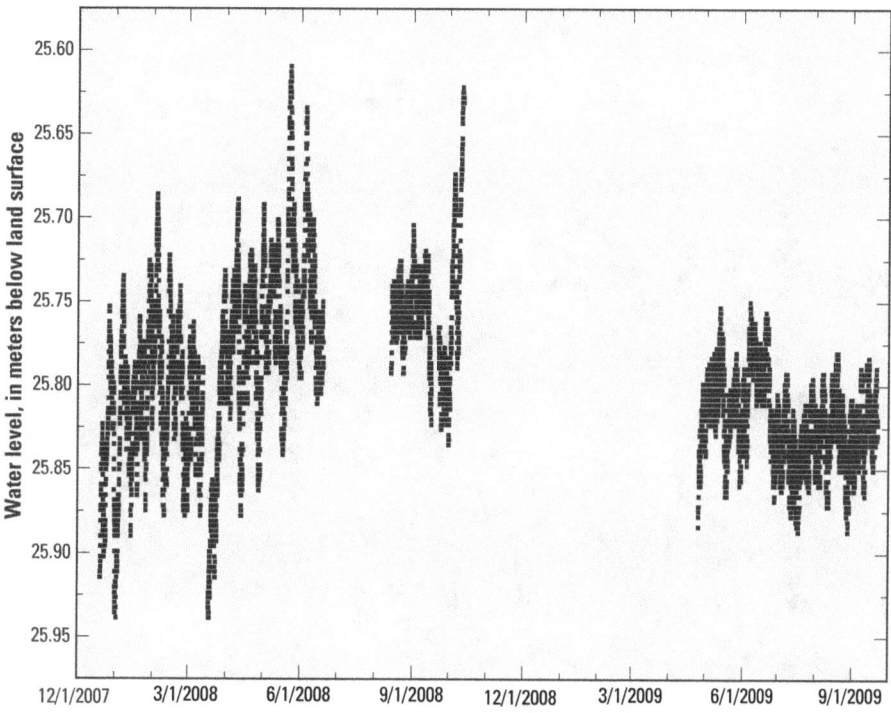

Figure 6. Hourly water levels measured in the West well from December 20, 2007, to September 22, 2009.

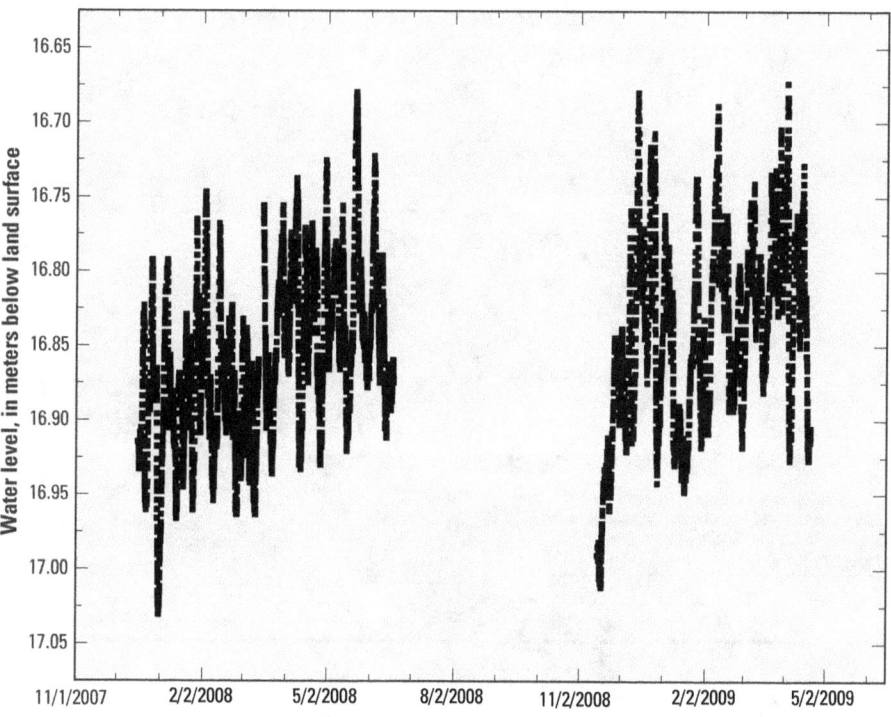

Figure 7. Hourly water levels measured in the East well from December 17, 2007, to April 21, 2009.

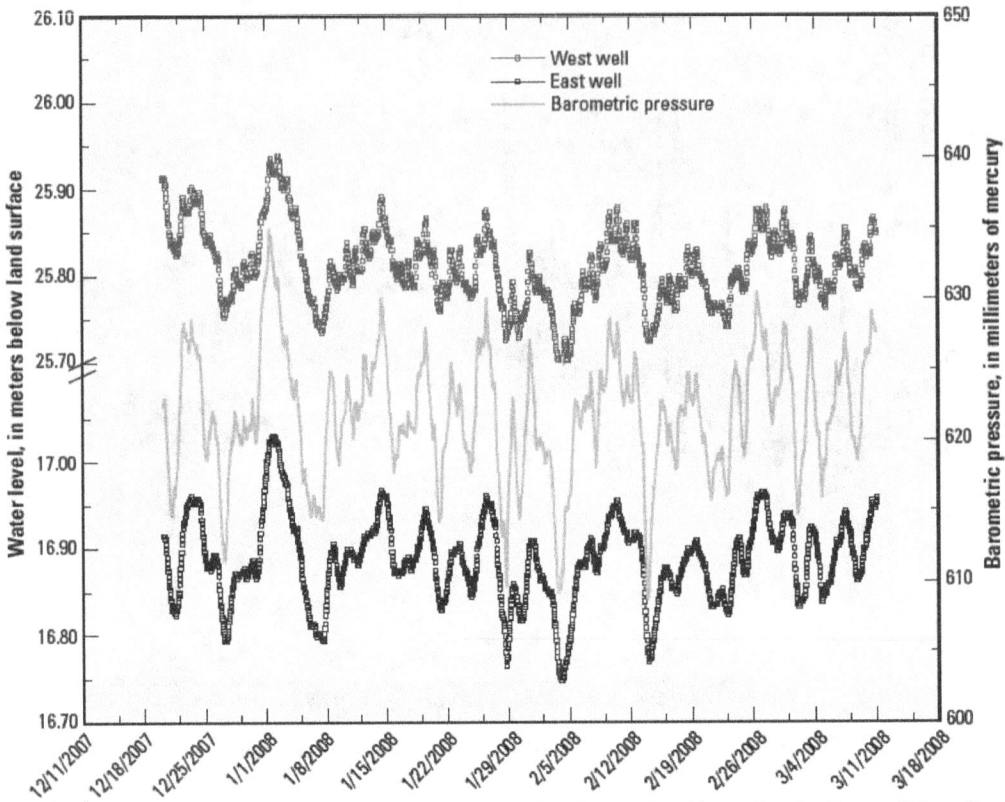

Figure 8. Water level and barometric pressure logged at the West well and water level logged at the East well from December 20, 2007, to March 11, 2008.

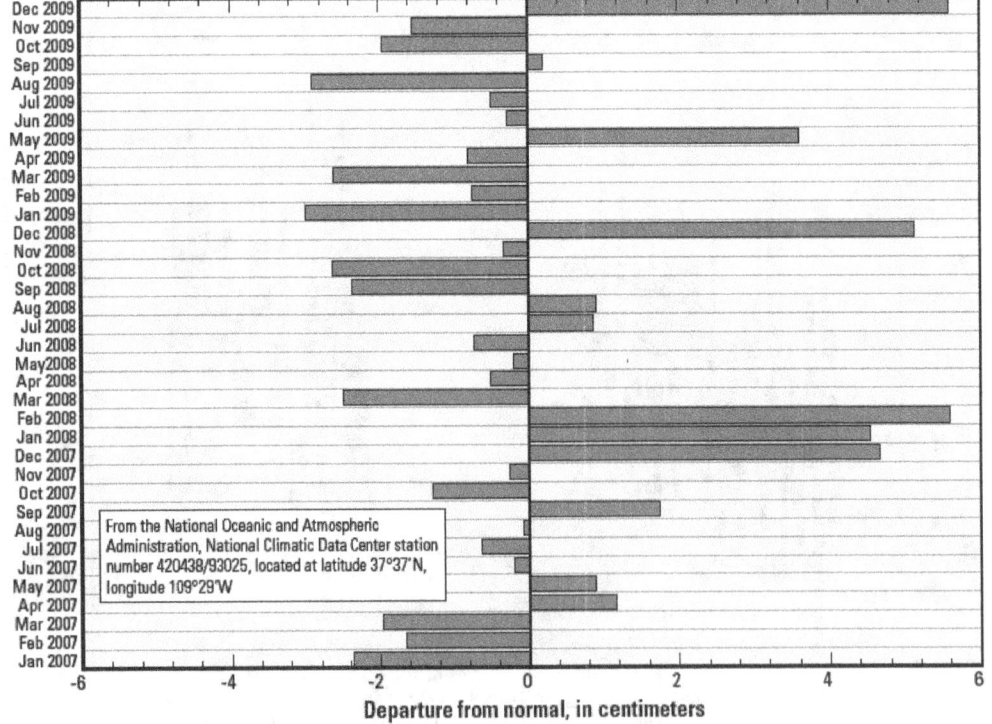

Figure 9. Monthly precipitation departure from normal, in inches, for Blanding, Utah, from January 2007 to December 2009.

Water-Rock Interaction

The mobility of U, if introduced from the mill into the unconfined aquifer in the Dakota Sandstone/Burro Canyon Formation, would be a function of the chemical composition of the groundwater in this aquifer. Therefore, in this section, a detailed analysis of the processes controlling the geochemistry of groundwater in this aquifer is undertaken using data presented in Appendix 1. In the "Uranium Mobility" section, the

information learned from this analysis is combined with the physical and chemical properties of U to evaluate the potential for U mobility in groundwater throughout the White Mesa.

Groundwater in the Dakota Sandstone/Burro Canyon Formation in the White Mesa is characterized by neutral pH, the presence of dissolved oxygen, and much greater spatial variability than temporal variability in the composition of major ions (figs. 10–13).

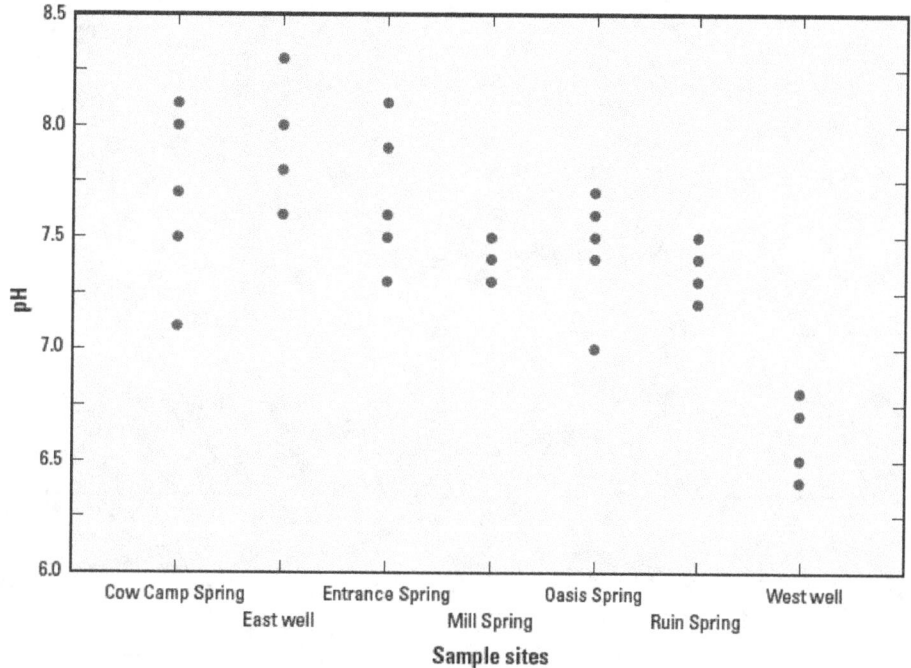

Figure 10.   pH of water samples collected from springs and wells in the vicinity of the White Mesa mill, San Juan County, Utah.

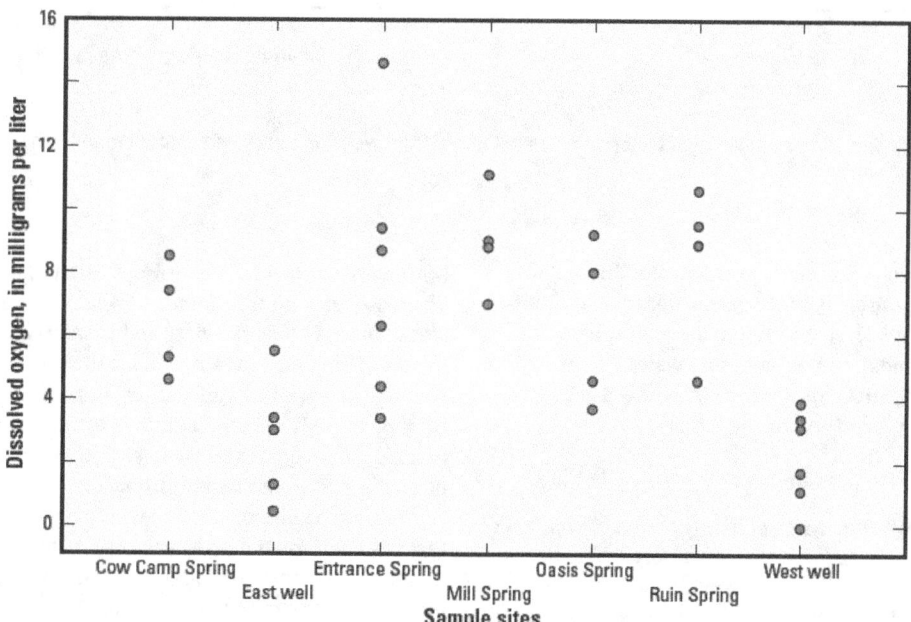

Figure 11.   Concentration of dissolved oxygen in water samples collected from springs and wells in the vicinity of the White Mesa mill, San Juan County, Utah.

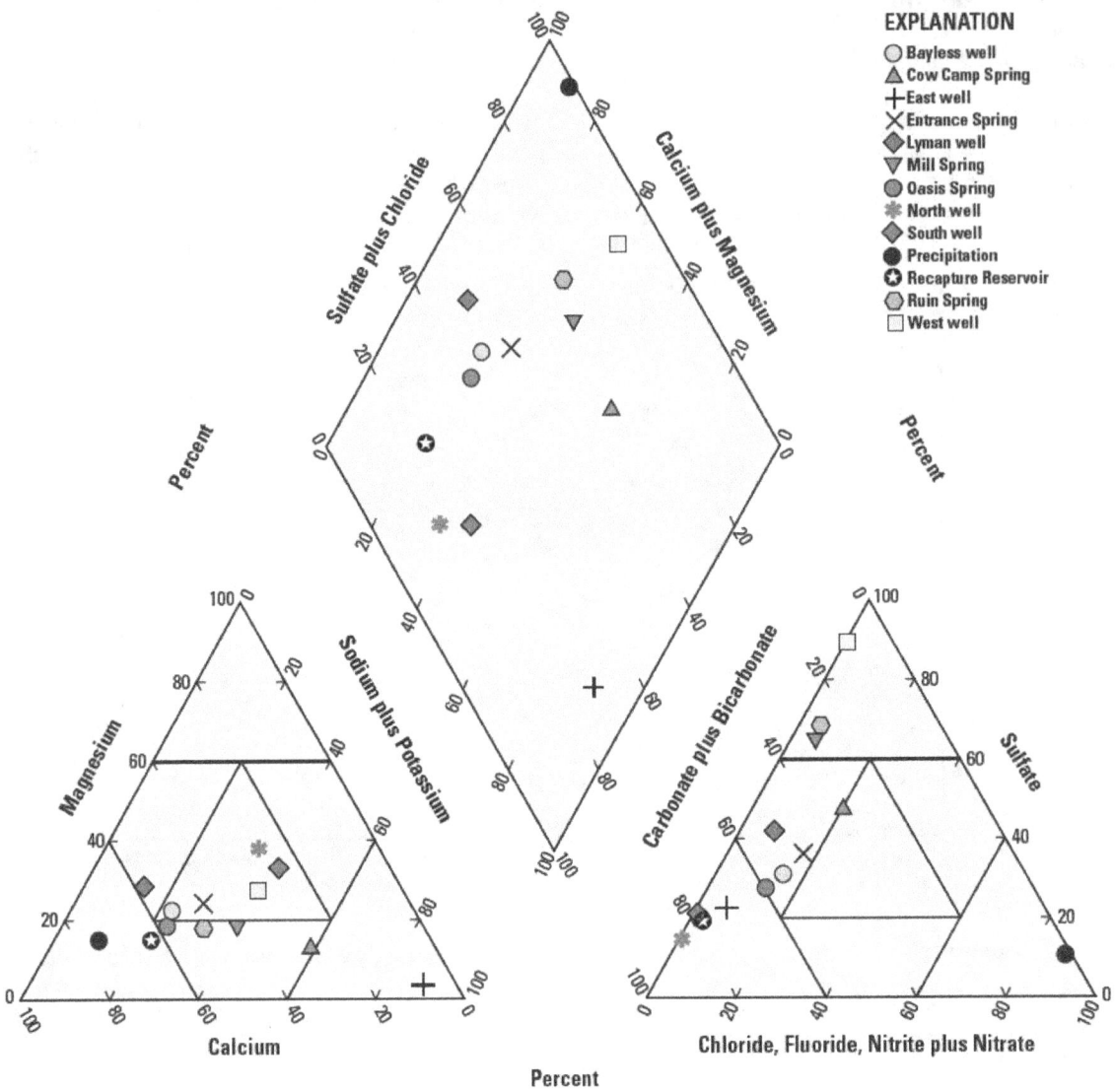

Figure 12. Average major-ion composition of water samples collected from wells and springs adjacent to the White Mesa mill, San Juan County, Utah.

Dissolved oxygen is present in groundwater throughout White Mesa because there is little organic matter in the soil (Hansen and Fish, 1993). As oxygen in the atmosphere infiltrates into the soil and dissolves in groundwater it comes into contact with soil organic matter, which is oxidized according to the following equation (Freeze and Cherry, 1979):

$$O_2(g) + CH_2O \text{ (simple carbohydrate)} = CO_2(g) + H_2O \quad (7)$$

Because soils on White Mesa contain so little organic matter, 0.5 to 2 percent, there is not enough organic matter to consume the oxygen present in the groundwater.

Piper Diagrams demonstrate that the major-ion compositions of water from the sampling sites form several groups (fig. 12). Samples from West well, Mill Spring, and Ruin

Spring are composed primarily of calcium and sulfate, whereas water from Entrance Spring, Oasis Spring, and the two domestic wells (Bayless and Lyman) are composed primarily of calcium, sulfate, and bicarbonate. Cow Camp Spring is composed primarily of sodium and sulfate. The two domestic supply wells are predominated bycomposed primarily of calcium and bicarbonate, whereas the East well is composed primarily of sodium and bicarbonate.

Average values of specific conductance (fig. 14) show that there is a great deal of variation in the concentration of dissolved ions that, for the most part, parallels the variation in the composition of major ions. For example, the two public supply wells (North and South wells) have the lowest values of specific conductance relative to the other sampling sites

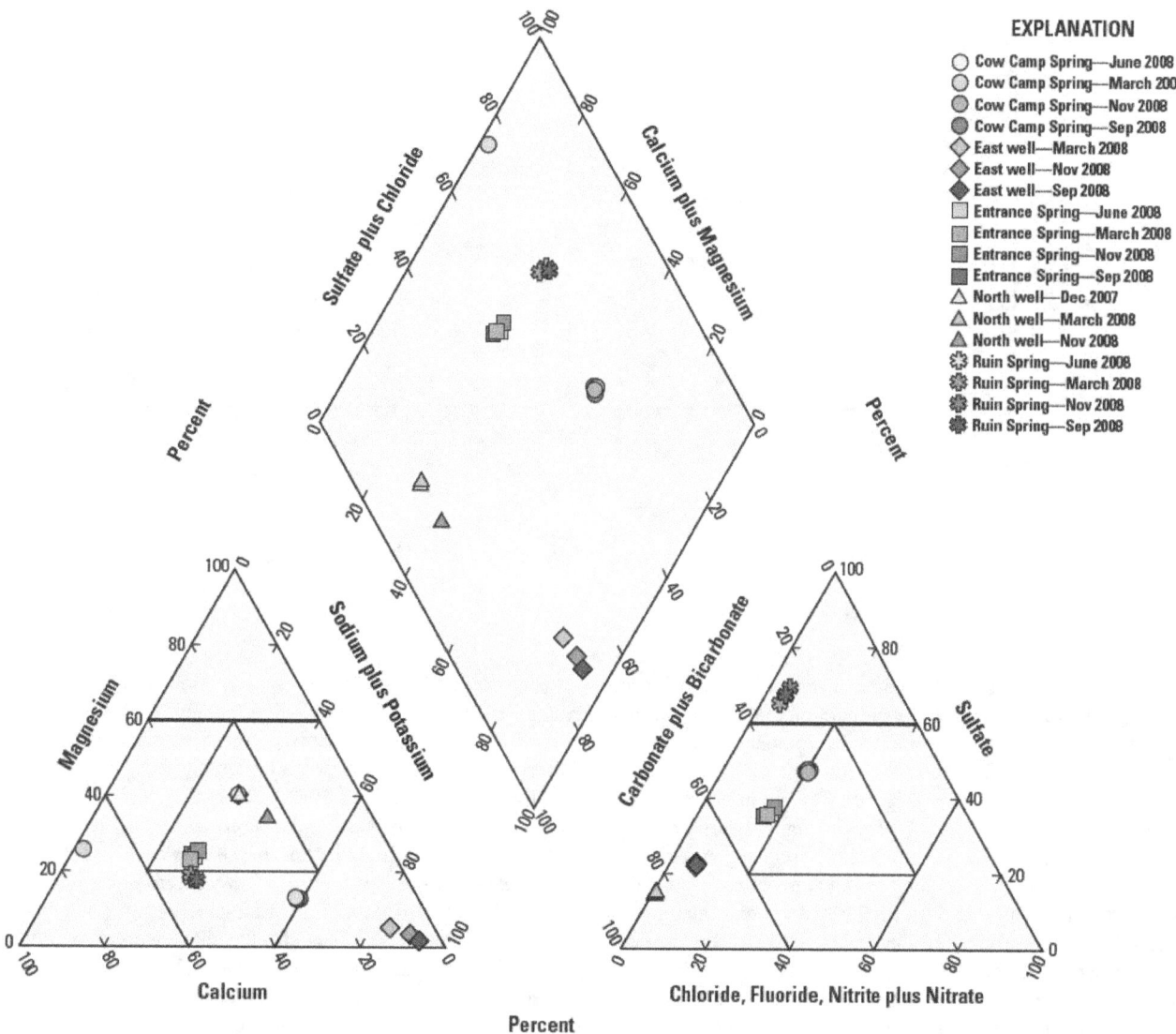

Figure 13. Seasonal changes in major-ion composition of water samples collected from wells and springs adjacent to the White Mesa mill, San Juan County, Utah.

(less than 500 microsiemens per centimeter (µS/cm)). The East well also has a relatively low value of specific conductance (624 µS/cm). Entrance Spring, Oasis Spring, and the two domestic wells (Bayless and Lyman wells) have very similar values of specific conductance and are also similar in major-ion composition. Mill and Ruin Springs have relatively high values of specific conductance, but the West well, while similar in major-ion composition to these two wells, has the highest average value of specific conductance measured in this study at 5,086 µS/cm. Cow Camp Spring also has a relatively high value of specific conductance (1,543 µS/cm) but falls between Mill Spring and Ruin Spring.

The process controlling the major ion chemistry of groundwater in the White Mesa can be a combination of (1) evaporative concentration due to the arid climate of the region and (2) weathering reactions between precipitation and the rocks composing the Dakota Sandstone and the Burro Canyon Formation. The effect of evaporation on the composition of water quality was evaluated by plotting the concentration of calcium, magnesium, sodium, bicarbonate, and sulfate as a function of the concentration of chloride (Kimball, 1981), and the effect of weathering reactions on groundwater chemistry was modeled using the Inverse Modeling function of the USGS Geochemical model PHREEQC (Parkhurst and Appelo, 2010).

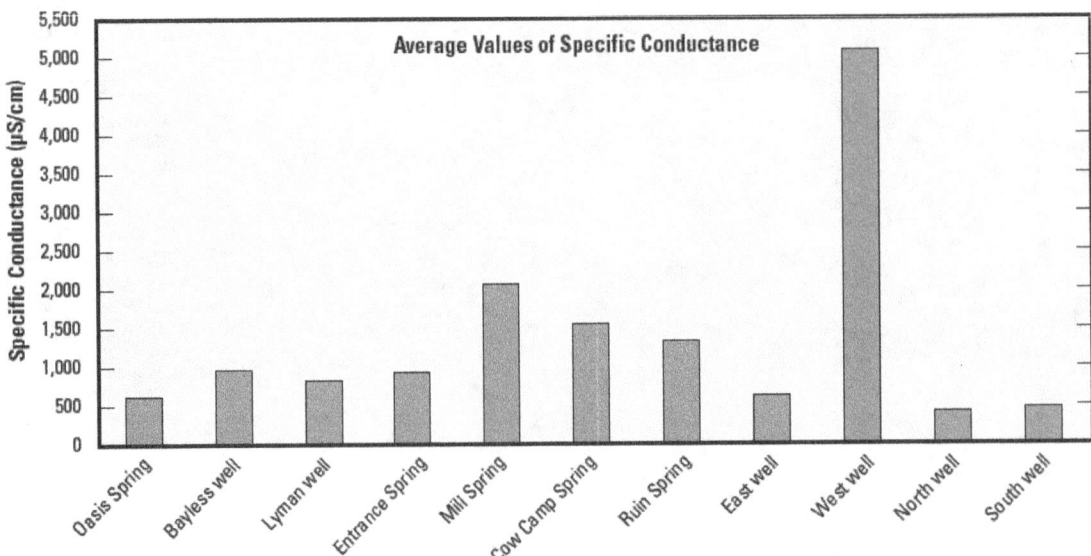

Figure 14.   Average values of specific conductance in water samples collected from springs and wells in the vicinity of the White Mesa mill, San Juan County, Utah.

The concentrations of calcium, magnesium, sodium, bicarbonate, and sulfate are plotted as a function of the concentration of chloride in figures 15 and 16. On these plots, the square represents the concentration of each ion in rainfall as measured at the National Atmospheric Deposition Program/National Trends Network (NADP/NTN) precipitation chemistry site in Canyonlands National Park in 2007 (http://nadp.sws.uiuc.edu/). The concentration of the bicarbonate ion was obtained from calculations used in the PHREEQC inverse modeling described below because the concentration of bicarbonate was not measured in the NADP/NTN analysis because of the low pH of the rainwater (5.21). A line of unit slope, or one-to-one concentration, is drawn from this point showing the path of simple evaporative concentration. If the major ions were not added or taken away from the groundwater, the concentration of the major ions would plot along this line (Kimball, 1981). Only results from the December 2007 and September 2008 sampling are shown in figures 15 and 16; however, similar results were obtained for each sampling event. Figures 15 and 16 show that the concentrations of magnesium, sodium, bicarbonate, and sulfate at all sites, and the concentrations of calcium and bicarbonate at all sites except the East well and Cow Camp Spring, exceed that expected from simple evaporative concentration of groundwater. These results indicate that mineral weathering reactions are the primary process controlling the major-ion composition of groundwater on White Mesa at all sites, with the possible exception of calcium at the East well and Cow Camp Spring.

The inverse modeling function of PHREEQC was used to quantify the weathering reactions controlling groundwater chemistry by allowing precipitation, as measured in 2007 from the NADP/NTN site at Canyonlands National Park,

to react with the minerals present in the Dakota Sandstone/Burro Canyon Formation at all sites, except for Bayless well, Lyman well, and Entrance Spring. For these 3 sites, Recapture Reservoir water was used instead of precipitation because, as discussed in the "Isotopes of Oxygen and Hydrogen" section, in this area of White Mesa, groundwater is recharged with water from Recapture Reservoir. In the area of the Bayless and Lyman wells, this is a result of water from Recapture Reservoir used for irrigation. At Entrance Spring, water from Recapture Reservoir is used to fill the wildlife ponds on mill property and leakage from those ponds recharges groundwater. On the basis of a literature review and the results of the mineralogical analyses of samples collected from several of the spring sampling sites, the following reactions were incorporated into the PHREEQC model:

Calcite Dissolution:

$$CaCO_3(s) + CO_2(g) + H_2O \leftrightarrow Ca^{2+} + 2HCO_3^- \qquad (8)$$

Dolomite Dissolution:

$$CaMg(CO_3)_2(s) + 2CO_2(g) + 2H_2O \leftrightarrow Ca^{2+} + Mg^{2+} + 4HCO_3^- \qquad (9)$$

Gypsum Dissolution:

$$CaSO_4 \cdot 2H_2O \leftrightarrow Ca^{2+} + SO_4^{2-} + 2H_2O \qquad (10)$$

Quartz Dissolution:

$$SiO_{2(quartz)} + 2H_2O = H_4SiO_4(aq) \qquad (11)$$

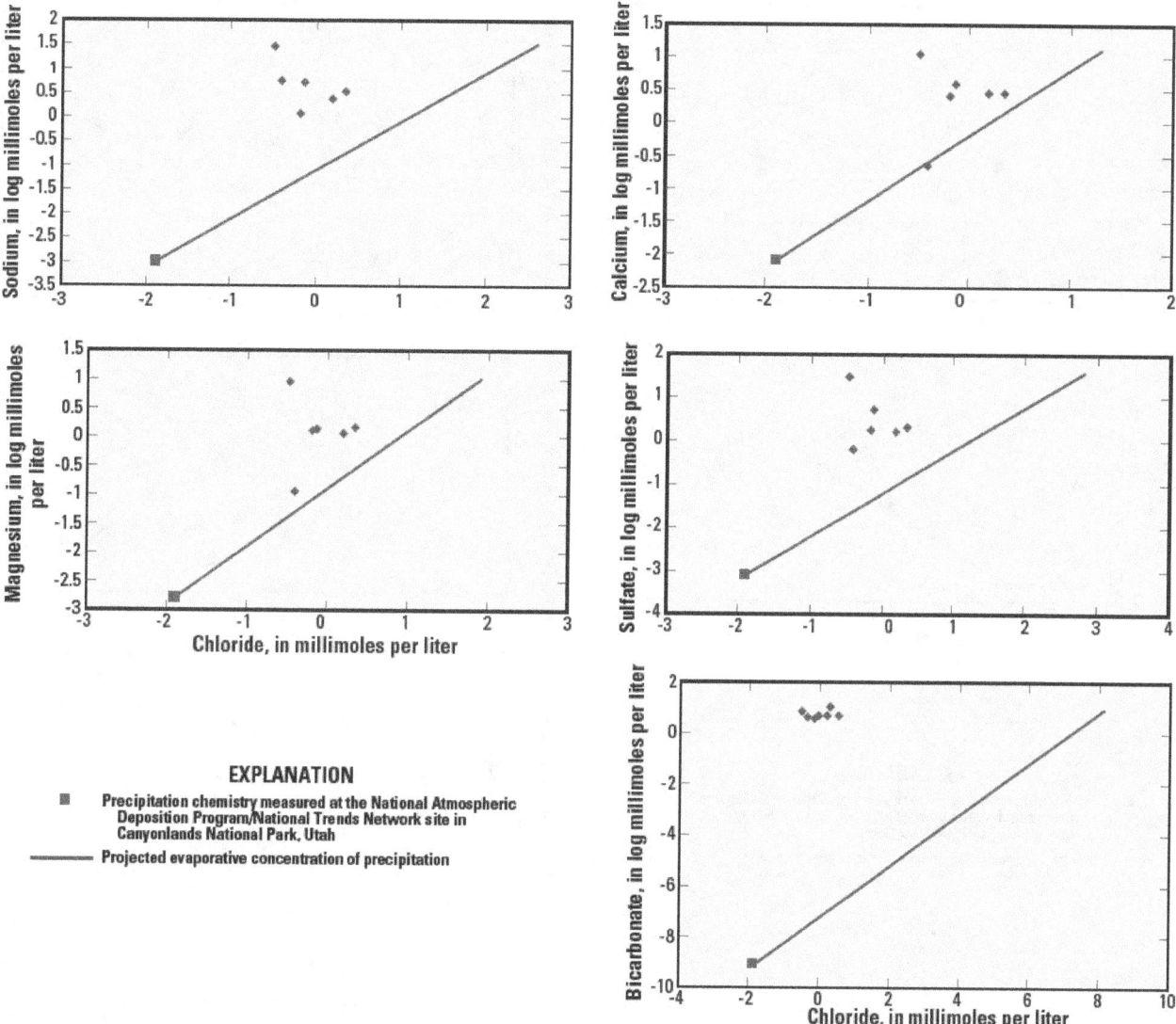

Figure 15. Concentration of sodium, calcium, magnesium, sulfate, and bicarbonate in water samples collected from springs and wells during December 2007 in the vicinity of the White Mesa mill, San Juan County, Utah, compared to evaporative concentration of precipitation.

Halite Dissolution:

$$NaCl \rightarrow Na^+ + Cl^- \qquad (12)$$

Incongruent Dissolution of Albite:

$$2NaAlSi_3O_8 + 2CO_2 + 11H_2O = 2Na^+ + 2HCO_3^- + 4H_4SiO_4 + Al_2Si_2O_5(OH)_4 \text{ (kaolinite)} \qquad (13)$$

Incongruent Dissolution of Orthoclase:

$$2KAlSi_3O_8 + 2CO_2 + 11H_2O = 2K^+ + 2HCO_3^- + 4H_4SiO_4 + Al_2Si_2O_5(OH)_4 \text{ (kaolinite)} \qquad (14)$$

Cation Exchange:

$$1/2Ca^{2+} + Na\text{-}X \rightarrow 1/2Ca\text{-}X_2 + Na^+ \qquad (15)$$

Quantifying the reactions to account for the difference in concentration between precipitation or Recapture Reservoir and groundwater at each sampling site requires a cumulative integration of all the reactions that occur as precipitation infiltrates to the water table and travels to the sampling site. So, for example, at an upgradient site like Oasis Spring, the results of the PHREEQC modeling show the kind and degree of weathering reactions that occur upgradient of Oasis Spring only. At a downgradient site like Ruin Spring, the PHREEQC models include the reactions occurring upgradient of Oasis Spring as well as reactions that occur between the two sites. This approach was considered to be an accurate representation of the reactions controlling groundwater chemistry in the unconfined aquifer for two reasons: first, precipitation

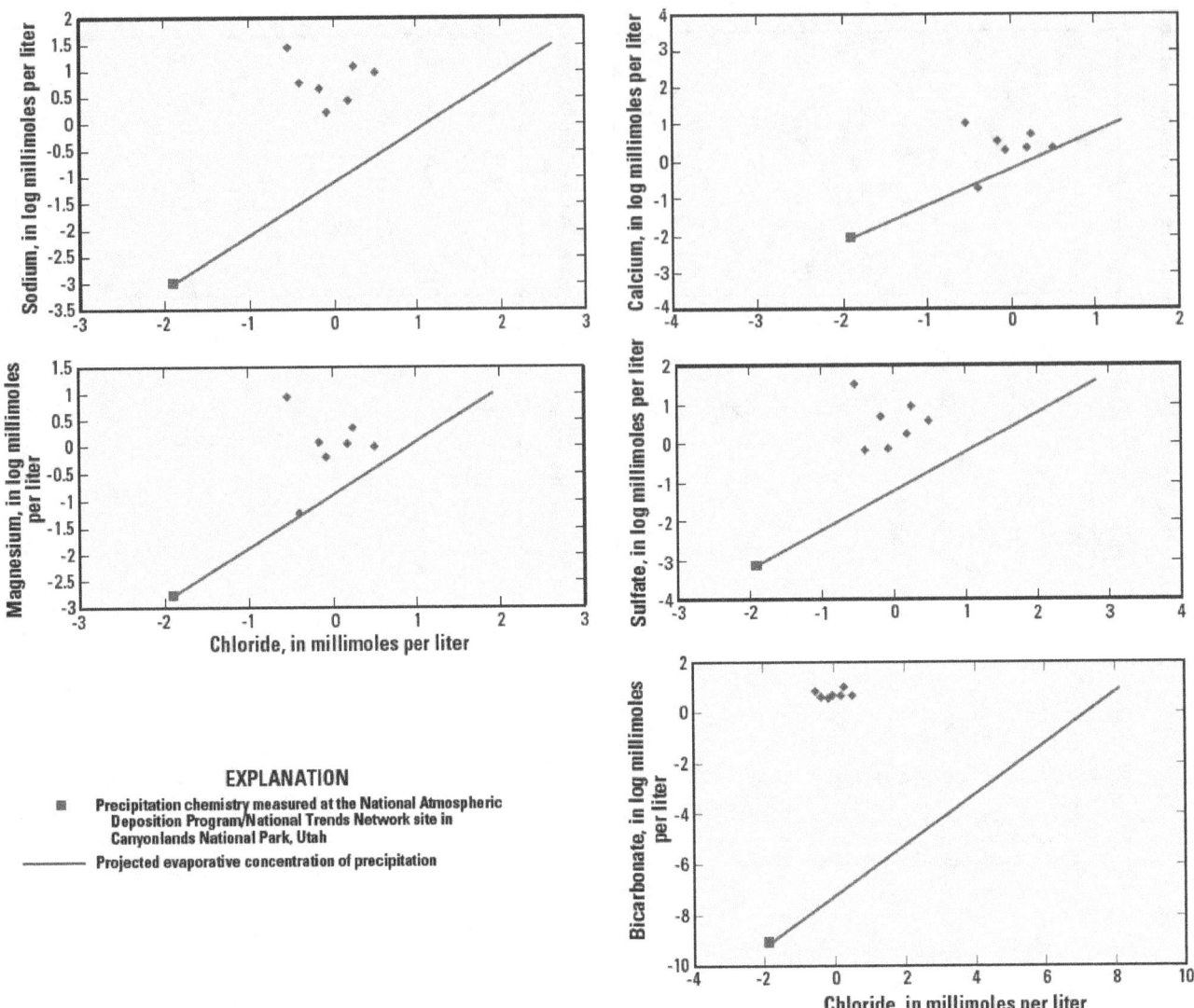

Figure 16. Concentration of sodium, calcium, magnesium, sulfate, and bicarbonate in water samples collected from springs and wells during September 2008 in the vicinity of the White Mesa mill, San Juan County, Utah, compared to evaporative concentration of precipitation.

and water from Recapture Reservoir are the only sources of groundwater recharge; second, a potentiometric surface map of the unconfined aquifer presented in Kirby (2008) shows that groundwater flows from the north end of the White Mesa toward the south. On a more local scale, potentiometric surface maps of mill property shown in unpublished quarterly monitoring reports prepared by the mill and in a consultant's report commissioned by the mill (Titan, 1994) show groundwater flowing from the mill south to the reservation. In the modeling, it was assumed the groundwater system is open to exchange with CO_2, and the partial pressure of CO_2 in equilibrium with precipitation was set at $10^{-3.5}$ atmosphere (atm).

The PHREEQC simulations compute several different models to account for the differences in chemistry between precipitation and groundwater at each site. It is up to the user to select the model which is the most valid on the basis of geochemical principles and knowledge of the geology and hydrology of the area. One factor in selecting the most appropriate model was the interpretation of saturation indices computed by PHREEQC for each mineral used in the inverse modeling function. Groundwater at all sites during each sampling event was saturated with respect to quartz and undersaturated with respect to halite, gypsum, albite, and orthoclase. The East well was undersaturated with respect to calcite and dolomite, but

Calcite

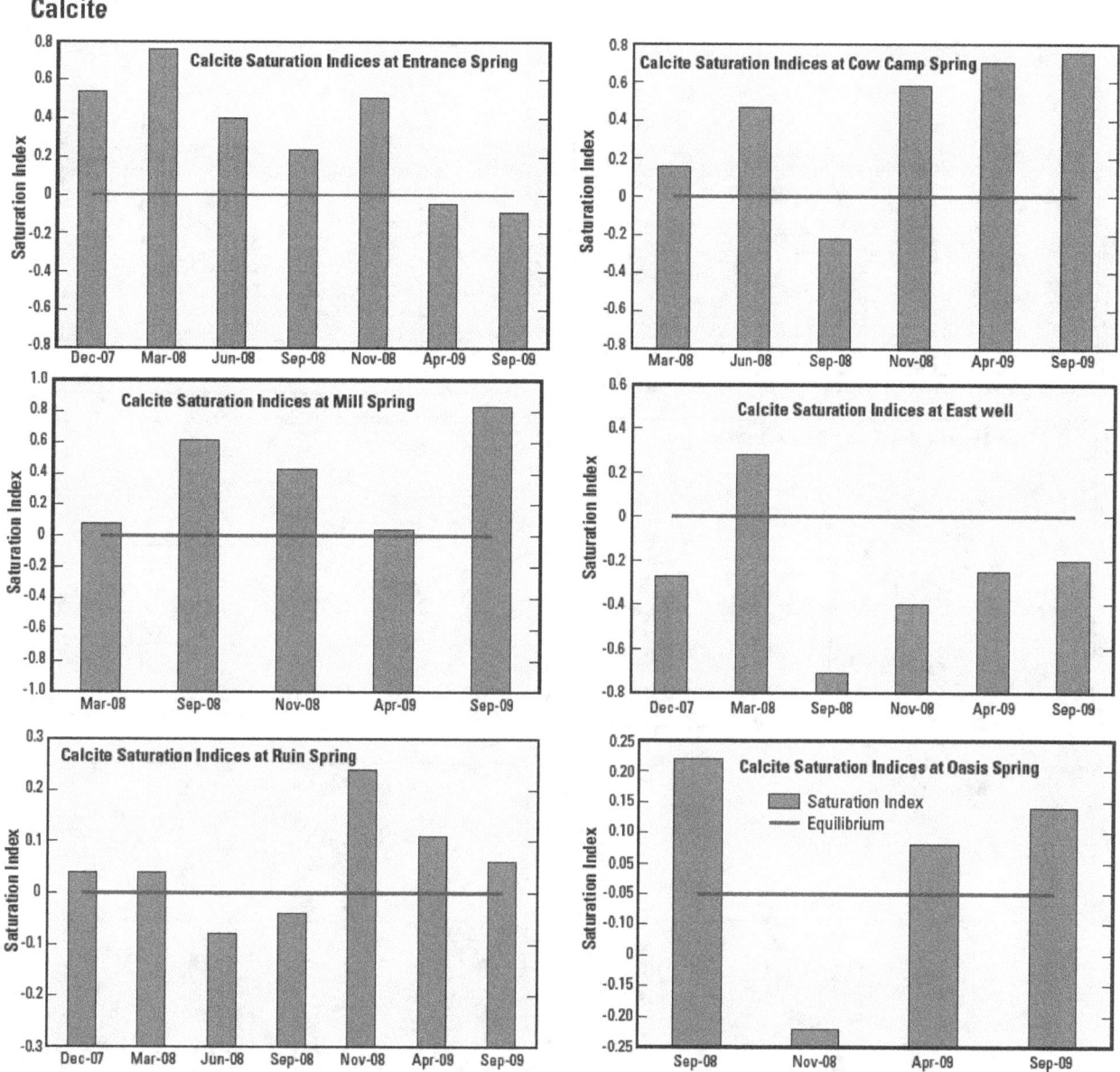

Figure 17. Saturation indices calculated for water samples from springs and wells surrounding the White Mesa mill, San Juan County, Utah, for calcite and dolomite.

the groundwater at all other sites was either in equilibrium, or saturated, with respect to calcite and dolomite. Recapture Reservoir water is saturated with respect to calcite, at equilibrium with respect to dolomite, and undersaturated with respect to the other minerals reacted in the PHREEQC Inverse Modeling. In evaluating the degree of saturation of the groundwater at each site with respect to calcite and dolomite, we assumed that values with the range of 0.0 ± 0.1 for calcite and 0.0 ± 0.4 for dolomite indicated equilibrium (David Parkhurst, USGS, personnel communication, 2010).

The temporal variability in the values of the saturation indices for calcite and dolomite at Oasis, Mill, Entrance, and Cow Camp Springs made determination of the degree

of saturation with respect to calcite and dolomite difficult (fig. 17). We suspect that the positive values of the calcite and dolomite saturation indices at these sites are a result of CO_2 degassing while the sample was collected, causing calcite precipitation, so that it appeared supersaturated when, in fact, the groundwater is undersaturated or in equilibrium with respect to calcite. We conclude this for several reasons. Calculations made by PHREEQC indicated that CO_2 concentrations in groundwater at all sites is an order of magnitude higher than atmospheric. Mill Spring, Entrance Spring, and Cow Camp Spring are sampled downstream from their source, and all flow as very shallow, slow rivulets of water, which would easily allow for the excess CO_2 to escape to the atmosphere. Oasis

Dolomite

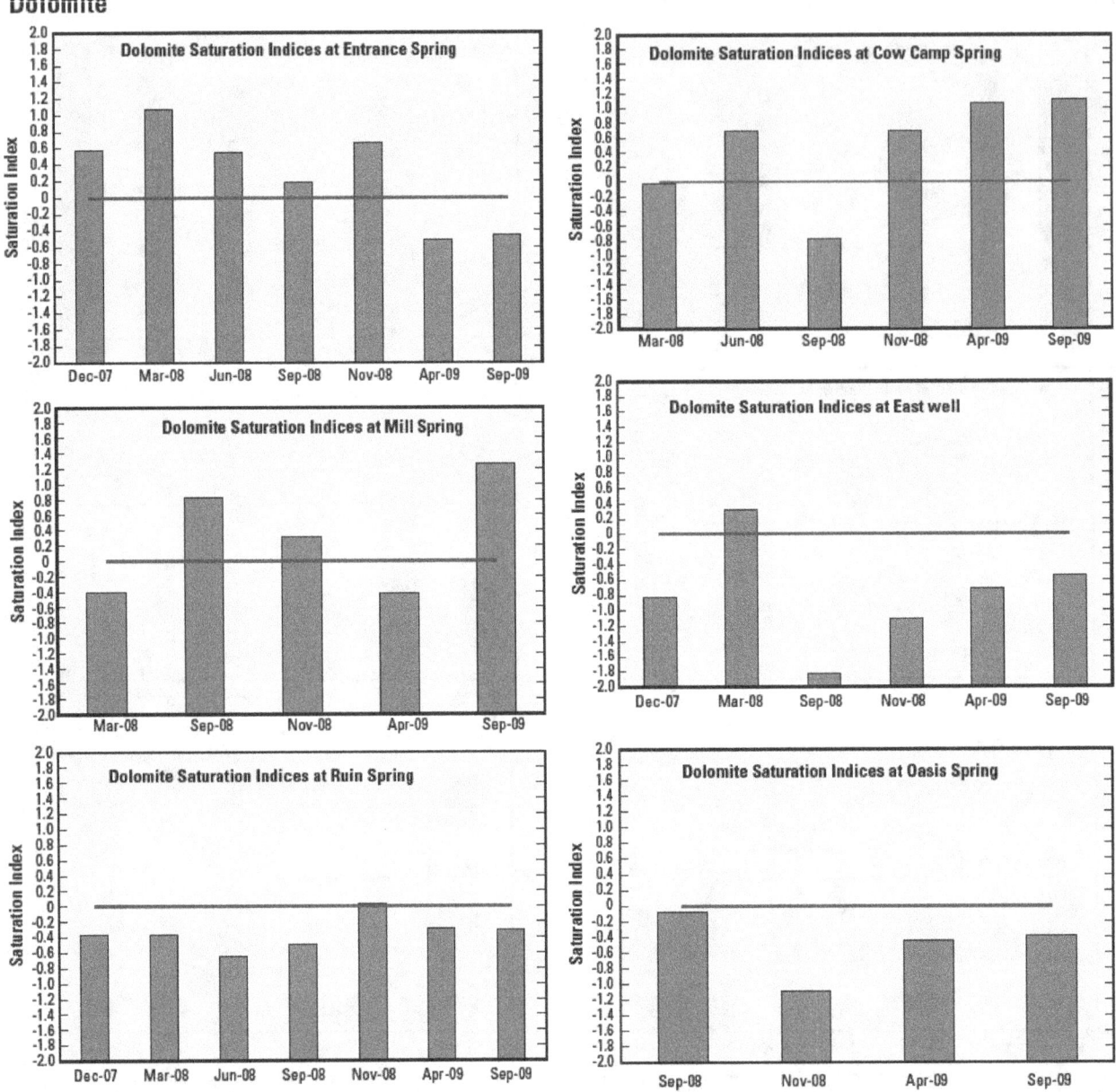

Figure 17. Saturation indices calculated for water samples from springs and wells surrounding the White Mesa mill, San Juan County, Utah, for calcite and dolomite.—Continued

Spring is sampled on a seepage face and usually was only dribbling out of the rock so that it took a long time to fill the sample bottle, which would allow for the escape of excess CO_2 to the atmosphere. In contrast, Ruin Spring is sampled directly from a pipe set into a basin that captures water from the spring discharge and may have minimized the opportunity for CO_2 degassing to occur relative to Mill Spring, Entrance Spring, and Cow Camp Springs. the rock that serves as the source of the spring, which is always flowing at a relatively quick velocity, so there is no chance for CO_2 to escape to the atmosphere. Therefore, on the basis of the calcite saturation indices at Ruin Spring, we assume that the positive calcite saturation values calculated at Oasis Spring, Mill Spring, Entrance Spring, and

Cow Camp are a result of CO_2 degassing to the atmosphere, and in the interpretation of the PHREEQC inverse models, we assume that these waters are in equilibrium with calcite. Similarly, on the basis of the saturation index values computed for Ruin Spring and the East well, we assume that the groundwater at all of our sampling sites is undersaturated with respect to dolomite.

Results of the PHREEQC inverse model calculations are given in table 11. The results indicate that variations in the degree of cation exchange and, perhaps, in the spatial distribution of gypsum are the cause of the spatial variability in the composition of the major ions. This analysis was not done for the West well because after the project began it was learned

Table 11. Transfer of minerals in groundwater (millimoles per liter).

[Negative values represent minerals removed from groundwater **Abbreviations**: mmol, millimoles; —, not available]

	Bayless domestic well, December 2007	Lyman domestic well, December 2007	Oasis Spring, November 2008	Entrance Spring, April 2009	Mill Spring, March 2008	Ruin Spring, September 2008	Cow Camp Spring, September 2008	East monitoring well, September 2008
Calcite	—	−0.418	0.220	—	—	−0.610	0.501	2.02
Dolomite	1.18	1.25	0.631	1.15	1.39	1.28	1.01	0.057
Halite	1.55	0.658	0.860	1.23	0.96	0.678	3.13	0.402
Gypsum	1.59	1.89	0.909	1.47	4.13	4.84	3.72	0.664
Orthoclase	0.068	0.048	0.032	0.100	0.050	0.084	0.145	0.030
Albite	0.744	0.110	0.624	0.028	0.069	0.011	—	0.028
Quartz	−1.38	—	−1.11	—	—	—	—	—
Kaolinite	−0.406	−0.079	−0.328	−0.064	−0.059	−0.048	−0.072	−0.029
CaX_2	—	−0.209	—	−0.627	−2.15	−1.96	−3.03	−2.58
NaX	—	0.419	—	1.25	4.31	3.93	6.06	5.45

that this well is screened in both the Dakota Sandstone and the Burro Canyon Formation and the Brushy Basin Formation; therefore, water is being withdrawn from two different formations, and the water chemistry is influenced by the minerals present in the Brushy Basin Formation as well as in the Dakota Sandstone and the Burro Canyon Formation. We are interested in the chemistry of the latter formation only, however, because this is the aquifer of concern to the Ute Mountain Ute Tribe. An interpretation of the results of the PHREEQC simulations to determine the source of the major ions for each distinctive major ion chemistry group identified on the Piper Diagram (fig. 12) is presented (fig.18).

Precipitation falling on the White Mesa is essentially a dilute solution of carbonic acid that is undersaturated with respect to all of the minerals present in the Dakota Sandstone and the Burro Canyon Formations. The chemical changes that occur as precipitation infiltrates into the Dakota Sandstone and the Burro Canyon Formation and subsequently flows as groundwater in these formations include an increase in specific conductance; equilibrium, or saturation, with calcite at many of the sites; and a varied major-ion composition at all of the groundwater sampling sites, which is distinctly different from that of precipitation. The processes responsible for these changes include (1) concentration due to evaporation, (2) dissolution of the aluminosilicate minerals orthoclase and albite, (3) cation exchange of Ca^{2+} for Na^+ on the surface of the clay mineral kaolinte, and (4) dissolution of the readily soluble minerals calcite, dolomite, gypsum, and halite. Of these processes, it is the dissolution of calcite, dolomite, gypsum, and halite that is primarily responsible for the evolution of groundwater chemistry. At all sites, dissolution of the aluminosilicate minerals, orthoclase and albite, occurs also, but it is about two orders of magnitude less than that of calcite, dolomite, and gypsum because the kinetics of dissolution of aluminosilicate minerals are much slower than for carbonate and sulfate minerals, such that the former contribute only a very minor amount of ions into solution. Similarly, whereas concentration

resulting from evaporation while precipitation percolates to the water table occurs, the increase in concentration due to this process is minor relative to the amount of solutes released into solution from the dissolution of carbonate and sulfate minerals. The cation exchange reaction can alter the major-ion composition and the degree of saturation with respect to certain minerals but does not affect specific conductance.

At the north, or upgradient, end of White Mesa, in the area of Oasis Spring, because precipitation is undersaturated with respect to all minerals present in the Dakota Sandstone/Burro Canyon Formation, the dissolution of all these minerals occurs to varying degrees. Cation-exchange reactions appear to be of minor importance in controlling groundwater chemistry here. At the northeastern end of the White Mesa, in the area of Bayless well, Lyman well, and Entrance Spring, calcite dissolution does not occur because water from Recapture Reservoir is saturated with respect to calcite. Nonetheless, the (1) predominance of gypsum dissolution, as at Oasis Spring, (2) dissolution of calcite at Oasis Spring, and (3) infiltration of water that is saturated with respect to calcite into the Dakota Sandstone and the Burro Canyon Formation in the northeastern section of the White Mesa results in these three sites having a similar major-ion composition to each other and to Oasis Spring.

As groundwater moves further south to Mill Spring and Ruin Spring, it attains equilibrium with respect to calcite and is greatly enriched with sulfate and slightly enriched with sodium relative to the upgradient sites. The groundwater is at equilibrium with respect to calcite most likely because Ca^{2+} released by the dissolution of gypsum suppresses the dissolution of calcite through a common ion effect. As a result, at some point between these sites and Oasis Spring, calcite dissolution ceases and calcite precipitation could occur. The shift in the anion composition of the groundwater to sulfate dominated water results from large amounts of gypsum dissolution. The shift in the cation composition of the groundwater to slight enrichment with sodium results from the exchange of Ca^{2+} for Na^+ on the surface of kaolinite. This cation-exchange reaction

---

Precipitation

Average pH is 5.2; average specific conductance is 7.2 µS/cm; calcium sulfate type water; undersaturated with respect to all minerals present in the White Mesa.

↓

Bayless and Lyman wells, Oasis Spring, and Entrance Spring

Neutral pH; average specific conductance is 619–963 µS/cm; dissolved oxygen present; calcium bicarbonate sulfate type water; dissolution of calcite, gypsum, and perhaps dolomite are the major controls on water quality.

↓

Mill Spring and Ruin Spring

Neutral pH; average specific conductance is 1,321–2,066 µS/cm; dissolved oxygen present; calcium sulfate bicarbonate water; dissolution of gypsum and cation exchange of Ca^{2+} for Na^+ on the surface of kaolinite are the major controls on water quality.

↓

Cow Camp Spring

Neutral pH; average specific conductance is 1,543 µS/cm; dissolved oxygen present; sodium calcium sulfate water; greater cation exchange than at Mill Spring and Ruin Spring and gypsum dissolution are the major controls on water quality.

↓

East well

Neutral pH; average specific conductance is 624 µS/cm; dissolved oxygen present; sodium bicarbonate type water; greater cation exchange than at Cow Camp Spring and calcite dissolution are the major controls on water quality.

↓

North and South wells

Neutral pH; average specific conductance is 409–467 µS/cm; dissolved oxygen absent; calcium bicarbonate type water; dissolution of calcite is the major control on water quality.

Figure 18.   Schematic describing the geochemical evolution of groundwater in the surficial aquifer, White Mesa, San Juan County, Utah.

occurs on the surface of the clay mineral kaolinite because the surfaces of clay minerals are charged, such that they engage in ion exchange to some degree (Drever, 1997). Thus, at some point along the groundwater flow path from Oasis Spring, the suppression of calcite dissolution and the initiation of cation exchange reactions begin to change the major-ion composition of the groundwater.

At Cow Camp Spring, an even greater amount of the exchange of Ca^{2+} for Na^+ results in the groundwater becoming enriched with Na^+ relative to the upgradient sites. The difference in water quality between Cow Camp Spring and the East well is related to a much greater amount of gypsum dissolution occurring at Cow Camp Spring relative to that at the East well. Even though the exchange of Ca^{2+} for Na^+ occurs to a greater degree here than at the East well, the greater amount of gypsum dissolution causes release of relatively large amounts of Ca^{2+} and SO_4^{2-} into solution and can explain the shift in composition at Cow Camp Spring to one in which Ca^{2+} and

SO_4^{2-} compose a greater percentage of the cation and anion composition, respectively, relative to the East well.

At the East well, the processes responsible for creating a sodium-bicarbonate water are the release of Ca^{2+} into groundwater from the dissolution of calcite, primarily, and of gypsum, secondarily, followed by cation exchange of Ca^{2+} for Na^+ and the release of HCO_3^- from the dissolution of calcite. The groundwater is undersaturated with respect to calcite, and the largest amount of calcite dissolution occurs here. The water, however, evolves to a sodium bicarbonate composition because a cation-exchange reaction removes Ca^{2+} from solution and introduces Na^+ into solution. A lack of gypsum dissolution limits the common-ion effect, and, thus, keeps the water undersaturated with respect to calcite, which allows for calcite dissolution to occur and furnishes Ca^{2+} for the cation exchange reaction. The small amount of gypsum dissolution relative to calcite dissolution allows for an increase in the concentration of HCO_3^- relative to SO_4^{2-}.

The PHREEQC modeling was not done for the two public supply wells in the Navajo sandstone. Given that the composition of the Navajo Sandstone is primarily quartz and calcite, and that the major-ion composition of the two wells plots well into the calcium bicarbonate region of the Piper Diagram (fig. 12), it is clear that dissolution of calcite is the dominate reaction controlling water chemistry in this aquifer. A major difference in the chemistry of the groundwater in the Navajo Formation relative to the groundwater in the Dakota Sandstone and the Burro Canyon Formation is the absence of oxygen in groundwater in the Navajo Formation. The presence of iron oxides, presumably hematite (Fe_2O_3) and/or goethite ($FeO(OH)$), as a film on the quartz grains provides a clue about how the oxygen in the groundwater could have been consumed. Groundwater in the Navajo Sandstone is very old, and as groundwater moved along the flow path from its place of recharge to the monitoring wells, enough time would have elapsed to allow reactions between iron-containing minerals and oxygen, which could have consumed all of the oxygen dissolved in the groundwater at the time of recharge. One example is the reaction of iron pyroxene with water and oxygen to form hematite:

$$2FeSiO_3 + 4H_2O + 2O_2 \rightarrow 2Fe_2O_3 + 2H_4SiO_4 \qquad (16)$$

Another example is the reaction of iron ions released into solution by the dissolution of minerals, such as pyrite, that can also react with oxygen to form hematite:

$$4Fe^{3+} + 3O_2 \rightarrow 2Fe_2O_3 \qquad (17)$$

We conclude that spatial variability in the major-ion composition of groundwater in the Dakota Sandstone and the Burro Canyon Formation results primarily from spatial variation in the extent of cation-exchange reactions and from spatial variation in the extent of gypsum dissolution. Why there is such variability in the relative importance of these reactions among our sampling sites is not known. One possibility could be the heterogeneous nature of the stream deposits that compose the Dakota Sandstone and the Burro Canyon Formations. Given the nature of these deposits, it is probably not unexpected that there would be spatial variability in the distribution of minerals composing the rocks in these formations. Another factor controlling groundwater chemistry is that the amount of time (residence time) in which the groundwater reacts with these minerals generally increases as it flows south within White Mesa. Thus, the major-ion composition and/or concentration of the groundwater will continue to evolve along the groundwater flow path until or if the groundwater becomes saturated with respect to the minerals present in the Dakota Sandstone/Burro Canyon Formations.

Trace-Element Geochemistry

Concentration data were compiled for selected chemical constituents analyzed in water samples from monitoring wells, springs, and pond/reservoir sites in the vicinity of the White Mesa mill site (fig. 19 and Appendix 1). Box plots were used to summarize data from each sample site containing at least three samples collected during the time period from September 2007 through September 2009 (fig. 20). The chemical constituents selected for display are generally associated with U deposits or are mobile under the chemically oxidizing and alkaline conditions present in selected ground- and surface-water resources adjacent to the White Mesa mill. Box plots were not used to summarize data from sample sites with less than three samples. When appropriate, the chemical constituent data were compared to U.S. Environmental Protection Agency maximum contaminant levels (MCL) and maximum contaminant level goals (MCLG; U.S. Environmental Protection Agency, 2009).

With the exception of arsenic, thallium, and uranium, the concentration of most trace elements in water samples collected during the study were below both the MCLs and MCLGs established by the U.S. EPA. Arsenic concentrations in unfiltered water samples are below the MCL of 10 µg/L at most sampling sites (fig. 20); however, public supply wells, South and North, contain median arsenic concentrations greater than 8 µg/L, which are well above those measured at the other sampling sites. Both of the public supply wells with elevated arsenic concentration were completed in the Navajo sandstone. Heilweil and Susong (2007) found elevated levels of arsenic, ranging from 2 to 44 µg/L, in groundwater samples collected from the Navajo Sandstone associated with an artificial recharge project in southwestern Utah.

Water samples from Entrance Spring had the highest median U concentration (26.6 µg/L, sample number [n] = 8) relative to water samples collected from the other sites (fig. 20). Water samples collected from both Entrance and Mill Springs exceeded the MCL for U in drinking water. Entrance Spring is located on the eastern boundary of the White Mesa uranium mill, and Mill Spring is located on the western boundary of the mill site (fig. 19). Thallium concentration in all water samples were below the MCL for drinking water; however, thallium levels in water samples from the Lyman well and West well did exceed the MCLG for thallium set by the US EPA at 0.5 µg/L.

The concentration of selenium is below the MCL for drinking water in all the water samples that were analyzed. Water samples from Entrance and Ruin Springs contain the highest selenium concentrations, with some samples exceeding 10 µg/L (fig. 20). Selenium is a common element associated with U deposits (Miesch, 1962; 1963).

The highest median concentration of vanadium (unfiltered; 6.8 µg/L) was found in water samples collected from Entrance Spring (fig. 20). Elevated concentrations of vanadium also were found in water samples collected from the South Mill (9.9 µg/L) and Anasazi pond (8.2 µg/L) sites. Vanadium is an element commonly associated with U deposits (Northrop and others, 1990). The occurrence of elevated concentrations of selenium, U, and vanadium in water samples from Entrance Spring could indicate contaminant migration from within the mill boundaries or contact with undiscovered and naturally occurring U ore bodies in the vicinity of the mill site.

Multiple passive diffusion bag samplers (Vroblesky and others, 2003) were deployed in monitoring wells MW3A, and West and MW 18 during December 2008 and October 2009 (fig. 21) to assess vertical variation of U in the Dakota Sandstone/Burro Canyon Formation (surficial aquifer). Vertical variation of dissolved-U concentrations during the December 2008 deployment period in all three wells was low and did not show any discernable trends. The U concentration in the three diffusion bag samples deployed in the West well ranged from 11.4 to 13.7 µg/L, which was slightly lower than the average U concentration of 16.0 µg/L determined from pumped samples collected during the study. Two diffusion sampling bags deployed in MW3A had U concentrations of 17.6 and 18.8 µg/L, compared to the U concentration of 19.9 µg/L determined from a pumped water sample collected and analyzed by Hurst and Solomon (2008). Three passive diffusion bag samples deployed in MW18 had U concentrations that ranged from 27.2 to 38.4 µg/L and were similar to the U concentration of 40.8 µg/L in a pumped water sample collected and analyzed by Denison Mines (writ. commun., 2008).

Vertical variation of U concentrations during the October 2009 deployment period was similar to the December 2008 results in the West and MW3A wells (fig. 21). Vertical variation of U concentration in MW18 was greater during the October 2009 deployment than in December, however, ranging from 20.2 µg/L in the shallowest diffusion bag to 44.5 µg/L in the deepest.

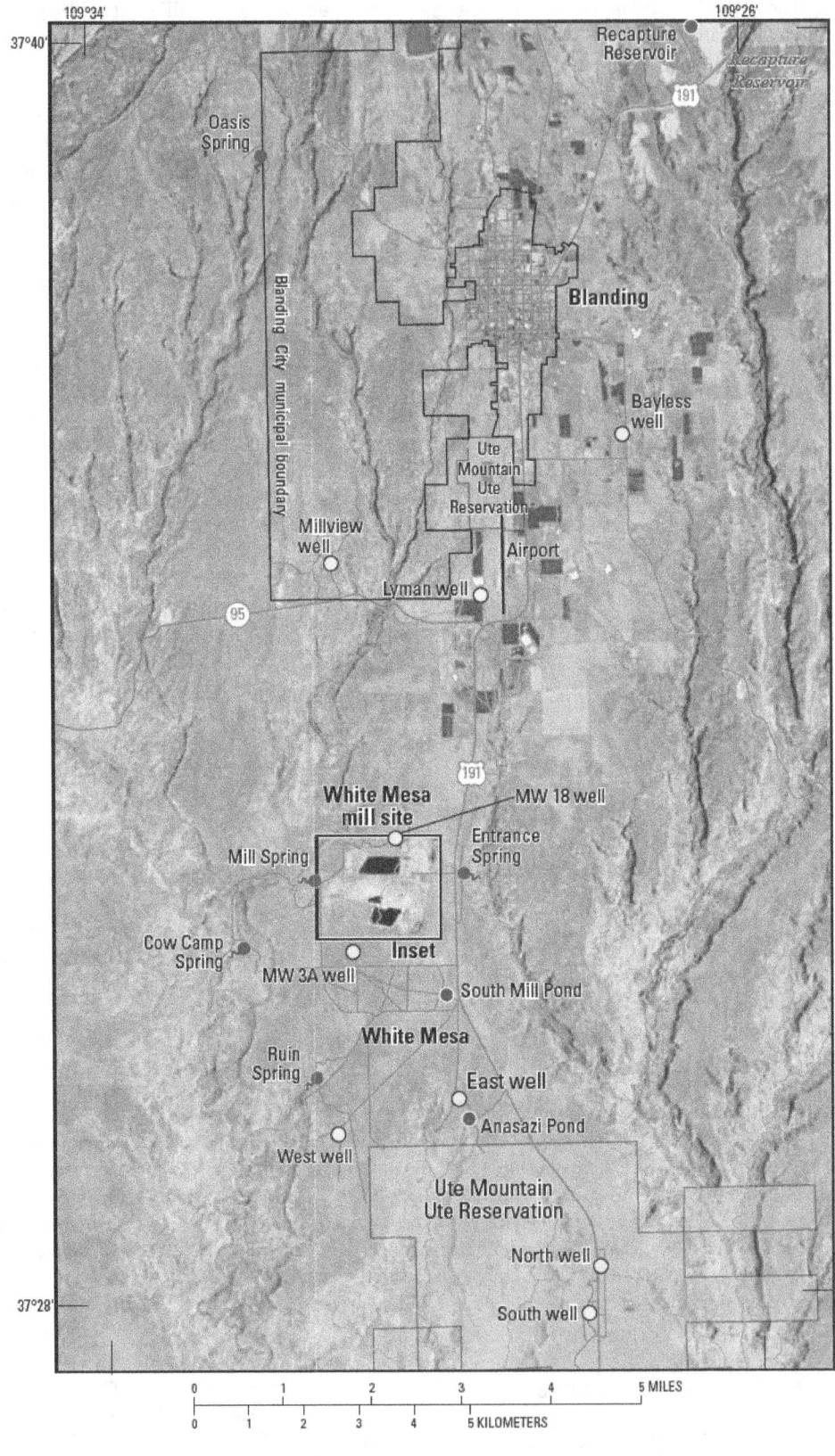

Figure 19.   Location of water-sampling sites in the vicinity of the White Mesa mill, San Juan County, Utah, that were sampled during the study period.

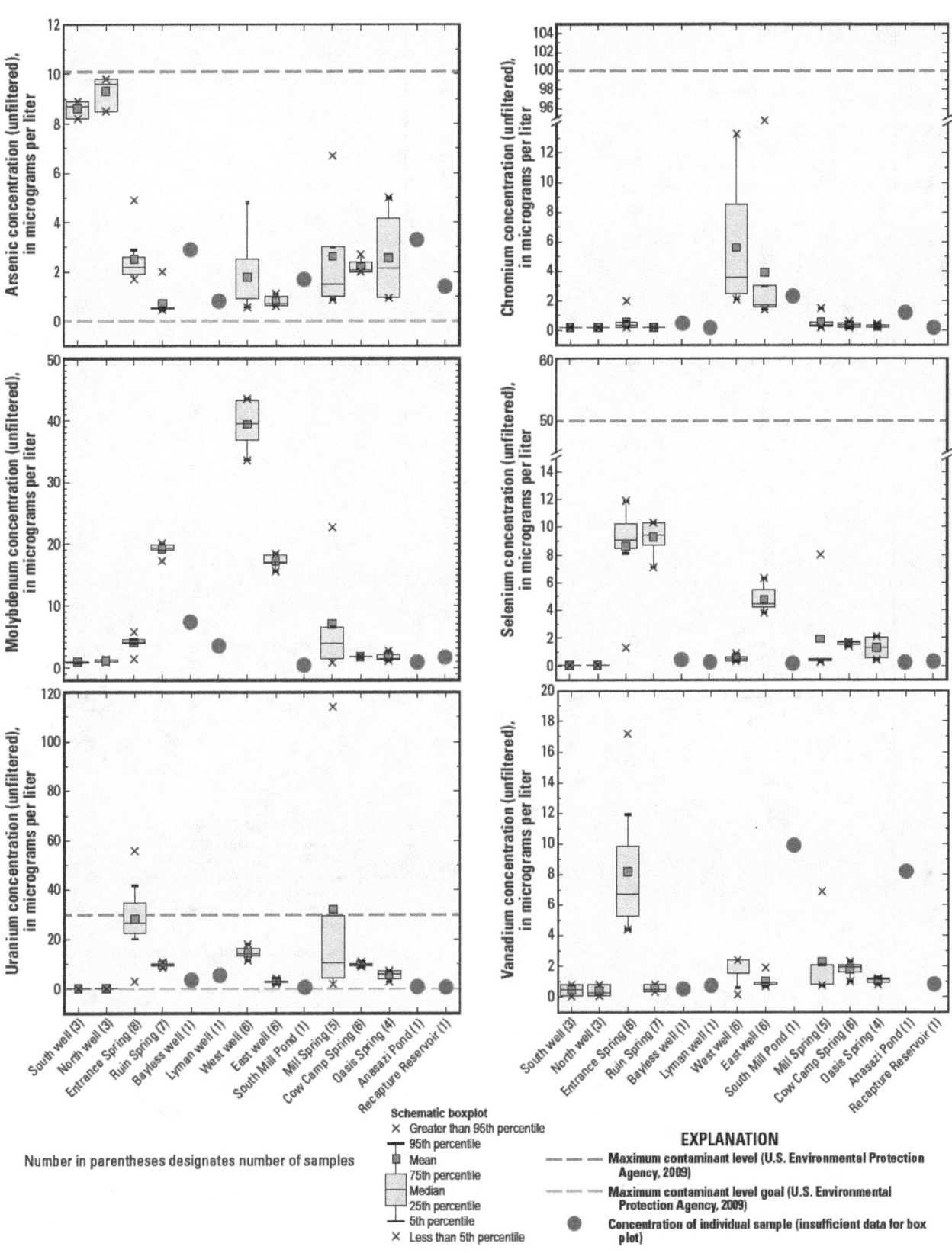

Figure 20.   Distribution of selected chemical constituents in unfiltered and filtered water samples collected from spring, monitoring well, and pond/reservoir sites near White Mesa uranium mill, San Juan County, Utah, compared to drinking water standards.

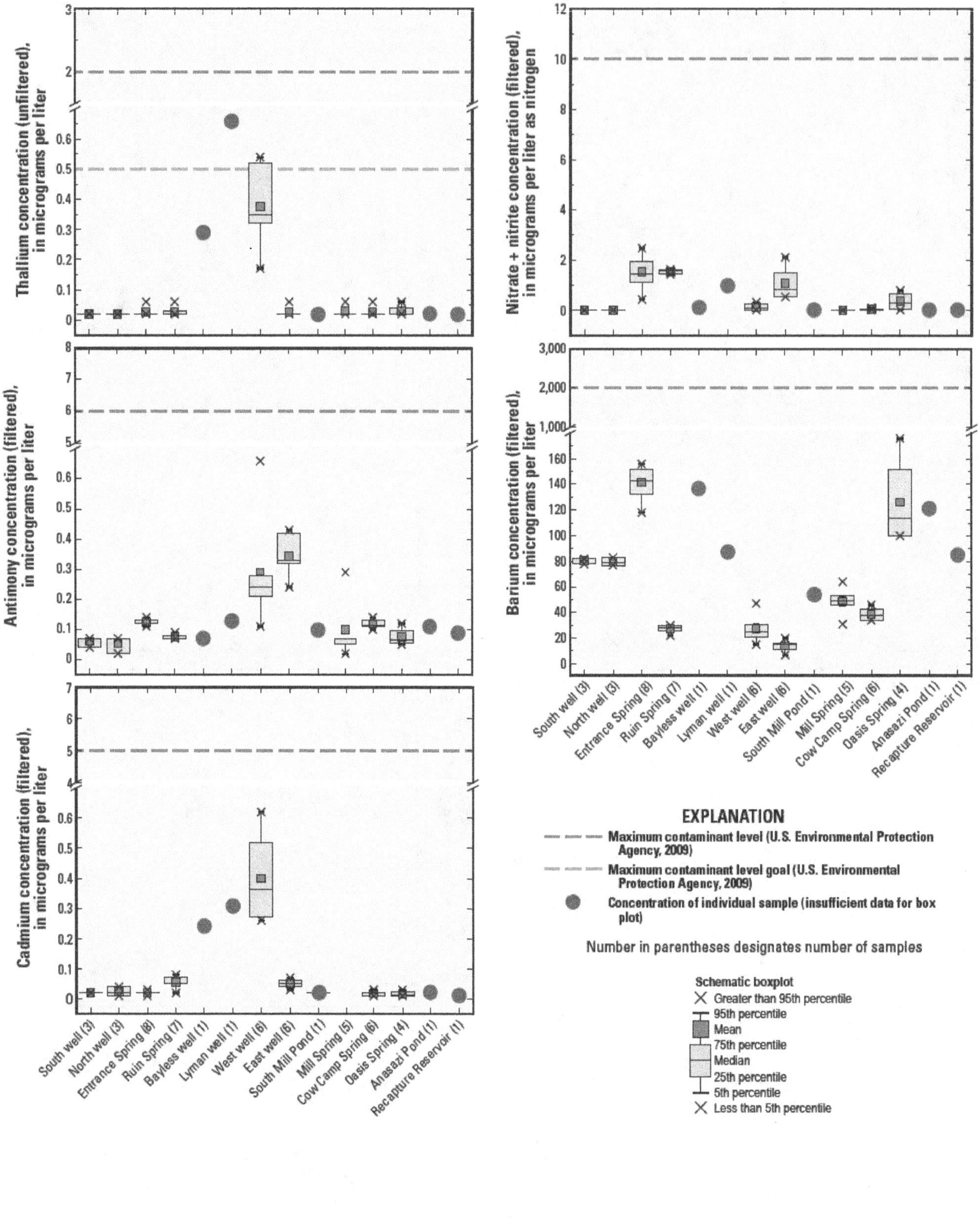

Figure 20.   Distribution of selected chemical constituents in unfiltered and filtered water samples collected from spring, monitoring well, and pond/reservoir sites near White Mesa uranium mill, San Juan County, Utah, compared to drinking water standards.— Continued

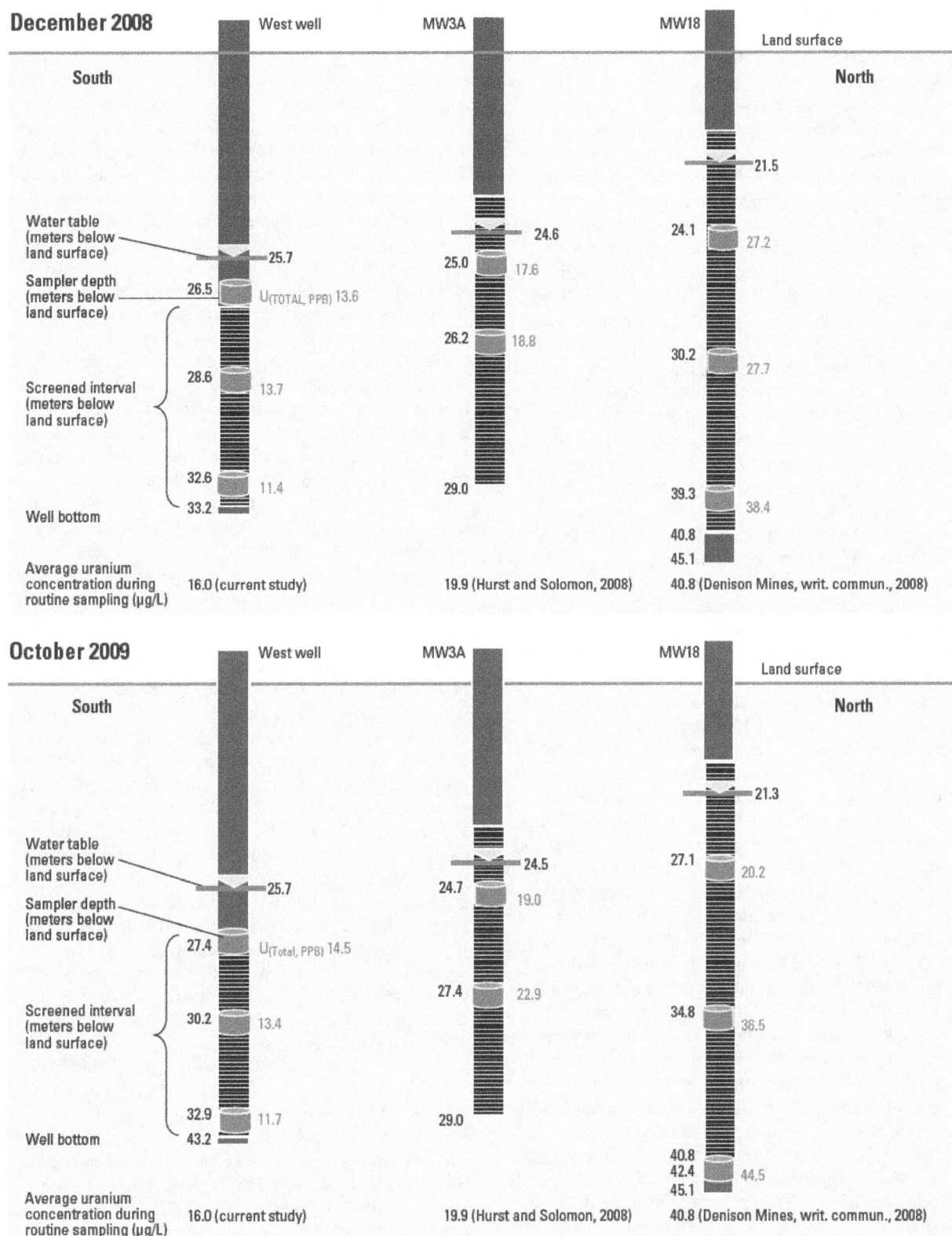

Figure 21. Schematic diagrams summarizing vertical variation in uranium concentration in passive diffusion bag samplers placed in three monitoring wells within and surrounding the White Mesa mill, San Juan County, Utah, during December 2008 and October 2009.

Uranium Mobility

An evaluation of the fate of U that could potentially be released from the mill into the aquifer in the Dakota Sandstone and the Burro Canyon Formation requires an understanding of the processes controlling the mobility of U in groundwater. The mobility of U in groundwater is determined by U solution-mineral equilibria and sorption reactions (Hsi and Langmuir, 1985). These properties are a function of pH, redox conditions, the presence of complexing agents, and the presence of other metals, such as vanadium, that can induce precipitation. The term mobile, in this report, means that the conditions in the unconfined aquifer favor U solubility and, thus, allow U to travel at rates nearly equal to groundwater movement. Uranium would not be considered mobile if the conditions in the unconfined aquifer retarded its movement in groundwater as a result of precipitation and/or sorption.

Uranium solubility is highly dependent on redox conditions. For example, under reducing conditions, U exists as U^{4+} and can form the insoluble compounds coffinite $(U(SiO_4)_{0.9}(OH)_{0.4})$ and uraninite (UO_2), and concentrations of dissolved U in groundwater would only be on the order of 0.06 µg/L (Sherman and others, 2007). Uranium can also precipitate out of solution as carnotite $(K_2(UO_2)_2(VO_4)_2)$ under all redox conditions over the pH range of 4–8 in the presence of dissolved vanadium in concentrations of 0.1 mg/L (Drever, 1997). Under oxidizing conditions, however, U is present as U^{6+} and is at least 10,000 times more soluble than U^{4+} (Sherman and others, 2007). In solution, U interacts strongly with carbonate (CO_3^{2-}) and phosphate (PO_4^{3-}) to form complexes, such as $UO_2(CO_3)_3^{4-}$, $UO_2(CO_3)_2^{2-}$, and $UO_2(HPO_4)_2^{2-}$. The formation of these complexes increases the solubility of U because, as Drever (1997) states, "The simplest process that might regulate the concentration of a trace element in solution is equilibrium with respect to a solid phase containing the element as a major component. The presence of ligands that can form complexes with U can increase the dissolved concentration of U above that expected on the basis of equilibrium with any U bearing mineral than it would be in water free of ligands." The formation of $UO_2(CO_3)_3^{4-}$, $UO_2(CO_3)_2^{2-}$, and $UO_2(HPO_4)_2^{2-}$ complexes also affects the capacity for U adsorption to clay minerals and iron oxides and, thus, influences the mobility of dissolved U in groundwater.

As stated previously, U tends to be most mobile in groundwater when it exists in solution as U^{6+} and forms soluble phosphate and uranyl-carbonate complexes in oxidizing alkaline water (Zielinski and others, 1997; Sherman and others, 2007). These conditions can occur in near-surface, unconfined aquifers that are open to exchange with the atmosphere and contain little organic matter (Zielinski and others, 1997). As discussed below, the conditions that favor U mobility in groundwater exist in the unconfined Dakota Sandstone and the Burro Canyon Formation aquifer despite the variability of the major-ion chemistry in this aquifer.

Within the unconfined aquifer, dissolved U was observed to be present at concentrations at or below 10 µg/L at all sites,

except in several samples at West Well, Entrance Spring and Mill Spring. The concentration of dissolved U was at or above the EPA Drinking Water MCL of 30 µg/L on several occasions at Entrance Spring and Mill Spring. Almost all of the U measured is in the aqueous phase, and the small concentrations of dissolved U result in groundwater being extremely undersaturated with respect to common U bearing minerals (fig. 22). The WATEQ database used in the PHREEQC modeling did not contain data for the mineral carnotite, so a saturation index for this mineral could not be calculated. Given that the highest concentration of dissolved vanadium measured at any of our sites is 6.5 µg/L, however, it is assumed that groundwater is also undersaturated with respect to this mineral.

Another factor that enhances the mobility of U in the groundwater in the White Mesa is the formation of uranyl-carbonate and uranyl-phosphate complexes. Dissolved U at all sampling sites does not exist as the free ion (U^{6+}), or as UO_2^{2+}, in solution but exists primarily as $UO_2(CO_3)_3^{4-}$ and secondarily as $UO_2(CO_3)_2^{2-}$ and $UO_2(HPO_4)_2^{2-}$, and there is spatial variation in the relative amount of these three complexes (fig. 23). These complexes decrease the adsorption of U to the surface of kaolinite in the Dakota Sandstone and the Burro Canyon Formation because of the low pH of the point of zero charge (pzc) of kaolinite. The pzc is the pH at which the surface charge on a solid, such as a clay mineral or iron oxide, submerged in an electrolyte is zero (Drever, 1997). In acid solutions, or when the pH of groundwater is less than the pzc, the surface of a solid will be positively charged and will attract anions and repel cations. In alkaline solutions, or when the pH of groundwater is greater than the pzc, the surface of a solid will be negatively charged and will attract cations (cation-exchange capacity is significant) and will repel anions (anion-exchange capacity will be small or zero; Drever, 1997). Since the pzc of kaolinite is 4.6 (Appelo and Postma, 2005), and the pH of the groundwater in the Dakota Sandstone and the Burro Canyon Formation is above 7, the negatively charged uranyl-carbonate and uranyl-phosphate complexes will not adsorb to kaolinite. The pzc of iron oxides, such as hematite (8.5), goethite (9.3), and $Fe(OH)_3$ (8.5; Appelo and Postma, 2005), suggests that adsorption to iron oxides is possible; however, because dissolved carbonate species (HCO_3^- and CO_3^{2-}) are preferentially adsorbed to soil surfaces compared to the uranyl-carbonate and uranyl-phosphate complexes, adsorption to iron oxides will not occur either (Duff and Amrhein, 1996; Echevarria and others, 2001).

The fact that groundwater in White Mesa contains dissolved oxygen, is extremely under-saturated with respect to common U bearing minerals, and contains enough CO_3^{2-} and PO_4^{3-} to completely complex dissolved U, leads to the conclusion that any solid phase U in contact with the groundwater would readily dissolve and any aqueous phase U would remain in solution. Thus, any U introduced into the unconfined aquifer in the Dakota Sandstone/Burro Canyon Formation from the mill, whether as dust blown off of the ore-storage pads, from trucks delivering ore to the mill, or as liquid from a leak in the tailings cells, would be mobile.

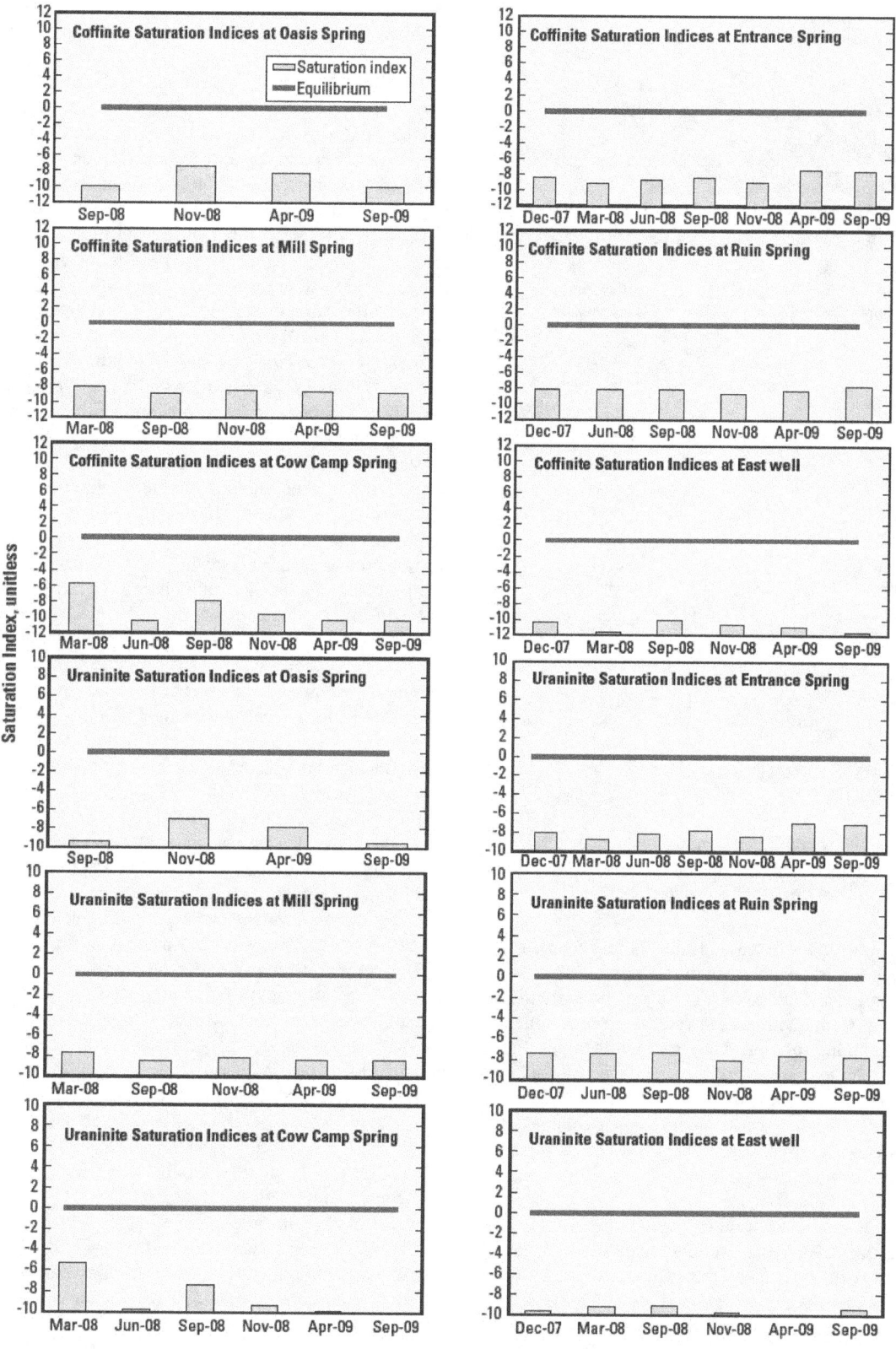

Figure 22. Saturation indices calculated for water samples collected from springs and wells surrounding the White Mesa mill, San Juan County, Utah, for coffinite and uraninite.

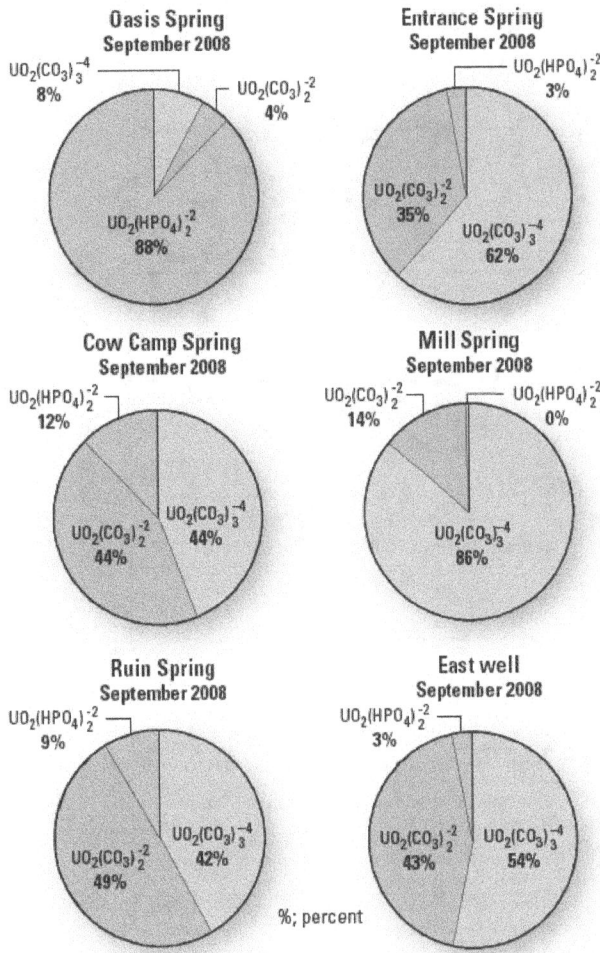

Figure 23. Pie charts showing dominant uranium complexes calculated for water samples collected from springs and wells surrounding the White Mesa mill, San Juan County, Utah.

The hypothesis that U released from a tailing cell as a result of a leak would be mobile in the unconfined aquifer was tested by using PHREEQC to mix, in varying proportions, the water in tailings Cell 1 with groundwater in the unconfined aquifer. The resulting solution, equilibrated with atmospheric concentrations of oxygen and carbon dioxide, was mixed with the groundwater composition measured at Oasis Spring in November 2008. The first scenario mixed equal volumes of tailing-cell water and groundwater, the second scenario mixed a solution of 90-percent groundwater and 10-percent tailing-cell water, and the third scenario mixed a solution of 70-percent groundwater and 30-percent tailing-cell water.

Under all simulations, the resulting mixed solution was very undersaturated with respect to coffinite and uraninite; thus, precipitation of U as these mineral phases would not occur. Also, in all mixed solutions, dissolved U existed as U^{6+}, but the type of complexes that formed differed. In the solution resulting from mixing equal volumes of tailing-cell water and groundwater, 27 percent of the dissolved U exists as UO_2^{2+}, 9 percent as $(UO_2)_2(OH)_2^{2+}$, 5 percent as $(UO_2)_4(OH)^{7+}$ and the

remainder of the dissolved U forms various positively and negatively charged and neutral complexes. In the solution resulting from mixing 90-percent groundwater with 10-percent tailing-cell water, 66 percent of the dissolved U exists as $UO_2(CO_3)_2^{2-}$, 13 percent as $UO_2(CO_3)_3^{4-}$, 9 percent as UO_2CO_3, and the remainder of the dissolved U forms various positively and negatively charged and neutral complexes. In the solution resulting from mixing 70-percent groundwater with 30-percent tailing-cell water, 19 percent of the dissolved U exists as $(UO_2)_4(OH)^{7+}$, 9 percent as UO_2CO_3, and 5 percent as $UO_2(CO^3)_2^{2-}$, and the remainder of the dissolved U forms various positively and negatively charged and neutral complexes.

The implication of this modeling is that under conditions in which small amounts of tailing-cell solution mixes with groundwater, the U would tend to remain in solution because U remains undersaturated with respect to common U-bearing minerals and forms predominantly negatively charged complexes, which limits adsorption to clay minerals and iron oxides. Under conditions in which the solution is composed of higher amounts of tailing cell water, it is possible that dissolved U would not be as mobile as the predominant complexes that form, which are positively charged and have the potential to adsorb to clay minerals and iron oxides. Thus, it appears that if a leak in a tailings cell occurred, dissolved U would tend to remain in solution, unless the proportion of tailing cell water that mixes with groundwater composes about 30 percent or greater of the resulting mixed solution. Whether U would precipitate out of solution as carnotite could not be determined because Hurst and Solomon (2008) did not measured the concentration of vanadium in tailing Cell 1. Since the pH of the mixed solution under the three scenarios described above ranged between 4.58 and 6.79, it is possible that U could precipitate as carnotite.

In the model, tailing-cell water also was mixed with water in one of the public supply wells in the Navajo Formation in the following proportions: 1-percent tailing-cell water and 99-percent groundwater, 10-percent tailing-cell water and 90-percent groundwater, and 50-percent tailing-cell water and 50-percent groundwater. Results were similar to those obtained with mixing the tailing-cell water with water in the unconfined aquifer. Under all simulations the mixed solution was very undersaturated with respect to coffinite and uraninite, and dissolved U existed as U^{6+}, but the type of complexes that formed differed. In the solution formed from mixing with 1-percent tailing-cell water, the dissolved-U concentration was 5.8 mg/L, with 63 percent of the dissolved U existing as $UO_2(CO_3)^{4-}$. In the solution formed from mixing with 10-percent tailing-cell water, the dissolved-U concentration was 58.1 mg/L, with 84 percent of the dissolved U existing as $UO_2(CO_3)^{4-}$ and 11 percent as $UO_2(CO_3)^{4-}$. In the solution formed from mixing with 50-percent tailing-cell water, the dissolved-U concentration was 290 mg/L, with 19 percent of the dissolved U existing as UO_2^{2+}, 10 percent as $(UO_2)_2(OH)_2^{2+}$, and 9 percent as $(UO_2)_3(OH)^{5+}$. The implication of this modeling is that under conditions in which tailing-cell water mixes with groundwater in the Navajo Formation in proportions of 10 percent or less of the total solution, U would be mobile because precipitation of U would not occur and predominately negatively charged complexes would form, which limit adsorption to clay minerals and iron oxides. When the tailing cell water composes a small amount of the solution, 1 percent or less, the concentration of U is less than the EPA MCL of 30 µg/L, however. Under conditions in which the solution is composed of 50 percent or more of tailing-cell water, the mobility of U could be limited because predominately positively charged complexes would form, which enhance adsorption of U to clay minerals and iron oxides.

Isotope Geochemistry

Uranium Isotope Geochemistry

After describing the controls on groundwater chemistry in an unconfined aquifer and its effect on the mobility of U, it is important to determine the source of U in the aquifer. Specifically, are the concentrations of U measured in this study, especially those at Entrance Spring, indicative of the range of natural or background concentrations, or is there evidence of contamination from the mill? Examining the spatial variation in dissolved-U concentrations can provide some insight. Hem (1989) states that U is present in concentrations between 0.1 and 10 µg/L in most natural water. In addition to the sites shown in figure 19, dissolved-U concentrations were measured at three other sites in the unconfined aquifer upgradient from the mill. The Lyman and Bayless domestic wells, sampled in December 2007 only, had dissolved-U concentrations of 5.36 µg/L and 3.1 µg/L, respectively. Reference Spring North (fig. 1), a very slow flowing seep on a hillslope 9 km northwest of the mill sampled in June 2007, had a dissolved-U concentration of 8.1 µg/L. Uranium concentrations at these three upgradient sites fall within the concentration range of most natural waters. All dissolved-U concentrations in groundwater at down-gradient sample sites sampled during this study, except for Entrance Spring and the September 2008 and September 2009 samples collected at Mill Spring, had dissolved-U concentrations in the range expected for naturally occurring U and that of upgradient sites. The fact that dissolved-U concentrations at Entrance and Mill Springs are elevated relative to the limited number of surrounding monitored sites does not, of itself, indicate that they are the result of a non-natural input of U to the White Mesa groundwater system. Work at Fry Canyon, about 50 miles to the west of the mill site, has shown that dissolved-U concentrations in groundwater at or above 40 µg/L are derived entirely from natural sources (Wilkowske and others, 2002). Concentration data for U alone cannot be used to identify the source of U in the groundwater of the unconfined aquifer.

In this study, U isotopes were used to help distinguish the source of U in the groundwater in the unconfined aquifer. All elements exist as a mixture of two or more isotopes. Uranium exists as three isotopes: on a mass basis, 99 percent of U exists as ^{238}U; 0.7 percent of U exists as ^{235}U; and 0.0054 percent of U exits as ^{234}U. Zielinski and others (1997) demonstrated that the $^{234}U/^{238}U$ alpha activity ratio (AR) can help to distinguish

between U derived from weathering and U derived from processing mills. They state that most natural groundwater has a $^{234}U/^{238}U$ alpha activity ratio greater than 1.0, with typical values in the range of 1 to 3, but values in excess of 10 can occur. By contrast, U in raffinate, a term used to describe the liquid waste generated by the processing of U ores, is derived from a mixture of materials with AR both above and below 1.0; considering the variety of U ores that are processed in a mill, a time-integrated average AR of 1.0 ± 0.1 is estimated for it.

Raffinate contains residual amounts of U originally brought into solution by reacting the U ore with strong oxidizing solutions of acid or alkali. The raffinate should retain the U-isotope composition of the processed ore because neither rapid, nearly complete dissolution of U from finely crushed ore samples for further chemical processing of the leachate to efficiently remove most U from solution by solvent exchange, sorption, or precipitation will promote isotopic fractionation (Zielinski and others, 1996). As a result, we assume that any solid-phase ore, such as that stored on the ore-storage pads at the mill, if blown offsite and deposited in water, will dissolve, and the uranium derived from this source will have an average AR of 1.0 ± 0.1 also.

The difference in the $^{234}U/^{238}U$ AR between U derived from raffinate and U derived from oxidative leaching by groundwater of soil and rocks is due to a process known as "alpha recoil" that occurs during radioactive decay of a ^{238}U atom (Sherman and others, 2007). Alpha recoil refers to the fractionation of ^{238}U and its daughter product ^{234}U during radioactive decay, which results from the displacement of a ^{234}U atom from the site of its parent ^{238}U atom. When ^{238}U decays to ^{234}Th (thorium) by alpha decay, the Th nucleus can be recoiled out of the mineral into the groundwater. The ^{234}Th decays via ^{234}Pa (protactinium) to ^{234}U, resulting in an excess of ^{234}U in the groundwater. By contrast, U ores that have not been subject to major oxidative leaching within the last million years approximate closed systems that are in radioactive (secular) equilibrium (Zielinski and others, 1996). In secular equilibrium, the rate of decay of ^{234}U is equal to the rate of decay of the ^{238}U parent, and if the isotopes are measured in terms of their alpha-emission rates, radioactive equilibrium between ^{238}U and ^{234}U represents a condition of equal alpha activity, where the $^{234}U/^{238}U$ AR is 1.0. The most likely reason that the AR is measured instead of absolute abundances of the two isotopes is that ^{234}U represents only 0.0054 percent of U by mass and there is a large difference in the half-life of the two isotopes: 4.47×10^9 years for ^{238}U and 2.44×10^5 years for ^{234}U.

A plot of AR values as a function of U concentration shows that $^{234}U/^{238}U$ AR values for U concentrations less than the EPA MCL of 30 µg/L fall within the range of 1.4 to 3.4, which indicate a natural source of U at these sites (fig. 24). For the three samples that had a dissolved-U concentration in excess of 30 µg/L, 33.2 and 48.4 µg/L at Entrance Spring and 75.6 µg/L at Mill Spring, the $^{234}U/^{238}U$ AR were 1.55, 1.26, and 2.29, respectively. While AR values for all samples collected at Entrance Spring fall within the range expected for U derived from natural sources, they showed a general decline with

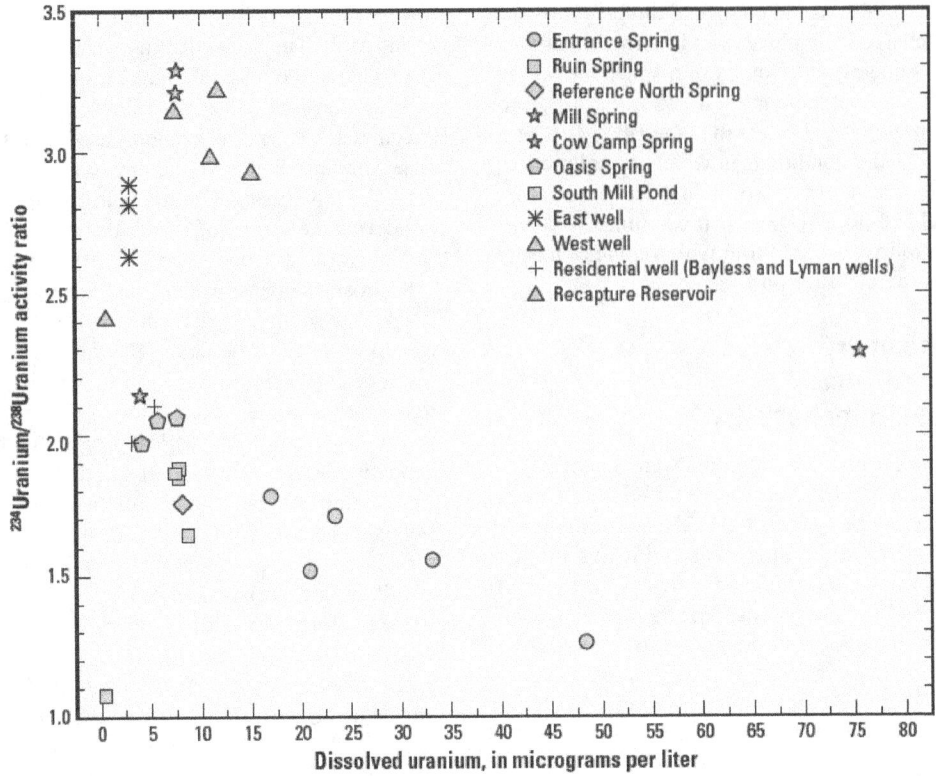

Figure 24.   Dissolved uranium and ^{234}U/^{238}U activity ratios measured in water samples collected from various sources near the White Mesa uranium mill, San Juan County, Utah.

increasing concentrations that approach the values expected for U derived from raffinate. The ^{234}U/^{238}U AR of 2.29, associated with a dissolved-U concentration of 75.6 µg/L, at Mill Spring indicates a natural source of U, but the AR of this sample is almost exactly the same (2.14) as that measured on another sample collected from Mill Spring that had a dissolved-U concentration of 3.98 µg/L. Is the pattern of AR values measured at Entrance Spring an indication of contamination by the mill? How can the large difference in concentration between two samples collected at Mill Spring, with virtually no difference in the value of the AR, be explained?

An attempt to answer these questions was made by plotting the ^{234}U/^{238}U AR for all of the sampling sites as a function of the reciprocal of dissolved-U concentration (fig. 25). The consistently low values of the AR at Entrance Spring, and the general decrease of these values with an increase in concentration fall on a mixing line (Zielinski and others, 1997) and suggest that perhaps there is some mixing of U derived from ore with groundwater at Entrance Spring. The two points for Mill Spring are displaced horizontally from one another, indicating a change in U concentration in the absence of isotopic changes. This same pattern can be seen for three samples collected at Oasis Spring. According to Zielinski (1997), such changes fall on a line indicating that evaporation or dilution is occurring. Thus, the increase in concentration at Mill Spring

from 3.98 to 75.6 µg/L is a result of evaporative concentration and is not evidence of contamination from the mill.

The ^{235}U/^{238}U ratio was determined for all of our samples, also, and is useful in distinguishing between anthropogenic and natural sources of U. The use of this isotope pair in this study is not as useful as ^{234}U/^{238}U AR, however, because the main source of anthropogenic ^{235}U is the manufacturing of atomic weapons and not U processing facilities such as the White Mesa mill. Therefore, this isotopic pair would be more appropriate for monitoring the effects of a weapons production facility, but enough work has been done with ^{235}U/^{238}U ratios to establish that the mass ratio of 0.0072 is indicative of naturally occurring U (Ketterer and others, 2000; Sherman and others, 2007). This ratio was 0.0072 in all of our samples, which supports the ^{234}U/^{238}U AR data that indicated the dissolved U at our sites is derived from natural sources.

The U isotope data indicate that the mill is not a source of U in the groundwater in the unconfined aquifer at any sites monitored during the study, with the possible exception of Entrance Spring. As defined previously, potential pathways of U transport from the mill to the groundwater system include (1) airborne dust from ore storage pads and emissions from the mill's drying ovens, with subsequent dissolution and seepage of contaminated water into the aquifer, and (2) direct leakage from the mill tailing ponds or seepage from tailings cells. If the elevated-U concentrations observed in Entrance Spring

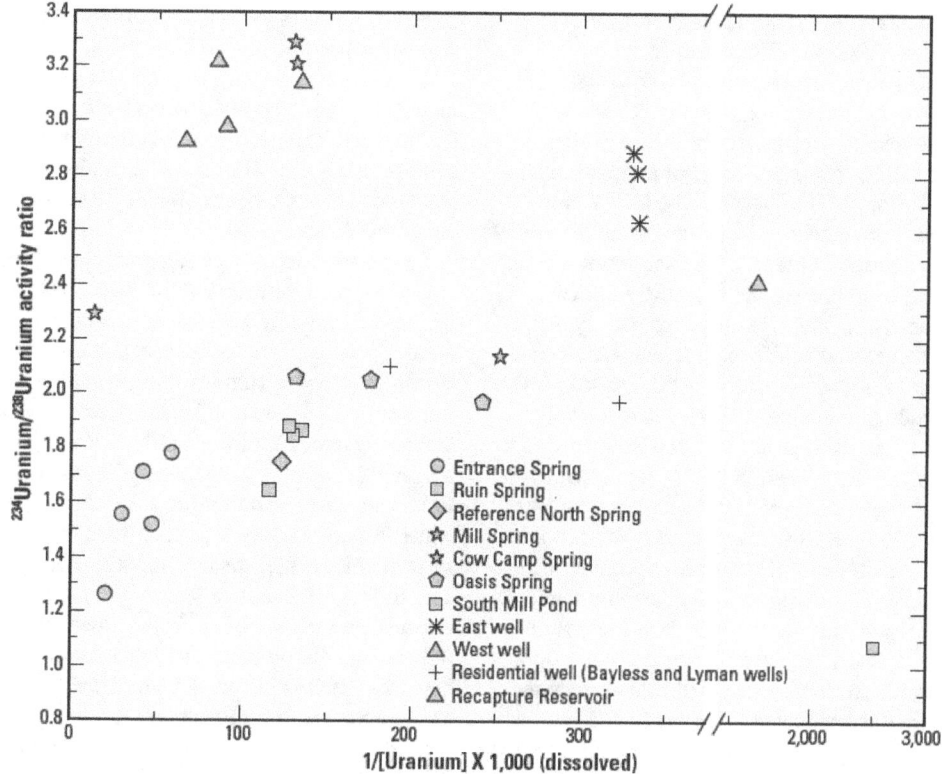

Figure 25. Transformed dissolved uranium (inverse concentration multiplied by 1,000) and $^{234}U/^{238}U$ activity ratios measured in water samples collected from various sources near the White Mesa uranium mill, San Juan County, Utah.

are not the result of natural sources, a possible pathway from the mill site to the spring is airborne transport with subsequent dissolution of the wind deposited material in the Entrance Spring drainage. This pathway is feasible for several reasons: (1) the ore to be processed in the mill is stored uncovered on ore storage pads directly across from Entrance Spring, and much of this material is fine grained, which easily can be transported by the wind; (2) starting approximately three years ago trucks delivering ore were covered, as stipulated Bureau of Land Management and Department of Transportation policies (Bureau of Land Management, 2011), but prior to that time trucks delivering ore were may have been uncovered and turned onto the mill from Highway 191, directly across from Entrance Spring; and (3) as discussed in the "Uranium Mobility" section, any solid phase U in contact with infiltrating water would dissolve readily, and any aqueous phase U would likely remain in solution. The tailings cells are not a likely source of U at Entrance Spring. An analysis of the groundwater flow paths on the White Mesa indicate that the prevailing groundwater flow direction is toward the south, and that any leakage from a tailings cell is unlikely to flow east toward Entrance Spring.

The evidence presented in this section, however, does not conclusively prove or disprove a hypothesis that the source of U in Entrance Spring is material from the ore storage pads deposited by wind into the drainage. We evaluated this

hypothesis further by collecting stream sediment and vegetation samples around the White Mesa. The results of this sampling are discussed in the "Sediment" and "Vegetation" sections.

Isotopes of Oxygen and Hydrogen

Water samples from selected springs, monitoring wells, domestic-supply wells, and surface-water sources (fig. 19) were analyzed for delta oxygen-18 ($\delta^{18}O$) and deuterium (δD) values by the USGS Stable Isotope Laboratory in Reston, Virginia. The $\delta^{18}O$ and δD values were compared to the global (Craig, 1961) and arid-zone (Welch and Preissler, 1986) meteoric water lines (fig. 26), and three distinct groupings of water samples were identified.

Group 1 includes water samples from the North and South wells that contain the isotopically lightest signature ($\delta^{18}O$ is less than -15.5 and δD is less than -115 permil) and plot directly on the global meteoric water line (fig. 26). Both of these wells are completed in the Navajo Sandstone, which represents a regional aquifer system that is recharged by higher elevation areas that include Comb Ridge to the west and the Abajo Mountains to the north (Freethey and Cordy, 1991; Naftz and others, 1997). The isotopic composition of water samples from the North and South wells is very similar to the isotopic composition of two snow samples (fig. 26) collected from the Abajo Mountains to the north of the study area (Spangler and others, 1996) that also plot in group 1.

The $\delta^{18}O$ and δD values for water samples in group 2 plot below the global meteoric water line (fig. 26) and are more aligned to the arid-zone meteoric water line. Wells and springs in group 2 include Oasis Spring, East and West wells, Mill spring, Ruin Spring, Mill View well, and Cow Camp Spring. The isotopic enrichment and deviation from the global meteoric water line indicate more localized and lower elevation recharge, which would be subject to isotopic enrichment through evaporation. These recharge characteristics typify the conditions in the surficial aquifer composed of the Dakota Sandstone and Burro Canyon Formation. As discussed previously, precipitation directly on the White Mesa is the only source of recharge to the surficial aquifer. The isotopic signature of recharge on the White Mesa is further supported by the $\delta^{18}O$ and δD values associated with water samples from two surface-water sites (South Mill and Anasazi Ponds, shown in fig. 19). Both of these sites collect localized precipitation characteristic of White Mesa that falls on lower elevations, and the isotopic composition of water samples from these sources is similar to the isotopically enriched composition of group 2 water samples collected from springs and wells associated with the surficial aquifer on White Mesa (fig. 26).

Water samples in group 3 plot below the arid-zone meteoric water line and represent the most isotopically enriched water samples collected from the study area. Group 3 sites include Entrance Spring, Lyman well, Bayless well, and Recapture Reservoir (fig. 26). A trend line through the $\delta^{18}O$ and δD values of group 3 water samples indicates an evaporative signature because the slope is lower than the meteoric water line (Drever, 1997).

Water from Recapture Reservoir is the primary water source for ore processing at the White Mesa mill and for irrigated agriculture in areas surrounding Blanding, Utah (Utah Division of Water Quality, 2006). The similar isotopic signature of water samples from Recapture Reservoir and Entrance Spring could indicate a linkage with mill runoff, seepage discharging from Entrance Spring, or the inputs from irrigated agriculture in the area utilizing water from Recapture Reservoir. The reason for the enriched isotopic signature of water from Recapture Reservoir likely is due to evaporation of snowmelt from the Abajo Mountains during reservoir storage. Inflow to Recapture Reservoir is entirely from ephemeral streams, and release of reservoir water to Recapture Creek only occurs during wet years when the reservoir reaches full capacity (Utah Division of Water Quality, 2006). Additional data are needed from Recapture Reservoir to better identify the seasonal variations in the isotopic composition.

The similar $\delta^{18}O$ and δD values in water samples from the Bayless well and Recapture Reservoir (fig. 26) could suggest the infiltration of irrigation water from Recapture Reservoir into the surficial aquifer. In addition, the enriched isotopic signature of the water sample from Lyman well is consistent with evapotranspiration of Recapture Reservoir water during irrigation and then subsequent recharge to the surficial aquifer. Both the Bayless and Lyman wells are in rural areas with irrigated agriculture.

The isotopic linkage between water samples from Entrance Spring and facilities water used at the mill site is further supported by $\delta^{18}O$ and δD values for water samples collected from the wildlife ponds that were published by Hurst and Solomon (2008). The wildlife ponds are unlined ponds on the eastern side of the mill site (fig. 1) and are filled with facilities water from Recapture Reservoir. The $\delta^{18}O$ and δD values of water samples from the wildlife ponds are enriched relative to the mill facilities water from Recapture Reservoir used to fill the ponds (fig. 27). This isotopic enrichment results from evaporation of the facilities

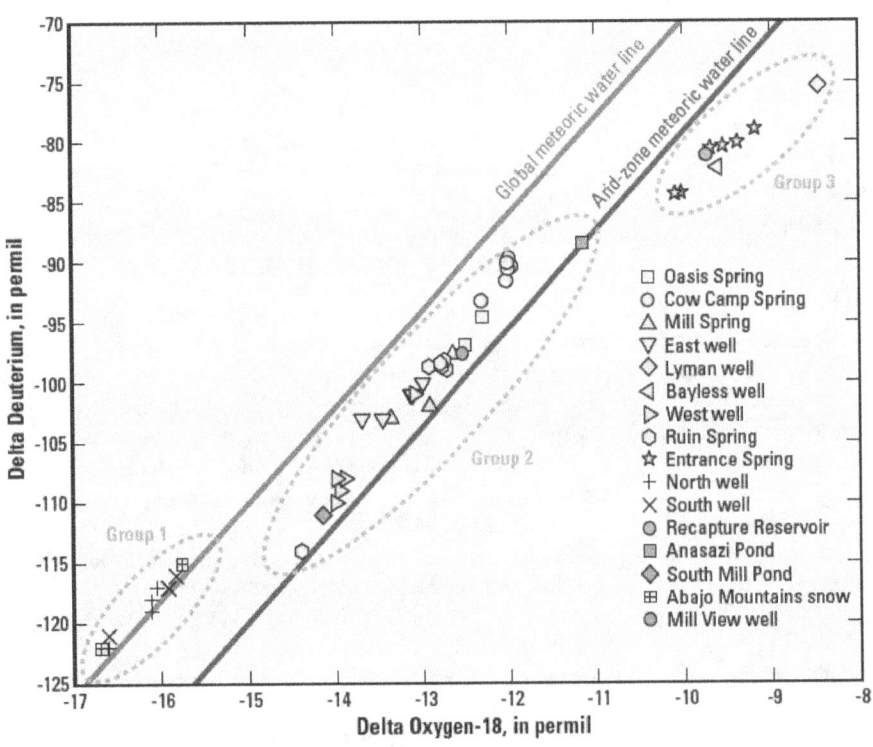

Figure 26.   Delta deuterium and delta oxygen-18 composition of water samples collected from the study area and comparison of sample groups 1, 2, and 3 to the global (Craig, 1961) and arid-zone (Welch and Preissler, 1986) meteoric water lines. Isotopic data from snow samples in the Abajo Mountains from Spangler and others (1996).

water. A mixing line constructed between the isotopic composition of water from Anasazi Pond and the wildlife ponds can assist in the depiction of likely water sources to Entrance Spring (fig. 27). Four of the six water samples collected from Entrance Spring are isotopically enriched relative to water from Recapture Reservoir (fig. 27). This isotopic enrichment can be explained by mixing with the isotopically enriched water from the unlined wildlife ponds or other ponded facilities water on the mill site that is subject to evaporation. The two water samples from Entrance Spring that are less isotopically enriched than the facilities water from Recapture Reservoir are likely the result of mixing between typical recharge water to the White Mesa (for example, water from Anasazi Pond) and facilities water from Recapture Reservoir and/or evaporated water from the wildlife ponds.

The $\delta^{18}O$ and δD data indicate that water discharging from Entrance Spring contains an isotopic fingerprint of water from Recapture Reservoir that also is used as facilities water on the mill site. In addition, water from Recapture Reservoir also is used to irrigate fields surrounding the town of Blanding. Infiltration of this irrigated water also could contribute to the enriched isotopic fingerprint observed for Entrance Spring. As noted in a previous report section, Entrance Spring also contains the highest median U concentration relative to the spring and groundwater sites that were sampled during the study period.

Isotopes of Sulfur and Oxygen in Sulfate

Filtered water samples from selected springs, monitoring wells, and domestic-supply wells (fig. 19) were analyzed for $\delta^{18}O$ in the sulfate ion ($\delta^{18}O_{sulfate}$) and delta sulfur-34 in the sulfate ion ($\delta^{34}S_{sulfate}$) by the USGS Stable Isotope Laboratory in Reston, Virginia. Because sulfuric acid is used in ore processing in the mill, the isotopic composition of both $\delta^{18}O_{sulfate}$ and $\delta^{34}S_{sulfate}$ can provide a unique isotopic fingerprint of groundwater contamination derived from mill sources. Hurst and Solomon (2008) determined the $\delta^{18}O_{sulfate}$ and $\delta^{34}S_{sulfate}$ values in water samples from multiple monitoring wells inside the mill property, as well as the tailings cells and wildlife ponds. The tailings cells were found to be enriched in $\delta^{18}O_{sulfate}$ (ranging from 3.9 to 4.5 permil) relative to other water samples on the mill property, and this isotopic enrichment was likely the result of evaporation of liquids in the tailing cells. In addition, the $\delta^{34}S_{sulfate}$ values in water from the tailings cells

had relatively consistent isotopic values that ranged from −1.04 to −0.89 permil (Hurst and Solomon, 2008) and is likely related to the $\delta^{34}S_{sulfate}$ isotopic signature of sulfuric acid used in ore processing.

The $\delta^{18}O_{sulfate}$ and $\delta^{34}S_{sulfate}$ values in water samples from wells and springs surrounding the mill site were compared to the isotopic composition of water from the tailings cells and wildlife ponds (fig. 28). The $\delta^{18}O_{sulfate}$ values in water samples from the tailings cells and wildlife ponds are isotopically enriched and likely reflect the evaporative processes that occur in these surface-water sites (Hurst and Solomon, 2008). Similarities in the $\delta^{34}S_{sulfate}$ values in water samples from the wildlife ponds and tailings cells indicate a potential linkage that may be related to eolian transport of aerosols from the tailings cells, surface runoff from the mill facility, and/or rainout of sulfuric acid released to the atmosphere from the mill site (Hurst and Solomon, 2008). None of the spring or monitoring well samples collected from areas surrounding the mill site contains $\delta^{18}O_{sulfate}$ and $\delta^{34}S_{sulfate}$ isotopic signatures that would indicate recharge from tailings cells within the mill boundary (fig. 28).

Figure 29 displays the relationship between sulfate concentrations and $\delta^{34}S_{sulfate}$ for water samples collected from the monitoring wells and spring sites adjacent to White Mesa mill, as well as water samples collected by Hurst and Solomon (2008) from the wildlife ponds and tailings cells at the mill site. With the exception of the water samples from the tailings

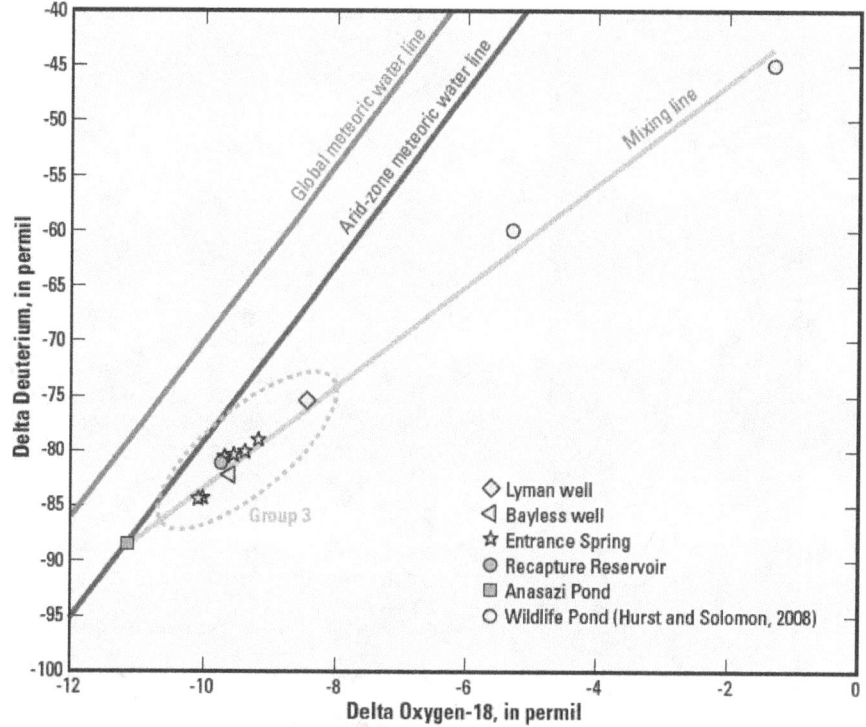

Figure 27. Delta deuterium and delta oxygen-18 composition of group 3 water samples compared to the isotopic composition of water samples from Anasazi Pond outside of the mill property and the wildlife ponds located within the mill site, San Juan County, Utah.

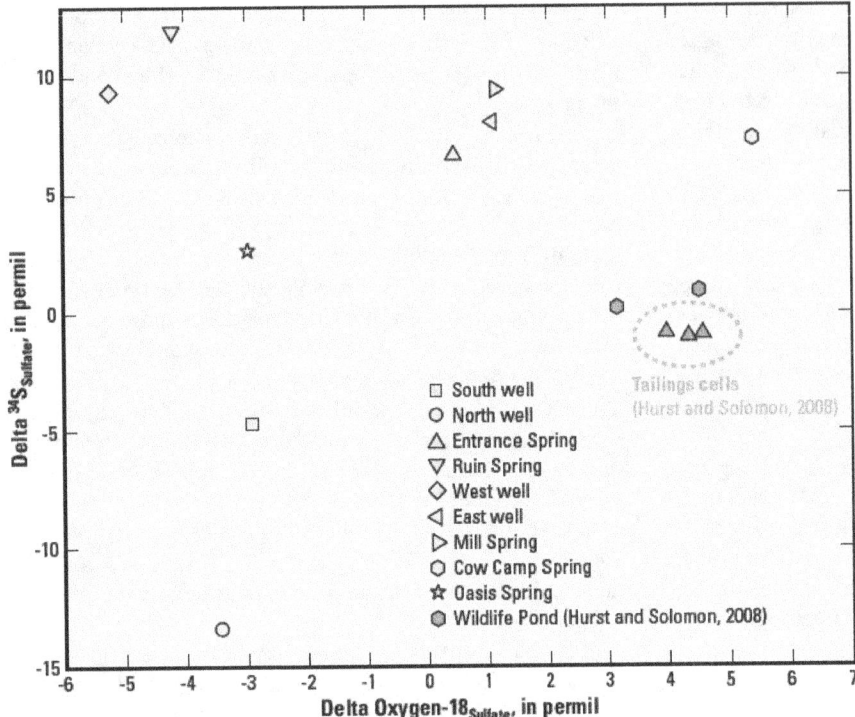

Figure 28.   Delta $^{18}O_{sulfate}$ and delta $^{34}S_{sulfate}$ composition of water samples collected from areas surrounding the White Mesa mill site compared to samples from the tailings cells and wildlife ponds located within the mill site, San Juan County, Utah.

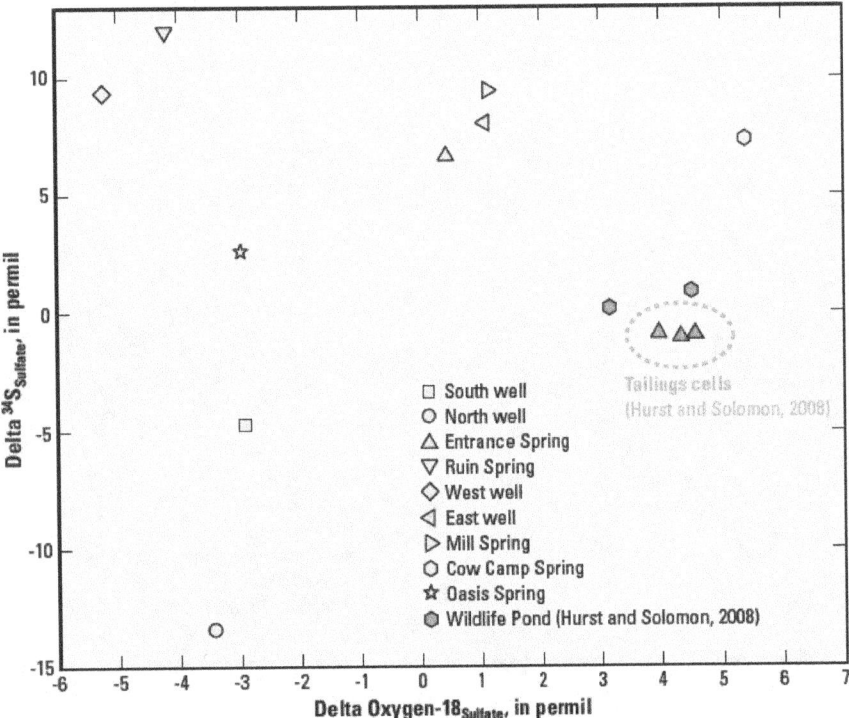

Figure 29.   Changes in delta $^{34}S_{sulfate}$ as a function of sulfate concentration in water samples collected from areas surrounding the White Mesa mill site compared to water samples from the tailings cells and wildlife ponds located within the mill site, San Juan County, Utah.

cells, increasing sulfate concentration tends to be associated with heavier $\delta^{34}S_{sulfate}$ values. The similarity in $\delta^{34}S_{sulfate}$ values for the tailings and wildlife ponds, and the difference between these values and those from other sites, provides a good fingerprint of water from these sources. To date (2010), the $\delta^{34}S_{sulfate}$ values measured in wells and springs surrounding the White Mesa mill site do not have an isotopic signature characteristic of the tailings cells. Because the wildlife ponds are actively leaking (Hurst and Solomon, 2008), it is likely that future groundwater samples from the surficial aquifer at sites within and adjacent to the mill site will exhibit decreasing trends in $\delta^{34}S_{sulfate}$ values; however, this potential decrease in $\delta^{34}S_{sulfate}$ values alone cannot be used to identify leakage from the tailings ponds exclusively.

Sediment

Trace-element geochemistry

Sediment samples from ephemeral drainages that could potentially receive and accumulate water and wind-blown material from the mill site were sampled during June 2008. Stream-sediment samples were collected from 28 sites in the ephemeral-stream channels draining the White Mesa uranium mill site (fig. 30). In addition, three stream-sediment samples were collected approximately five kilometers (km) north of the mill site (fig. 31) to represent local background conditions. The fine-grained fraction (−200 mesh) of each sediment sample underwent a multi-acid, total digestion and was analyzed for 42 major and trace elements (Appendix 2), including U.

Two standard reference materials obtained from the USGS (Green River Shale, SGR-1B, and Mica Schist, SDC-1) were submitted blindly with the routine stream sediment samples collected from the drainages surrounding the mill site. Analytical results from the standard reference materials were generally

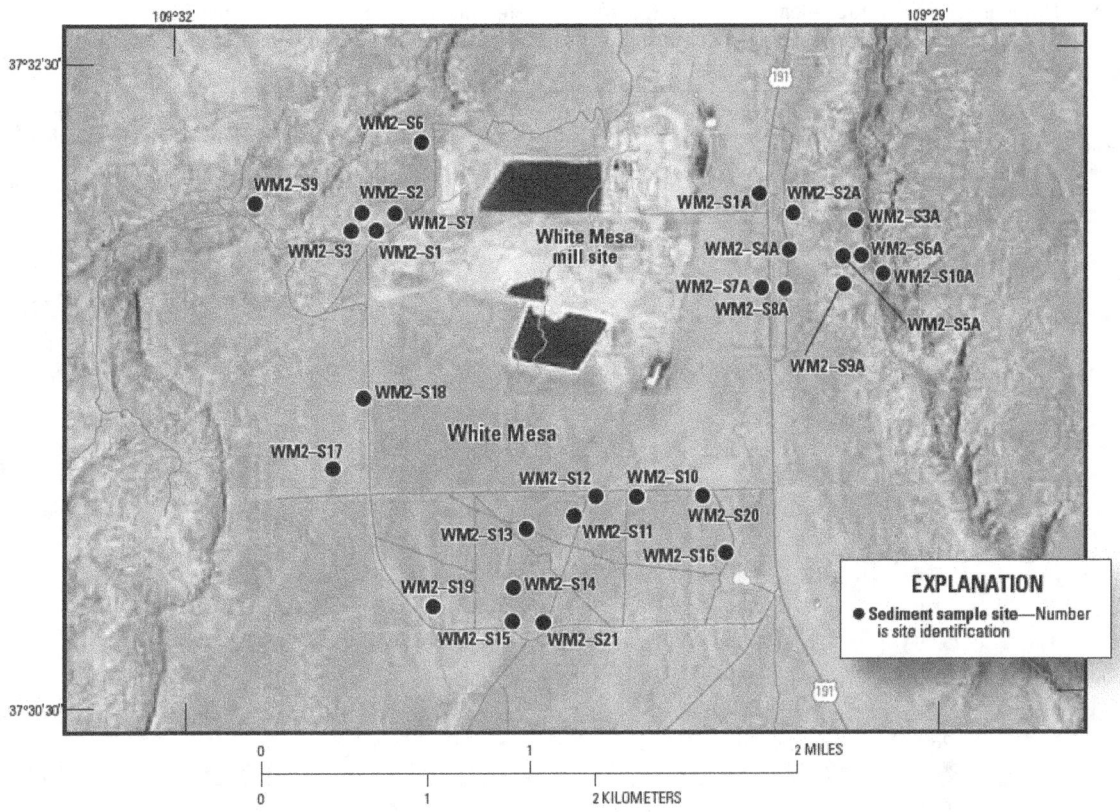

Figure 30.   Sites where sediment samples were collected in ephemeral drainages in close proximity to the White Mesa uranium mill, San Juan County, Utah, during June 2008.

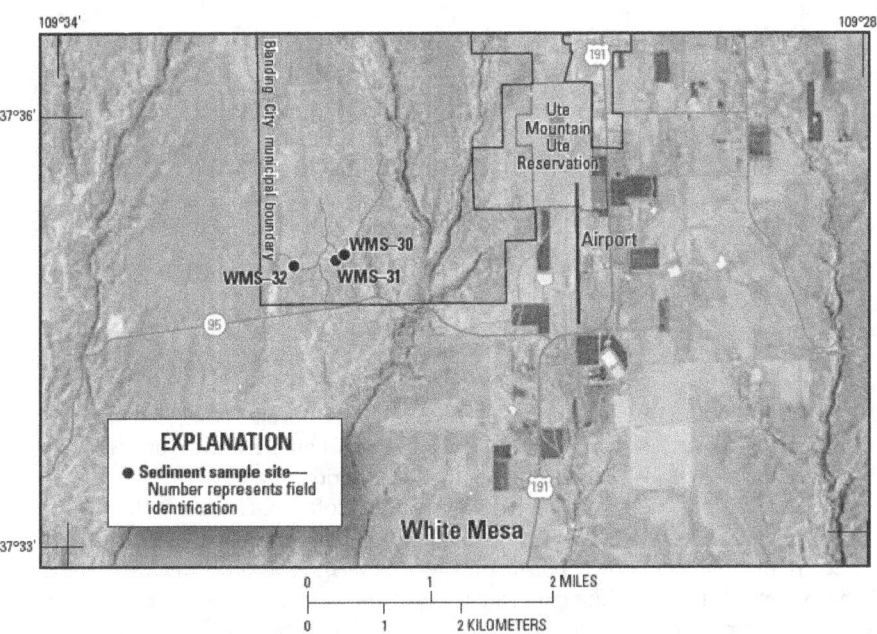

Figure 31.   Sites where background sediment samples were collected in ephemeral drainages approximately 5 kilometers north of the White Mesa uranium mill, San Juan County, Utah, during June 2008.

within acceptable limits and averaged within 12.4 percent for Green River Shale (SGR-1B) and 10.3 percent for Mica Schist (SDC-1; table 12).

The U concentration from the stream-sediment samples ranged from 1.5 to 16.2 parts per million (ppm). The highest U concentration measured in the local background samples (fig. 31), which ranged from 1.8 to 3.6 ppm, was equaled or exceeded in 8 of the 28 stream sediment samples. The stream-sediment data also were compared to the median concentration of stream-sediment samples collected in southeastern Utah (latitude range: 37.003 to 37.650 decimal degrees; longitude range: 109.044 to 110.779 decimal degrees) during the 1970s as part of the National Uranium Resource Evaluation (NURE) program (U.S. Geological Survey, 2010c). The median U

Table 12. Measurement errors for trace elements calculated from two reference materials that were submitted and analyzed with sediment samples collected from ephemeral drainages surrounding the White Mesa mill site, Utah, during June 2008.

[**Abbreviations**: ND, not determined; μg/g, micrograms per gram; <, less than lower reporting limit]

Chemical constituent	Green River Shale (SGR–1b) reference material, measured value, (μg/g)	Green River Shale (SGR–1b) reference material, expected value, (μg/g)	Green River Shale (SGR–1b) measurement error, (percent)	Mica Schist (SDC–1) reference material, measured value, (μg/g)	Mica Schist (SDC–1) reference material, expected value, (μg/g)	Mica Schist (SDC–1) measurement error, (percent)
Arsenic	64.0	67.0	−4.5	<1	0.2	ND
Barium	294.0	290.0	1.4	681.0	630.0	8.1
Beryllium	1.5	ND	ND	3.7	3.0	23.3
Cadmium	1.1	0.9	22.2	<0.1	ND	ND
Cerium	35.5	36.0	−1.4	90.8	93.0	−2.4
Cobalt	11.9	12.0	−0.8	18.3	18.0	1.7
Chromium	28.0	30.0	−6.7	60.0	64.0	−6.3
Cesium	5.0	5.2	−3.8	<5	4.0	ND
Copper	60.8	66.0	−7.9	26.0	30.0	−13.3
Gallium	9.5	12.0	−21.2	24.3	21.0	15.7
Lanthanum	19.2	20.0	−4.0	42.1	42.0	0.2
Lithium	128.0	147.0	−12.9	33.0	34.0	−2.9
Manganese	233.0	267.0	−12.7	839.0	880.0	−4.7
Molybdenum	35.2	35.0	0.6	0.2	ND	ND
Niobium	4.9	5.2	−5.8	15.7	21.0	−25.2
Nickel	26.5	29.0	−8.6	29.6	38.0	−22.1
Lead	40.2	38.0	5.8	21.2	25.0	−15.2
Rubidium	82.9	ND	ND	126.0	127.0	−0.8
Antimony	2.7	3.4	−22.1	0.5	0.5	−3.7
Selenium	5.0	4.6	8.7	16.0	17.0	−5.9
Tin	0.7	1.9	−63.2	2.9	3.0	−3.3
Strontium	377.0	420.0	−10.2	170.0	180.0	−5.6
Thorium	4.3	4.8	−10.4	11.0	12.0	−8.3
Thallium	0.5	ND	ND	0.5	0.7	−28.6
Uranium	5.4	5.4	0.0	2.7	3.1	−12.9
Vanadium	146.0	130.0	12.3	113.0	102.0	10.8
Tungsten	1.9	2.6	−26.9	0.6	0.8	−25.0
Yttrium	9.1	13.0	−30.0	31.7	ND	ND
Zinc	72.0	74.0	−2.7	102.0	103.0	−1.0
Selenium	2.5	3.5	−28.6	<0.2	ND	ND
Median			8.6			7.2
Mean			12.4			10.3

concentration in the NURE data set for southeastern Utah was 2.0 ppm (n = 627), and 27 of the 28 sediment samples collected in close proximity to the mill site exceeded the median value (fig. 32).

Figure 33 shows the location of the eight sediment samples that exceeded the maximum U concentration from the three local background samples. With the exception of site WM2-S21, sediment samples with elevated-U concentration cluster in the three ephemeral drainages east of the eastern mill boundary. In general, this area is downwind from the uncovered ore materials that are stockpiled at the mill and are in the same general area as Entrance Spring, which had the highest median U concentration of all the water monitoring sites sampled during the study period.

The USGS StreamStats software (Ries and others, 2008) was used to delineate the watershed for each of the three ephemeral drainages east of the mill site that were found to contain elevated-U concentrations in stream sediments (fig. 34). Because of the elevated-U found in the three ephemeral channels, it is likely that each of the designated watersheds could receive wind-blown dust with elevated-U concentrations from within the mill boundaries (for example,

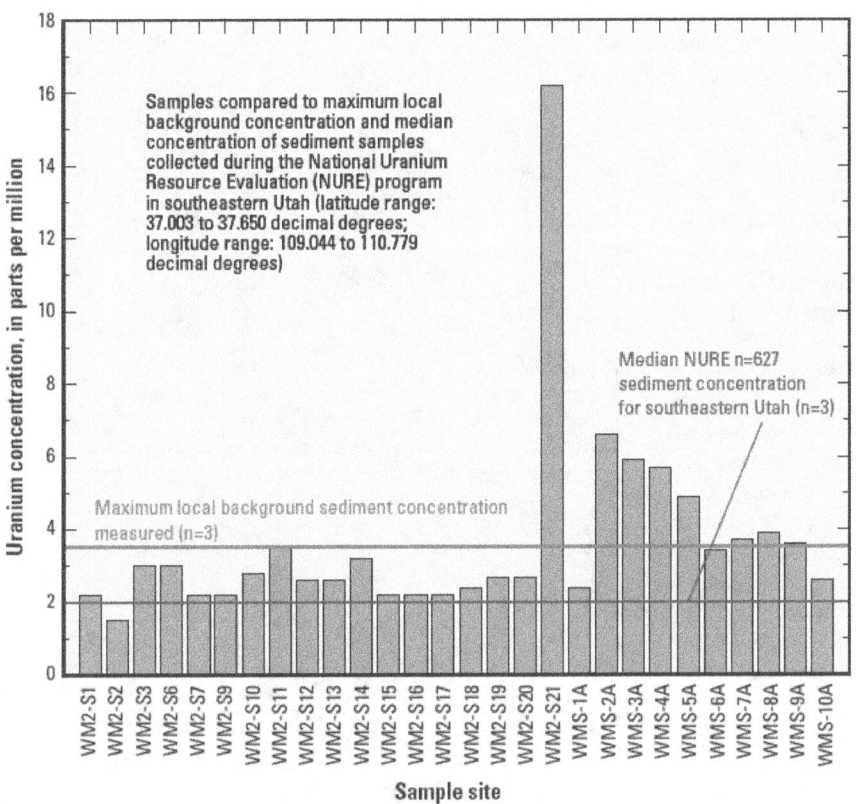

Samples compared to maximum local background concentration and median concentration of sediment samples collected during the National Uranium Resource Evaluation (NURE) program in southeastern Utah (latitude range: 37.003 to 37.650 decimal degrees; longitude range: 109.044 to 110.779 decimal degrees)

Median NURE n=627 sediment concentration for southeastern Utah (n=3)

Maximum local background sediment concentration measured (n=3)

Figure 32. Uranium concentration in sediment samples collected in ephemeral drainages in close proximity to the White Mesa uranium mill, San Juan County, Utah.

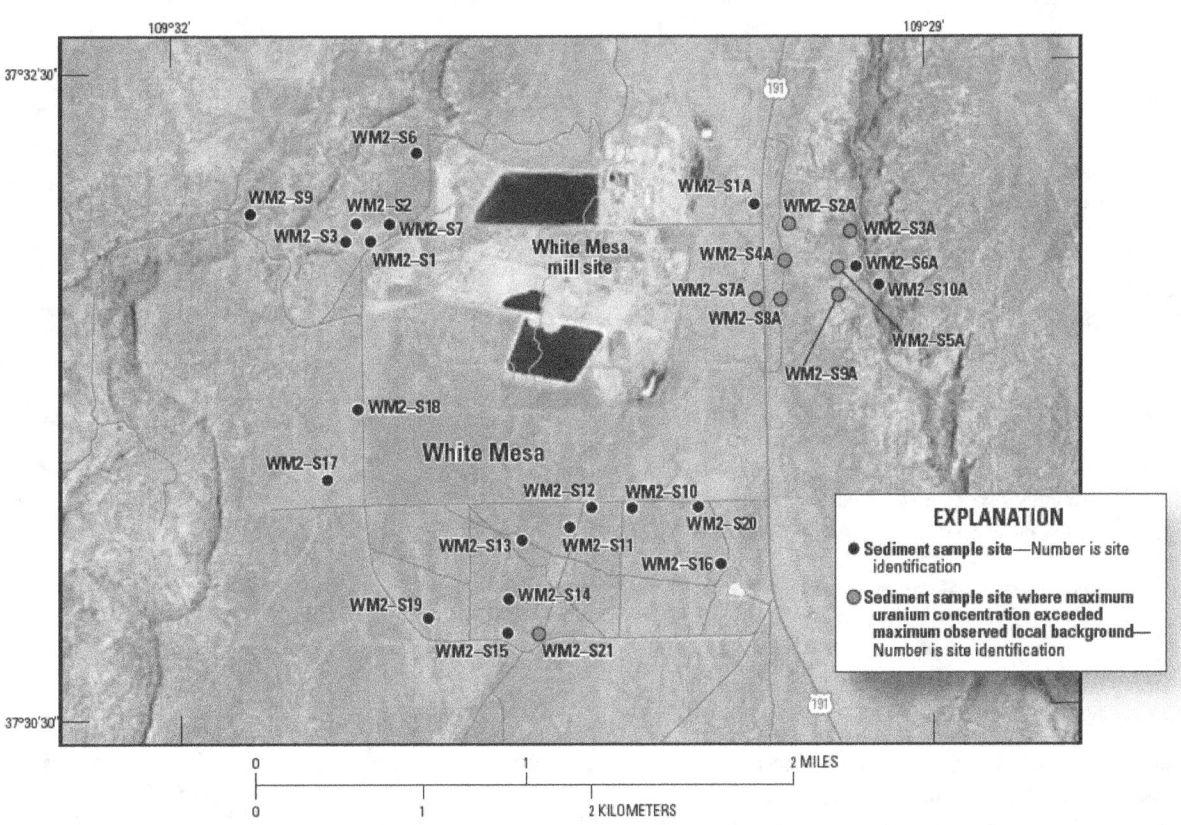

Figure 33. Sites where the measured uranium concentration in sediment samples exceeded the maximum uranium concentration observed in local background samples compared to sites where it did not during June 2008, San Juan County, Utah.

uncovered ore-storage piles or possible runoff from within the mill boundaries during rain and snowmelt events). Future assessments of offsite migration of ore material also should collect sediment samples from the two remaining unsampled ephemeral watersheds directly east of the mill site.

Sample site WM2-S21, located approximately 1.2 km south of the mill site in an ephemeral drainage originating within the mill boundaries, contained the highest U concentration (greater than 16 ppm) measured in any of the sediment samples. The elevated-U concentration in this sample was confirmed by two additional analyses of the stream-sediment sample (reanalysis 1 = 16.2 ppm and reanalysis 2 = 15.0 ppm). The U concentration in this sample was more than 8 times the median U concentration in the NURE data collected from southeastern Utah and likely is associated with transport of ore-grade material during a runoff event that was capable of transporting sediment down the ephemeral stream channel. The USGS StreamStats program (Ries and others, 2008) was used to delineate the watershed above sediment sample site WM2-S21 (fig. 35) also. The watershed boundaries delineated by the StreamStats program did not include the White Mesa mill site. Because of low surface gradients in this area, it is possible that the watershed boundaries estimated by the StreamStats program are not representative of actual conditions, which could include areas of the mill site. Additional data collected upstream of sample site WM2-S21 could help to determine the likely source(s) of the elevated-U concentration that was observed and to better delineate the watershed above the sample site.

Geochemical fingerprinting

In addition to U, the concentration of 41 other chemical constituents was determined in the 31 sediment samples collected from the ephemeral drainages surrounding the White Mesa mill site. Pattern-recognition modeling techniques were applied to this multivariate database to identify multi-element "geochemical fingerprints" that can be used to differentiate natural weathering of sediments from ore material and to use this information to identify areas that likely have received offsite migration of ore material through air or water transport.

Pattern-recognition modeling techniques have been used in a variety of environmental applications where multivariate chemical databases needed to be interpreted in the context of multiple environmental processes (for example, differentiating natural vs. anthropogenic trace-metal signatures). Naftz (1996a and 1996b) applied pattern-recognition modeling techniques to a large, chemical data base generated from the U.S. Department of the Interior's (DOI) National Irrigation Water Quality Program (NIWQP) to identify water that could pose a selenium hazard to waterfowl. Pattern-recognition techniques have been used for geochemical interpretation of organic biomarker signals (Christie and others, 1984). Archeological studies have used pattern-recognition techniques to discriminate marble sources (Mello and others, 1988) and classify ancient ceramics using major- and trace-element data (Heydorn and Thuesen, 1989). Pyrolysis-mass

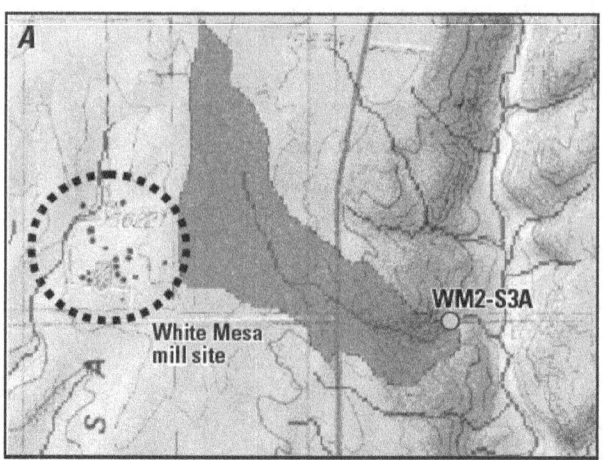

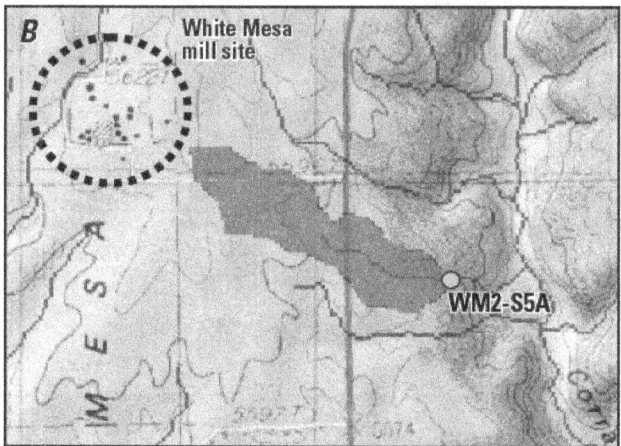

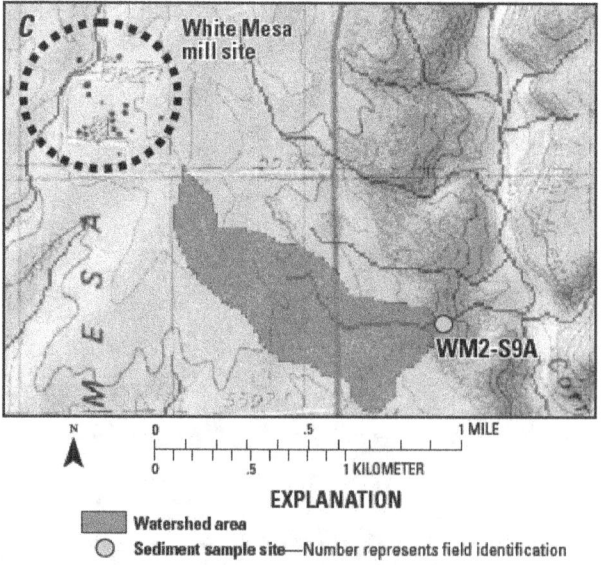

EXPLANATION

Watershed area

Sediment sample site—Number represents field identification

Figure 34. Location of sediment sample sites with elevated uranium and their corresponding watershed boundaries as estimated by the USGS StreamStats program (Ries and others, 2008) relative to the location of the White Mesa mill site, San Juan County, Utah: *A*, WM2-S3A; *B*, WM2-S5A; and *C*, WM2-S9A.

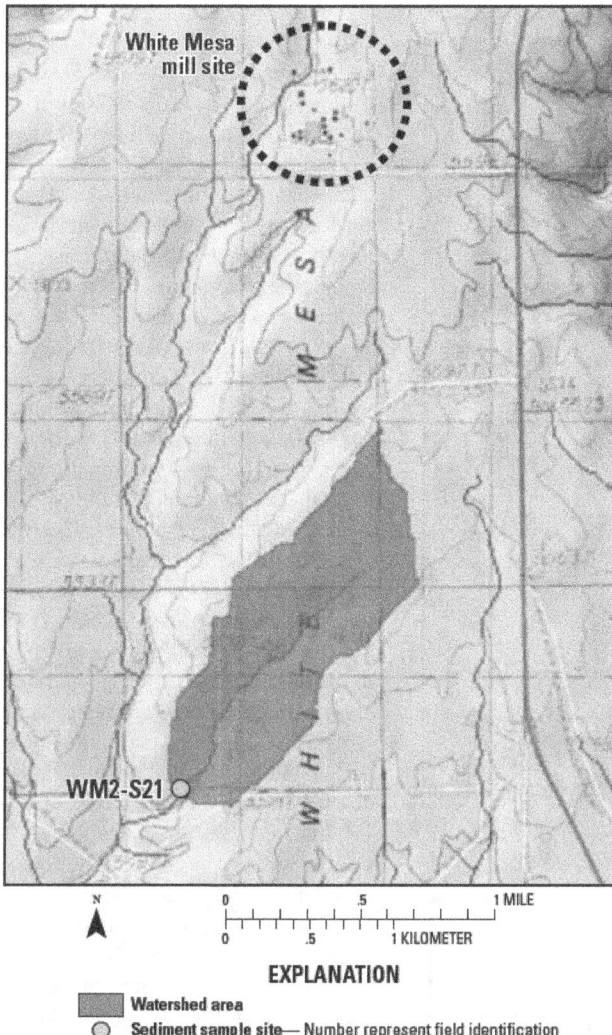

Figure 35. Location of sediment sample site WM2-S21 and the watershed boundary estimated by the USGS StreamStats program (Ries and others, 2008) relative to the location of the White Mesa mill site, San Juan County, Utah.

EXPLANATION

Watershed area

Sediment sample site— Number represent field identification

spectrometry analyses coupled with pattern-recognition techniques were useful in differentiating the origin of smoke aerosols (Voorhees and Tsao, 1985) and humic materials (MacCarthy and others, 1985). Also, pattern-recognition techniques applied to the elemental composition of oils have been used to determine spill-source identification in an oceanic setting (Duewer and others, 1975). In a hydrologic application, pattern-recognition techniques have been used to optimize multi-element groundwater quality monitoring programs at an oil-shale retort site (Meglen and Erickson, 1983).

Principal component analysis (PCA) was applied to the multi-element stream-sediment database to differentiate natural weathering from U ore "geochemical fingerprints." Two chemical constituents (cesium and tellurium) were not used in the PCA because the measured values consistently were below the lower reporting limit. Three factors were

found to account for 76 percent of the total variance of the multi-element stream-sediment database. The rotated loadings for the first two factors are shown in figure 36, with loading values (unitless) greater than 0.2 or less than –0.2 considered significant. Significant loadings associated with factor 1 include the elements Mg, Fe, Cr, K, Ti, and Y. The chemical elements associated with factor 1 were interpreted to be associated with the weathering of surficial geological units, predominantly the Burro Canyon Formation, surrounding the mill site. The Burro Canyon Formation consists primarily of sandstone, and the dominant minerals are quartz with small amounts of microcline and chert (Witkind, 1964). Calcite is the dominant cement; however, small amounts of silica and iron oxide cements were observed as well. Eolian sand deposits have been mapped by Haynes and others (1972) in the surficial materials east of the mill site, which are composed primarily of quartz grains covered by a thin film or iron oxide. Mineralogical analyses of rock samples collected during the study from areas surrounding White Mesa mill contained calcite, kaolinite, quartz, rutile, gypsum, orthoclase, anhydrite, and albite (Appendix 3).

The high loading for potassium (K) in factor 1 likely is explained by the presence of microcline and orthoclase, both K-containing feldspars, in the sediments. The high loading for iron (Fe) in factor 1 likely is explained by Fe oxide cement and coatings in the surficial geologic units, and the high loadings for chromium (Cr) and yttrium (Y) in factor 1 could be associated with trace elements in the Fe oxide coatings. The high factor 1 loading for titanium (Ti) could be associated with the mineral rutile, a Ti oxide that was detected in one of the mineralogical samples (Appendix 3). Finally, the high loading in factor 1 for magnesium (Mg) could be explained by the presence of calcite cements and the common substitution of magnesium for calcium in the mineral structure.

Significant loadings associated with factor 2 include Mo, As, S, Se, U, W, and Sb. These elements were interpreted to be associated with U-ore material contained within the White Mesa mill site. The elements, molybdenum (Mo), arsenic (As), and selenium (Se), are commonly associated with U deposits in the Salt Wash Member of the Morrison Formation (Miesch, 1962; 1963), sandstone-hosted U deposits in west-central Utah (Miller and others, 1984), as well as other U deposits in the western United States (Rose and others, 1979). Research by Miesch (1961 and 1963) found that antimony (Sb) was intrinsically related to U deposits in the Colorado Plateau. The high loading in factor 2 for sulfur (S) likely is related to the abundant amount of sulfide found in ores associated with U deposits (Miesch 1963).

The rotated scores for the first two factors were plotted (fig. 37) to evaluate the occurrence of distinct clusters in the data that could indicate common geochemical processes controlling the multi-element sediment chemistry observed among the ephemeral channel sampling sites in the study area. The rotated factor scores for the 31 sediment samples are grouped into two distinct clusters, identified as an ore migration and natural weathering grouping (fig. 37). The boundaries

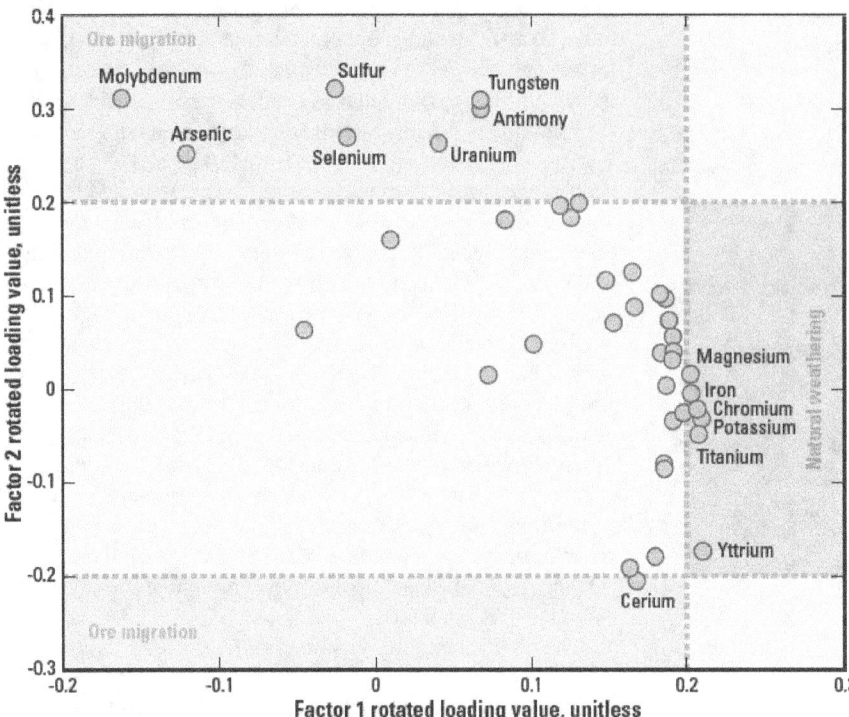

Figure 36. Loading values for principal components analysis factors 1 and 2 and chemical constituents with significant values for stream-sediment samples collected during June 2008 in the vicinity of the White Mesa mill site, San Juan County, Utah.

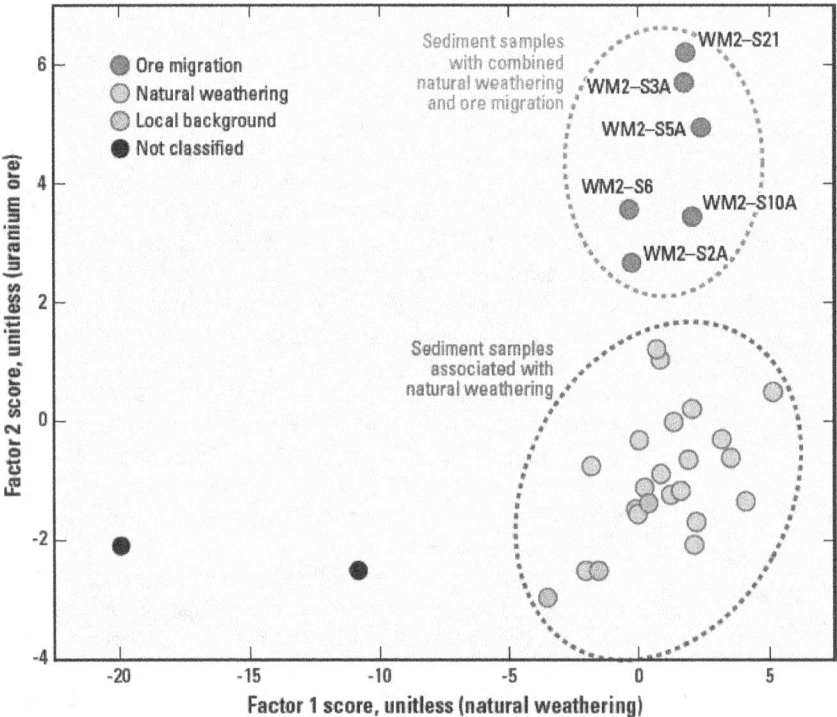

Figure 37. Scatter plot comparing factor 1 and factor 2 scores determined by principal components analysis of 31 stream-sediment samples collected from ephemeral drainages surrounding the White Mesa mill site, San Juan County, Utah, during June 2008.

drawn around the clusters of factor scores are not definitive, but aid in the visualization of the data and confirm possible commonalities in geochemical processes indicated by the variations in multi-element sediment chemistry for each of the score clusters and the sample-site locations identified in each of the clusters. Samples with high ore-migration factor scores also contain high scores associated with natural weathering. This combination of high scores with respect to both the ore-migration and natural-weathering factors is consistent with an ore-migration imprint in drainages containing naturally weathered stream sediments. The locations of the six samples with high ore-migration scores are shown in figure 38 and are located primarily in the ephemeral drainages directly east of the mill site. These are the same areas with elevated-U concentrations in the ephemeral drainage watersheds designated by StreamStats, which are downwind from the uncovered ore materials that are stockpiled at the mill site. The two remaining sediment samples with elevated ore-migration scores are located south and directly west of the mill site (fig. 38).

The three background samples are not shown in figure 38, but they contain low ore-migration scores and high natural-weathering scores and plot within the natural-weathering score cluster (fig. 37). Two of the 31 sediment samples are outside of both the natural-weathering and ore-migration score clusters. It is unclear why these two samples do not plot within the two score clusters and could represent an anomalous lithology or other unique set of geochemical characteristics.

Vegetation

Big sagebrush

Big sagebrush (*Artemisia tridentata*) is one of the most widely distributed and easily recognized shrubs in the western United States and has been used to establish

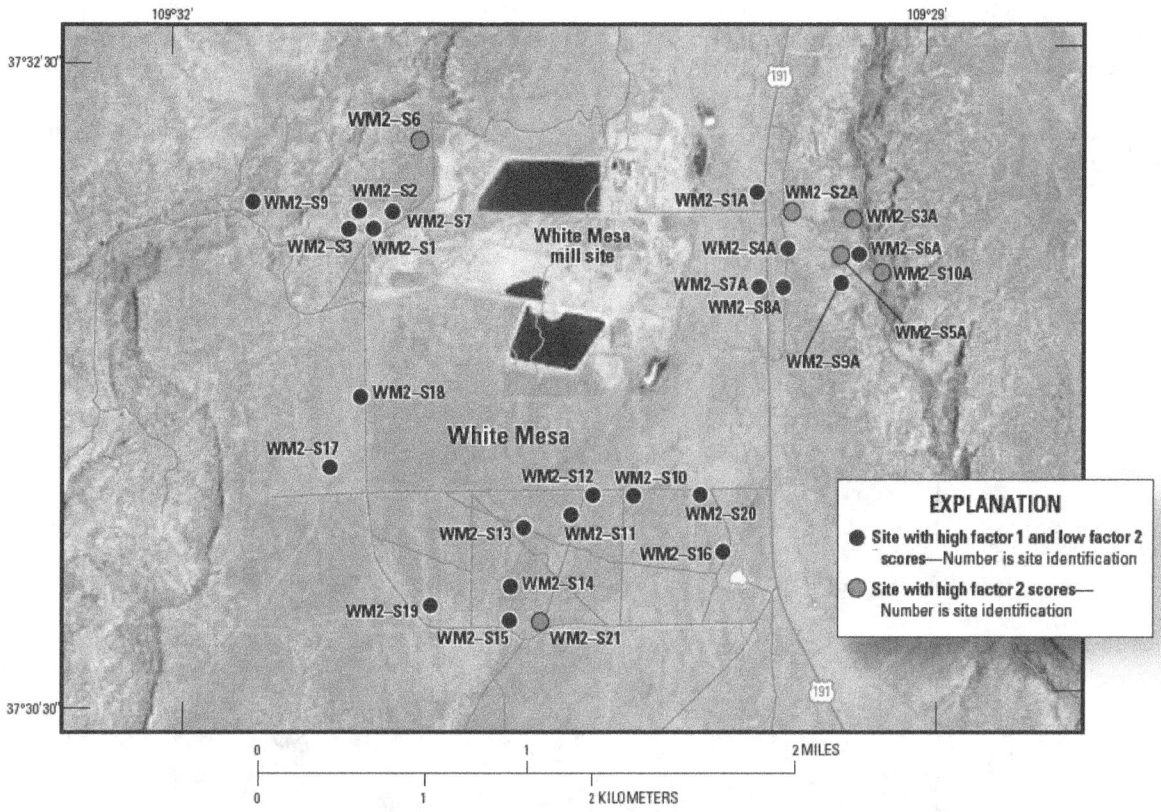

Figure 38. Location of sediment-sampling sites with high factor 2 scores (ore migration) compared to the location of sites with high factor 1 scores (natural weathering) and low factor 2 scores (ore migration), San Juan County, Utah, during June 2008.

geochemical baselines for selected chemical constituents since the late 1970s (Gough and Erdman, 1980; 1983). Big sagebrush develops an extensive root system and can accumulate trace-chemical constituents from soil water and groundwater containing mobile ions associated with ore deposits (Stewart and McKown, 1995). Because of the rough surface texture and resins on the leaf surfaces, sagebrush has been found to be very efficient at trapping dust (Wilt and others, 1992; Cutter and Guyette, 1993). Dust trapping on leaf surfaces was utilized in previous work to identify eolian transport of gold from a mill site (Smith and Kretschmer, 1992) and to detect ore spillage (Busche, 1989). Tissue samples were collected from big sagebrush in areas surrounding the White Mesa mill site during September 2009 (fig. 39) to determine areas of offsite migration of ore and associated material from eolian transport.

Tissue samples of new growth from plants growing within a 15-m radius from the center of each sample grid were composited and submitted for chemical analyses without surface rinsing to preserve any dust deposition geochemical signal (Appendix 4). Analytical results from the laboratory were verified by blindly submitting a certified standard reference material (National Institute of Standards plant reference material

1573a, tomato leaves) with the routine samples. On average, the laboratory results were within 13.1 percent of the accepted value and ranged from 3.4 to 33.4 percent (table 13).

In addition to the routine plant tissue samples collected from the center of each grid cell, additional samples were collected to determine the analytical and within-grid-cell variance. Six samples were collected from sample splits taken prior to laboratory analysis to assess analytical variance (table 14). Selected chemical constituents in the splits were consistently below the lower reporting limit (Ag, Cs, In, Te, Tl) or above the upper reporting limit (P) and could not be used to determine a mean percent difference between the analytical splits. The mean percent difference for the remaining 37 major and trace constituents was small, averaging 7.3 percent and ranging from 1.0 percent for strontium to 39.4 percent for chromium (table 14). In addition to the analytical sample splits, an additional plant composite sample was collected 200 m in a random direction from the routine sample site in the center of 10 of the grid cells (fig. 39) and used to qualitatively assess the within-grid-cell variance. The analytical results for the 10 sample pairs are shown in table 15 and generally indicate similar concentrations for the paired samples at this smaller, within-grid-cell geographic scale.

The U concentration in the plant-tissue samples from sagebrush ranged from 1.3 to 171 ppm (dry weight). The highest concentrations of U were found in plant tissue samples collected from regions north, south, and east of the mill site, and the lowest U concentrations were found west, northwest, and southwest of the mill site (fig. 40). Wind data collected from 2000 to 2008 at the Blanding airport (National Oceanic

and Atmospheric Administration, 2010), located about 6 km north of the mill, offers insight into the likely U source for the observed spatial distribution of U in the plant tissue samples (fig. 40). The predominant wind direction during the nine-year monitoring period was from the south-southwest (SSW) at an

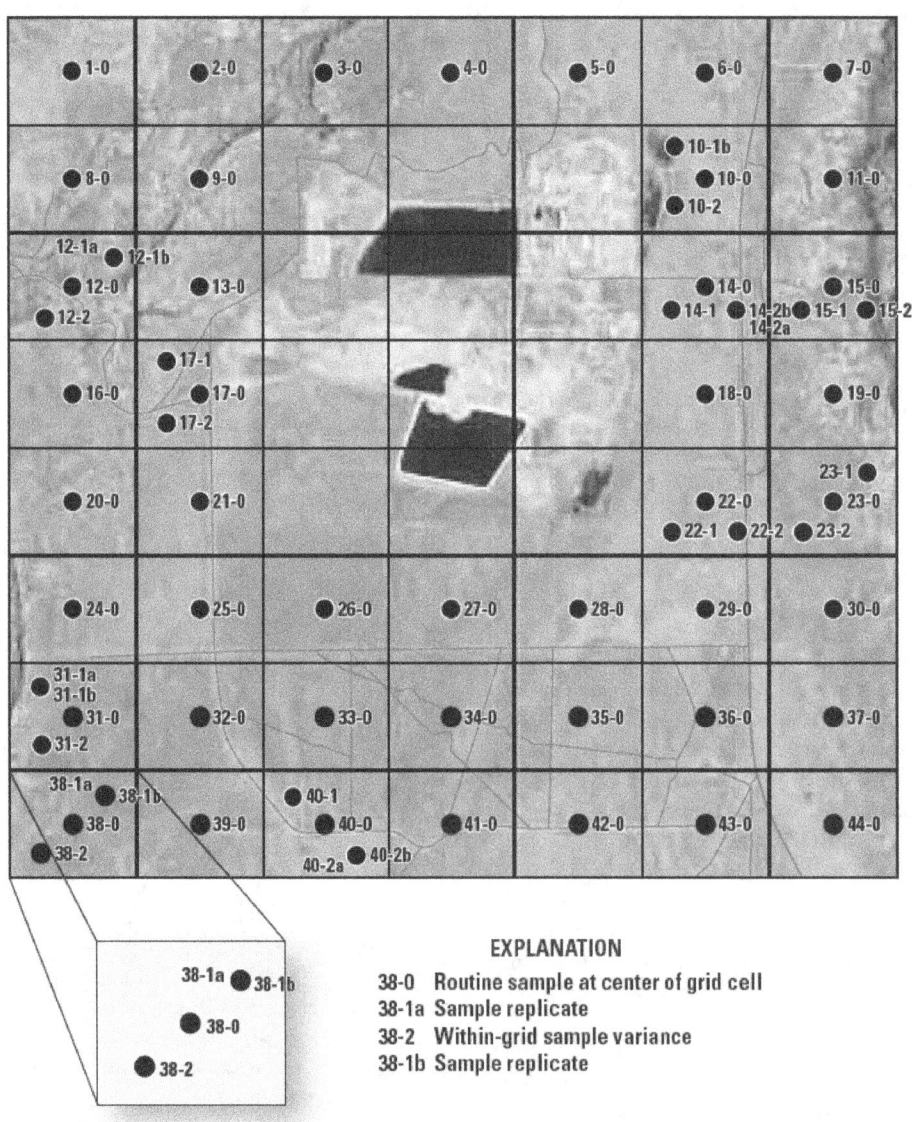

EXPLANATION

38-0   Routine sample at center of grid cell
38-1a   Sample replicate
38-2   Within-grid sample variance
38-1b   Sample replicate

Figure 39.   Sites where plant-tissue samples were collected from big sagebrush (*Artemisia tridentata*) in grid cell areas surrounding the White Mesa uranium mill site, San Juan County, Utah, during September 2009.

Table 13. Measurement errors calculated for National Institute of Standards and Technology (NIST) reference material that was submitted and analyzed with vegetation samples collected from areas surrounding the White Mesa mill site, Utah, during September 2009.

[**Abbreviations**: mg/kg, milligrams per kilogram; NIST, National Institute of Standards; ND, not determined; %, percent; *, element concentration determined but not NIST certified]

Chemical constituent and concentration units	NIST reference material (1573a, tomato leaves), measured value 1, concentration as specified	NIST reference material (1573a, tomato leaves), measured value 2, concentration as specified	NIST reference material (1573a, tomato leaves), measured value 3, concentration as specified	NIST reference material (1573a, tomato leaves), expected value, concentration as specified	NIST reference material (1573a, tomato leaves) average measurement error, (percent)
Aluminum (mg/kg)	560.0000	570.0000	560.0000	598.0000	5.8
Antimony (mg/kg)	0.0600	0.0640	0.0660	0.0630	3.7
Arsenic (mg/kg)	ND	ND	ND	0.1120	ND
Boron (mg/kg)	ND	ND	ND	33.3000	ND
Cadmium (mg/kg)	1.3000	1.5000	1.5000	1.5200	5.7
Chromium (mg/kg)	1.9000	1.8000	1.6000	1.9900	11.2
Cobalt (mg/kg)	0.5000	0.6000	0.6000	0.5700	7.6
Copper (mg/kg)	4.3000	4.4000	4.3000	4.7000	7.8
Iron (mg/kg)	350.0000	350.0000	330.0000	368.0000	6.7
Manganese (mg/kg)	221.5000	223.5000	220.4000	246.0000	9.8
Mercury (mg/kg)	ND	ND	ND	0.0340	ND
Nickel (mg/kg)	1.3000	1.2000	1.3000	1.5900	20.3
Rubidium (mg/kg)	10.3910	10.1890	9.1880	14.8900	33.4
Selenium (mg/kg)	ND	ND	ND	0.0540	ND
Sodium (mg/kg)	140.0000	180.0000	140.0000	136.0000	12.7
Vanadium (mg/kg)	0.6000	0.6000	0.6000	0.8350	28.1
Zinc (mg/kg)	26.1000	26.0000	26.0000	30.9000	15.7
*Magnesium (%)	0.9670	0.9760	0.9640	1.2000	19.3
*Sulfur (%)	0.9190	0.9350	0.9290	0.9600	3.4
*Barium (mg/kg)	58.8000	59.2000	58.1000	63.0000	6.8
*Bromine (mg/kg)	ND	ND	ND	1,300.0000	ND
*Cerium (mg/kg)	1.4000	1.5000	1.5000	2.0000	26.7
*Cesium (mg/kg)	ND	ND	ND	0.0530	ND
*Gadolinium (mg/kg)	ND	ND	ND	0.1700	ND
*Hatnium (mg/kg)	ND	ND	ND	0.1400	ND
*Lanthanum (mg/kg)	2.0000	2.2000	2.3000	2.3000	5.8
*Molybdenum (mg/kg)	0.4000	0.4000	0.4000	0.4600	13.0
*Tin (mg/kg)	0.0830	0.1230	0.1440	0.1000	28.0
*Silver (mg/kg)	ND	ND	ND	0.0170	ND
*Strontium (mg/kg)	79.5000	79.3000	78.7000	85.0000	6.9
*Thorium (mg/kg)	0.1000	0.1200	0.1200	0.1200	5.6
*Uranium (mg/kg)	0.0410	0.0410	0.0410	0.0350	17.1

Table 14. Comparison of analytical results from laboratory splits of sagebrush samples collected from areas surrounding the White Mesa mill site, Utah, during September 2009.

[**Abbreviations**: <, less than lower reporting limit; >, greater than upper reporting limit; ND, not determined; *, mean percent difference calculated with at least one missing value; ppm, parts per million]

Site ID	Aluminum, (percent dry weight)	Calcium, (percent dry weight)	Iron, (percent dry weight)	Potassium, (percent dry weight)	Magnesium, (percent dry weight)	Sodium, (percent dry weight)	Sulfur, (percent dry weight)	Titanium, (percent dry weight)	Silver, (ppm dry weight)	Barium, (ppm dry weight)
10–1a	0 67	10 7	0 33	12 3	4 98	0 18	3 97	0 03	<1	262
10–1b	0 61	10 4	0 3	13 1	4 9	0 17	4	0 03	<1	258
Percent difference	9.4	2.8	9.5	6.3	1.6	5.7	0.8	0.0	ND	1.5
12–1a	0 43	9 94	0 23	>15	2 18	0 11	3 25	0 02	<1	370
12–1b	0 44	9 89	0 23	13 7	2 19	0 1	3 28	0 02	<1	369
Percent difference	2.3	0.5	0.0	ND	0.5	9.5	0.9	0.0	ND	0.3
14–2a	1 16	10 4	0 58	12 3	2 97	0 24	3 18	0 06	<1	388
14–2b	1 04	10 6	0 53	12 5	3 02	0 22	3 22	0 05	<1	377
Percent difference	10.9	1.9	9.0	1.6	1.7	8.7	1.3	18.2	ND	2.9
31–1a	0 68	9 75	0 35	10 9	3 27	0 16	3 07	0 03	<1	356
31–1b	0 65	9 68	0 34	11	3 27	0 16	3 09	0 03	<1	357
Percent difference	4.5	0.7	2.9	0.9	0.0	0.0	0.6	0.0	ND	0.3
38–1a	0 81	9 55	0 41	9 81	3 01	0 15	3 14	0 02	<1	335
38–1b	0 79	9 37	0 4	11 4	3	0 15	3 02	0 02	<1	347
Percent difference	2.5	1.9	2.5	15.0	0.3	0.0	3.9	0.0	ND	3.5
40–2a	0 87	7 91	0 45	>15	2 38	0 13	2 3	0 03	<1	313
40–2b	0 82	7 84	0 43	15	2 31	0 12	2 23	0 02	<1	304
Percent difference	5.9	0.9	4.5		3.0	8.0	3.1	40.0		2.9
Mean percent difference (+/–)	5.9	1.5	4.7	6.0*	1.2	5.3	1.8	9.7	ND	1.9

Site ID	Beryllium, (ppm dry weight)	Bismuth, (ppm dry weight)	Cadmium, (ppm dry weight)	Cerium, (ppm dry weight)	Cobalt, (ppm dry weight)	Chromium, (ppm dry weight)	Cesium, (ppm dry weight)	Copper, (ppm dry weight)	Gallium, (ppm dry weight)	Indium, (ppm dry weight)	Lanthanium, (ppm dry weight)
10–1a	0 4	0 66	1 2	11	3 7	7	<5	207	2 05	<0 02	7 4
10–1b	0 3	0 63	1 1	10 5	3 6	8	<5	202	1 9	<0 02	7
Percent difference	28.6	4.7	8.7	4.7	2.7	13.3	ND	2.4	7.6	ND	5.6
12–1a	<0 1	0 12	1 2	5 06	1 5	5	<5	203	1 42	<0 02	2 6
12–1b	<0 1	0 11	1 1	4 99	1 4	4	<5	206	1 38	<0 02	2 6
Percent difference	ND	8.7	8.7	1.4	6.9	22.2	ND	1.5	2.9	ND	0.0
14–2a	0 3	0 45	1 6	13 7	4 4	15	<5	196	3 11	0 03	9 5
14–2b	0 3	0 43	1 6	12 8	4 2	8	<5	195	3 04	0 09	9 4
Percent difference	0.0	4.5	0.0	6.8	4.7	60.9	ND	0.5	2.3	100.0	1.1
31–1a	0 2	0 12	1 5	7 58	3 3	9	<5	151	1 83	<0 02	4 7
31–1b	0 2	0 11	1 5	7 24	3 2	6	<5	145	1 77	<0 02	4 5
Percent difference	0.0	8.7	0.0	4.6	3.1	40.0	ND	4.1	3.3	ND	4.3
38–1a	0 2	0 09	1 1	7 95	2 6	7	<5	191	1 96	<0 02	5 2
38–1b	0 2	0 1	1 2	8 14	2 7	7	<5	187	2 1	<0 02	5 3
Percent difference	0.0	10.5	8.7	2.4	3.8	0.0	ND	2.1	6.9	ND	1.9
40–2a	0 2	0 08	1 7	9 69	2 8	24	<5	141	2 27	<0 02	6
40–2b	0 2	0 06	1 6	9 05	2 6	8	<5	136	2 07	<0 02	5 6
Percent difference	0.0	28.6	6.1	6.8	7.4	100.0	ND	3.6	9.2	ND	6.9
Mean percent difference (+/–)	5.7*	10.9	5.4	4.4	4.8	39.4	ND	2.4	5.4	ND	3.3

Table 14. Comparison of analytical results from laboratory splits of sagebrush samples collected from areas surrounding the White Mesa mill site, Utah, during September 2009.—Continued

[**Abbreviations**: <, less than lower reporting limit; >, greater than upper reporting limit; ND, not determined; *, mean percent difference calculated with at least one missing value; ppm, parts per million]

Site ID	Lithium, (ppm dry weight)	Manganese, (ppm dry weight)	Molybdium, (ppm dry weight)	Niobium, (ppm dry weight)	Nickel, (ppm dry weight)	Phosphorous, (ppm dry weight)	Lead, (ppm dry weight)	Rubidium, (ppm dry weight)	Antimony, (ppm dry weight)	Scandium, (ppm dry weight)	Tin, (ppm dry weight)
10–1a	11	780	26 9	2 9	28 8	>10,000	13 8	26 8	0 37	1 7	8 1
10–1b	12	761	27 1	2 1	28 6	>10,000	12 4	23 9	0 29	1 5	2 1
Percent difference	8.7	2.5	0.7	32.0	0.7	ND	10.7	11.4	24.2	12.5	117.6
12–1a	58	755	21 7	0 9	9 7	>10,000	3 7	41 2	0 25	0 9	0 3
12–1b	60	764	20 9	0 9	9 6	>10,000	4 4	38	0 28	0 8	0 4
Percent difference	3.4	1.2	3.8	0.0	1.0	ND	17.3	8.1	11.3	11.8	28.6
14–2a	17	944	27 6	2 1	18 9	>10,000	12 3	34 8	0 33	2 2	1 7
14–2b	18	979	31	2 2	17 7	>10,000	11 4	34 9	0 31	2 2	1 7
Percent difference	5.7	3.6	11.6	4.7	6.6	ND	7.6	0.3	6.3	0.0	0.0
31–1a	19	649	17 4	1 5	23 3	>10,000	4 6	23	0 36	1 2	0 3
31–1b	16	647	17 1	1 5	23 2	>10,000	4 2	22 5	0 43	1 2	0 4
Percent difference	17.1	0.3	1.7	0.0	0.4	ND	9.1	2.2	17.7	0.0	28.6
38–1a	16	731	13 2	1 2	13 8	>10,000	4 4	25 1	0 29	1 3	0 3
38–1b	17	724	14 8	1 5	13 7	>10,000	4 7	30 4	0 39	1 5	0 3
Percent difference	6.1	1.0	11.4	22.2	0.7	ND	6.6	19.1	29.4	14.3	0.0
40–2a	10	789	11 6	1 2	21 3	>10,000	5 3	34 2	0 32	1 4	0 4
40–2b	9	773	11 2	1 1	20 6	>10,000	4 8	31 4	0 33	1 3	0 4
Percent difference	10.5	2.0	3.5	8.7	3.3	ND	9.9	8.5	3.1	7.4	0.0
Mean percent difference (+/–)	8.6	1.8	5.5	11.3	2.1	ND	10.2	8.3	15.3	7.7	29.1

Site ID	Strontium, (ppm dry weight)	Tellurium, (ppm dry weight)	Thorium, (ppm dry weight)	Thallium, (ppm dry weight)	Uranium, (ppm dry weight)	Vanadium, (ppm dry weight)	Tungsten, (ppm dry weight)	Yttrium, (ppm dry weight)	Zinc, (ppm dry weight)	Arsenic, (ppm dry weight)	Selenium, (ppm dry weight)
10–1a	1,220	<0 1	1 8	<0 1	56 8	250	2 7	3 2	447	1 1	0 4
10–1b	1,210	<0 1	1 7	<0 1	49 5	229	2 4	2 9	443	0 9	0 3
Percent difference	0.8	ND	5.7	ND	13.7	8.8	11.8	9.8	0.9	20.0	28.6
12–1a	1,360	<0 1	0 8	<0 1	2 3	14	0 2	1 7	556	2	0 5
12–1b	1,380	<0 1	0 7	<0 1	2 2	14	0 2	1 7	563	<0 6	0 5
Percent difference	1.5	ND	13.3	ND	4.4	0.0	0.0	0.0	1.3	ND	0.0
14–2a	1,100	<0 1	2 2	<0 1	44 9	165	2 1	4 8	340	0 8	0 7
14–2b	1,110	<0 1	2	<0 1	40 6	150	2	4 6	329	0 9	0 7
Percent difference	0.9	ND	9.5	ND	10.1	9.5	4.9	4.3	3.3	11.8	0.0
31–1a	1,030	<0 1	1 3	<0 1	15 3	61	0 3	2 9	271	<0 6	0 2
31–1b	1,020	<0 1	1 2	<0 1	14 9	59	0 3	2 7	268	1 5	0 2
Percent difference	1.0	ND	8.0	ND	2.6	3.3	0.0	7.1	1.1	ND	0.0
38–1a	1,090	<0 1	1 3	<0 1	8 1	40	0 4	2 9	329	0 7	0 2
38–1b	1,110	<0 1	1 3	<0 1	8 4	39	0 4	3 2	329	0 7	<0 2
Percent difference	1.8	ND	0.0	ND	3.6	2.5	0.0	9.8	0.0	0.0	ND
40–2a	604	<0 1	1 6	<0 1	7 6	31	0 4	3 4	261	<0 6	0 4
40–2b	603	<0 1	1 5	<0 1	6 7	29	0 4	3 1	253	<0 6	0 4
Percent difference	0.2	ND	6.5	ND	12.6	6.7	0.0	9.2	3.1	ND	0.0
Mean percent difference (+/–)	1.0	ND	7.2	ND	7.9	5.1	2.8	6.7	1.6	10.6*	5.7*

Table 15. Comparison of analytical results from sagebrush samples collected within each sample grid (200-meter separation distance) from areas surrounding the White Mesa mill site, Utah, during September 2009.

[ppm, parts per million; <, less than lower reporting limit; >, greater than upper reporting limit]

Site ID	Aluminum, in percent dry weight	Calcium, in percent dry weight	Iron, in percent dry weight	Potassium, in percent dry weight	Magnesium, in percent dry weight	Sodium, in percent dry weight	Sulfur, in percent dry weight	Titanium, in percent dry weight	Silver, in ppm dry weight	Barium, in ppm dry weight	Beryllium, in ppm dry weight	Bismuth, in ppm dry weight
10-0	1 31	9 1	0 62	12 9	3 11	0 22	2 68	0 04	<1	370	0 6	3
10-2	0 59	9 94	0 28	14 1	3 87	0 15	3 83	0 02	<1	174	0 3	1 46
12-0	0 68	11 3	0 36	13 4	2 89	0 13	2 94	0 04	<1	573	0 2	0 16
12-2	0 32	9 66	0 18	>15	2 58	0 07	2 29	0 02	<1	185	<0 1	0 12
14-0	1 34	9 7	0 69	11 8	2 67	0 26	2 75	0 05	<1	356	0 4	0 81
14-2a	1 16	10 4	0 58	12 3	2 97	0 24	3 18	0 06	<1	388	0 3	0 45
15-0	0 58	9 69	0 31	12 6	2 87	0 11	3 09	0 03	<1	277	0 1	0 29
15-2	0 3	7 43	0 18	>15	3 88	0 67	>5	0 02	<1	143	<0 1	0 16
17-0	0 73	8 44	0 38	12 2	2 77	0 17	3 02	0 03	<1	270	0 2	0 16
17-2	0 52	10 3	0 28	11 4	2 83	0 11	3 2	0 02	<1	330	0 1	0 08
22-0	0 98	9	0 57	12 5	2 79	0 24	1 99	0 02	<1	303	0 3	0 32
22-2	1 02	10 1	0 54	11 4	2 9	0 23	2 55	0 03	<1	359	0 2	0 36
23-0	0 89	9 66	0 45	12 6	2 95	0 15	2 96	0 04	<1	287	0 2	0 17
23-2	0 53	10 9	0 28	11 5	2 99	0 11	3 14	0 02	<1	376	<0 1	0 18
31-0	0 64	10 6	0 33	11 5	2 37	0 13	2 89	0 03	<1	259	0 2	0 09
31-2	0 9	8	0 45	12 2	3 01	0 17	2 9	0 03	<1	266	0 2	0 12
38-0	0 75	10	0 37	8 92	2 88	0 13	3 02	0 02	<1	325	0 2	0 09
38-2	0 78	9 84	0 39	10 8	3 26	0 14	3	0 03	<1	377	0 2	0 1
40-0	0 56	11 2	0 29	11 9	3	0 09	3 12	0 02	5	324	<0 1	0 07
40-2a	0 87	7 91	0 45	>15	2 38	0 13	2 3	0 03	<1	313	0 2	0 08

Site ID	Cadmium, in ppm dry weight	Cerium, in ppm dry weight	Cobalt, in ppm dry weight	Chromium, in ppm dry weight	Cesium, in ppm dry weight	Copper, in ppm dry weight	Gallium, in ppm dry weight	Indium, in ppm dry weight	Lanthanium, in ppm dry weight	Lithium, in ppm dry weight	Manganese, in ppm dry weight	Molybdium, in ppm dry weight
10-0	1 9	25	6 7	11	<5	166	3 64	0 83	16 1	25	700	35
10-2	1	11 7	3 7	17	<5	264	1 96	0 03	7 9	27	760	42
12-0	1 4	7 57	2 1	4	<5	192	2 03	<0 02	4	16	941	21 7
12-2	1 7	3 91	1 3	4	<5	171	1 1	<0 02	2 1	12	851	12 7
14-0	1	16 3	5 6	10	<5	250	3 49	0 03	12 6	20	798	45 3
14-2a	1 6	13 7	4 4	15	<5	196	3 11	0 03	9 5	17	944	27 6
15-0	5	7 34	6 7	6	<5	235	1 63	<0 02	4 2	43	678	10 7
15-2	1 5	3 23	1 4	4	<5	246	1 12	<0 02	1 8	134	570	7 5
17-0	1 2	7 98	3 1	5	<5	185	2 07	<0 02	4 4	12	723	15
17-2	2	6 04	2 3	6	<5	175	1 56	<0 02	3 4	8	707	12 3
22-0	2 3	12 5	4	14	<5	199	2 64	<0 02	7 5	15	869	18 7
22-2	2 3	13 3	3 9	10	<5	176	2 82	<0 02	7 6	14	665	43 6
23-0	1 1	10 5	2 5	10	<5	157	2 37	<0 02	6 3	19	587	15 8
23-2	1 5	6 77	2	5	<5	158	1 46	<0 02	4 2	9	741	10 2
31-0	1 4	7 96	2	6	<5	168	1 71	<0 02	4 7	12	599	12 1
31-2	1 3	9 91	2 7	8	<5	163	2 29	<0 02	5 9	13	920	13 1
38-0	1	8 15	2 5	8	<5	151	1 92	<0 02	5	14	679	13 6
38-2	1 3	8 29	2 6	7	<5	155	2	<0 02	5 1	14	837	15 3
40-0	1 6	6 15	2 2	6	<5	155	1 63	<0 02	3 8	9	834	20 3
40-2a	1 7	9 69	2 8	24	<5	141	2 27	<0 02	6	10	789	11 6

Table 15. Comparison of analytical results from sagebrush samples collected within each sample grid (200-meter separation distance) from areas surrounding the White Mesa mill site, Utah, during September 2009.—Continued

[ppm, parts per million; <, less than lower reporting limit; >, greater than upper reporting limit]

Site ID	Niobium, in ppm dry weight	Nickel, in ppm dry weight	Phosphorous, in ppm dry weight	Lead, in ppm dry weight	Rubidium, in ppm dry weight	Antimony, in ppm dry weight	Scandium, in ppm dry weight	Tin, in ppm dry weight	Strontium, in ppm dry weight	Tellurium, in ppm dry weight	Thorium, in ppm dry weight	Thallium, in ppm dry weight
10-0	19 2	28 9	>10,000	33 3	34	1 44	3 7	84	872	<0 1	5 1	0 1
10-2	2 4	27 6	>10,000	15 4	41 4	0 31	1 9	4 2	1,790	<0 1	2 4	<0 1
12-0	1 3	11 9	>10,000	3 7	30 6	0 29	1 3	0 4	1,560	<0 1	1 1	<0 1
12-2	0 7	17 7	>10,000	2 1	37 7	0 3	0 7	0 2	734	<0 1	0 6	<0 1
14-0	1 9	23 3	>10,000	17 7	26 7	0 35	2 7	2 7	1,080	<0 1	2 8	0 1
14-2a	2 1	18 9	>10,000	12 3	34 8	0 33	2 2	1 7	1,100	<0 1	2 2	<0 1
15-0	1 5	40	>10,000	5 9	36 7	0 28	1 2	3	1,290	<0 1	1 2	<0 1
15-2	0 6	23 1	>10,000	2 7	51 1	0 2	0 7	0 4	1,010	<0 1	0 5	<0 1
17-0	1 5	23 6	>10,000	4 5	31 7	0 37	1 4	0 4	1,080	<0 1	1 3	<0 1
17-2	1 2	21 9	>10,000	3 2	28 8	0 3	1	0 3	1,280	<0 1	1	<0 1
22-0	1 3	49 8	>10,000	8 7	43 8	0 29	1 7	2 9	800	<0 1	2	<0 1
22-2	1 5	44 7	>10,000	9 1	33 7	0 43	1 9	0 9	983	<0 1	2 1	<0 1
23-0	1 6	20 7	>10,000	6 6	26 2	0 51	1 5	0 7	901	<0 1	1 7	<0 1
23-2	1 4	15 4	>10,000	5	36 7	0 53	1	0 7	1,050	<0 1	1	<0 1
31-0	1 3	14 3	>10,000	3 9	25 3	0 2	1	0 3	865	<0 1	1 3	<0 1
31-2	1 7	13 7	>10,000	5 3	22 4	0 33	1 5	0 4	646	<0 1	1 7	<0 1
38-0	1 5	16 6	>10,000	4 1	28	0 29	1 3	0 5	1,050	<0 1	1 4	<0 1
38-2	1 5	14 8	>10,000	4 8	26 8	0 26	1 4	0 3	1,190	<0 1	1 4	0 3
40-0	1	22	>10,000	5 2	44 5	0 22	1	0 2	1,040	<0 1	1	<0 1
40-2a	1 2	21 3	>10,000	5 3	34 2	0 32	1 4	0 4	604	<0 1	1 6	<0 1

Site ID	Uranium, in ppm dry weight	Vanadium, in ppm dry weight	Tungsten, in ppm dry weight	Yttrium, in ppm dry weight	Zinc, in ppm dry weight	Arsenic, in ppm dry weight	Selenium, in ppm dry weight
10-0	171	582	11 5	7	515	1 2	0 6
10-2	74	220	3 1	3 5	474	0 8	0 5
12-0	3	19	0 3	2 7	421	<0 6	3 3
12-2	1 3	9	0 2	1 3	712	<0 6	<0 2
14-0	72 8	278	3 9	5 9	352	1 5	1
14-2a	44 9	165	2 1	4 8	340	0 8	0 7
15-0	15 7	55	1	2 4	615	1 7	0 2
15-2	5	15	0 3	1 1	679	<0 6	0 7
17-0	17 8	54	0 4	2 8	317	<0 6	0 2
17-2	9 4	31	0 3	2 2	285	1	0 4
22-0	41 9	91	1 4	3 9	286	0 9	0 3
22-2	40 5	80	1 5	4	237	1 6	0 4
23-0	15 3	45	0 7	3 5	294	0 7	<0 2
23-2	13 4	41	0 8	2 1	240	<0 6	<0 2
31-0	6 6	31	0 3	2 7	390	<0 6	<0 2
31-2	9 9	44	0 4	3 5	329	<0 6	<0 2
38-0	7 3	31	0 3	2 8	262	<0 6	0 4
38-2	7 1	32	0 3	2 9	281	<0 6	0 2
40-0	7	22	0 3	2 2	229	<0 6	0 4
40-2a	7 6	31	0 4	3 4	261	<0 6	0 4

azimuth of about 200 degrees (fig. 41). This could explain the anomalous U concentrations detected in plant tissue samples collected to the north and northeast of the mill site. Furthermore, some of the highest wind speeds, exceeding 4 meters per second (m/s) were from westerly directions (azimuth 200 to 340 degrees), providing an explanation for the anomalous U concentrations east of the mill site with the predominant direction from the SSW (205 degrees).

The second most predominant wind direction observed at the Blanding airport was from the north at an azimuth of 360 degrees (fig. 41). Wind originating from this direction likely can be responsible for the anomalous-U concentrations detected in plant tissue samples collected to the south of the mill site (fig. 40).

Elevated levels of vanadium (V) also would be present in ore material delivered to the White Mesa mill from mines operating in the Colorado Plateau. According to Northrop and others (1990), tabular-type V-U deposits occur in fluvial sandstones of the Salt Wash Member of the Morrison Formation in the Henry structural basin of southeastern Utah, and are characteristic of Salt Wash-hosted tabular V-U deposits throughout the Colorado Plateau. The V concentration in the plant tissue samples ranged from 9 to 582 ppm (dry weight), and its spatial distribution in the plant tissue samples was similar to the U distribution (fig. 42). Plant samples with elevated V concentrations consistently were found north-northeast,

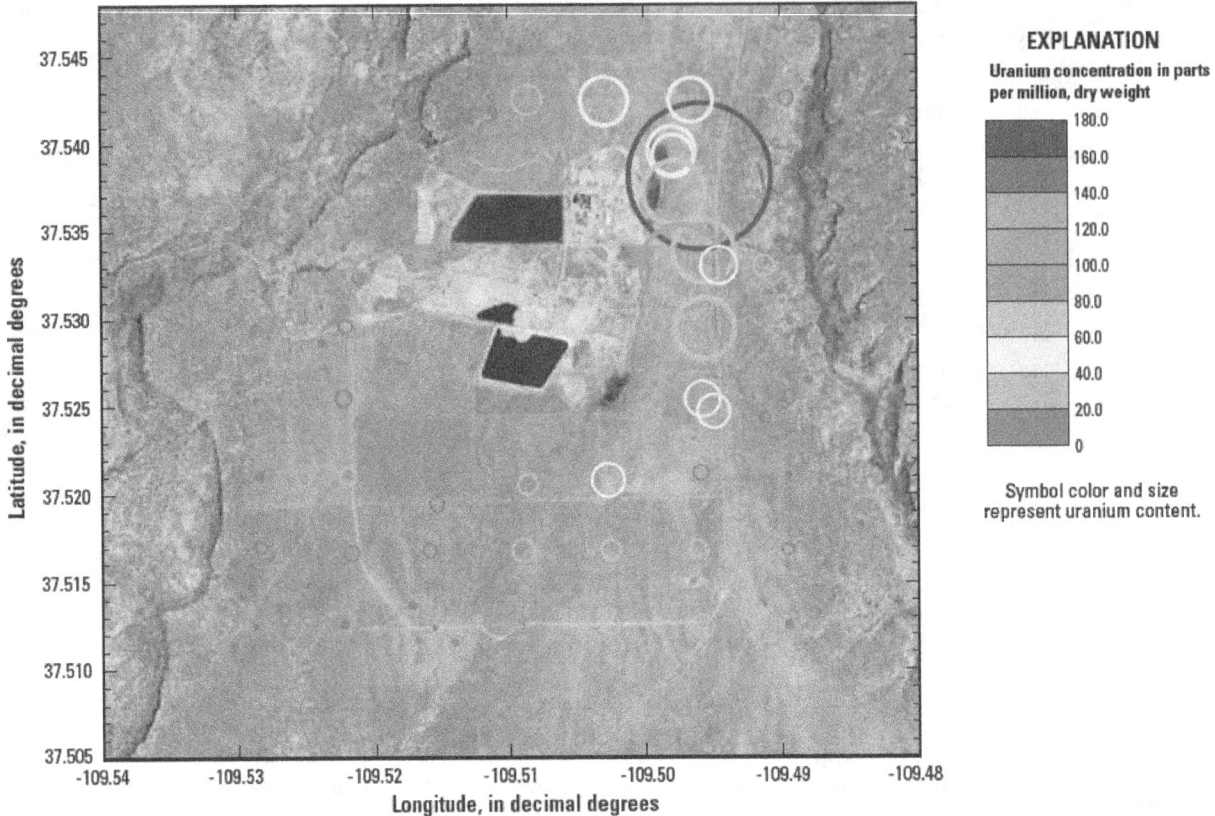

Figure 40. Uranium concentration in plant-tissue samples collected from big sagebrush (*Artemisia tridentata*) in areas surrounding and within the White Mesa uranium mill, San Juan County, Utah, during September 2009.

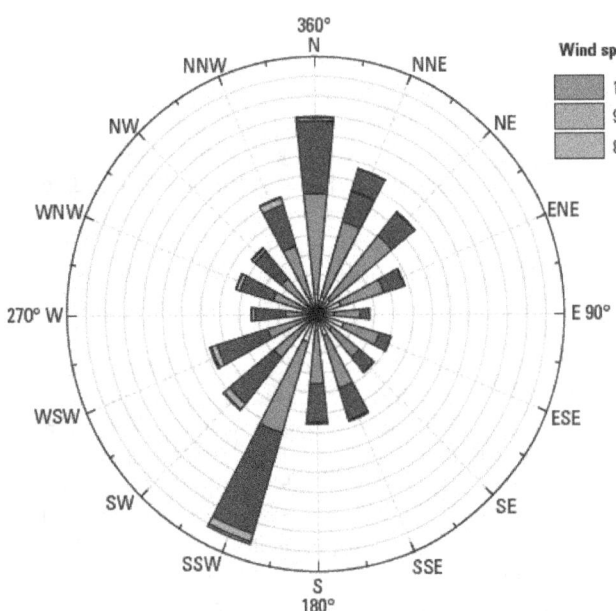

Figure 41. Rose diagram compiled from wind monitoring data collected at the Blanding airport, San Juan County, Utah, from January 2000 through May 2008 (National Oceanic and Atmospheric Administration, 2010).

east, and south of the mill site, indicating offsite transport in the predominant wind directions. The V concentration in plant samples collected west of the mill site was low (consistently less than 100 ppm, dry weight).

The spatial distribution of a non-ore related element, calcium, in plant tissue samples was investigated to substantiate the eolian transport of ore-material to areas surrounding the mill site. As noted in a previous section, calcite ($CaCO_3$) is the dominant cement in the Burro Canyon Formation and has been identified in rock samples collected from the study area (Appendix 3). Because calcium is present in the soil and rock material surrounding the mill site and not enriched in the ore material transported to the site, the spatial distribution of calcium concentration in plant tissue samples would not be elevated in the leeward areas surrounding the mill site. The calcium concentration in the plant-tissue samples from

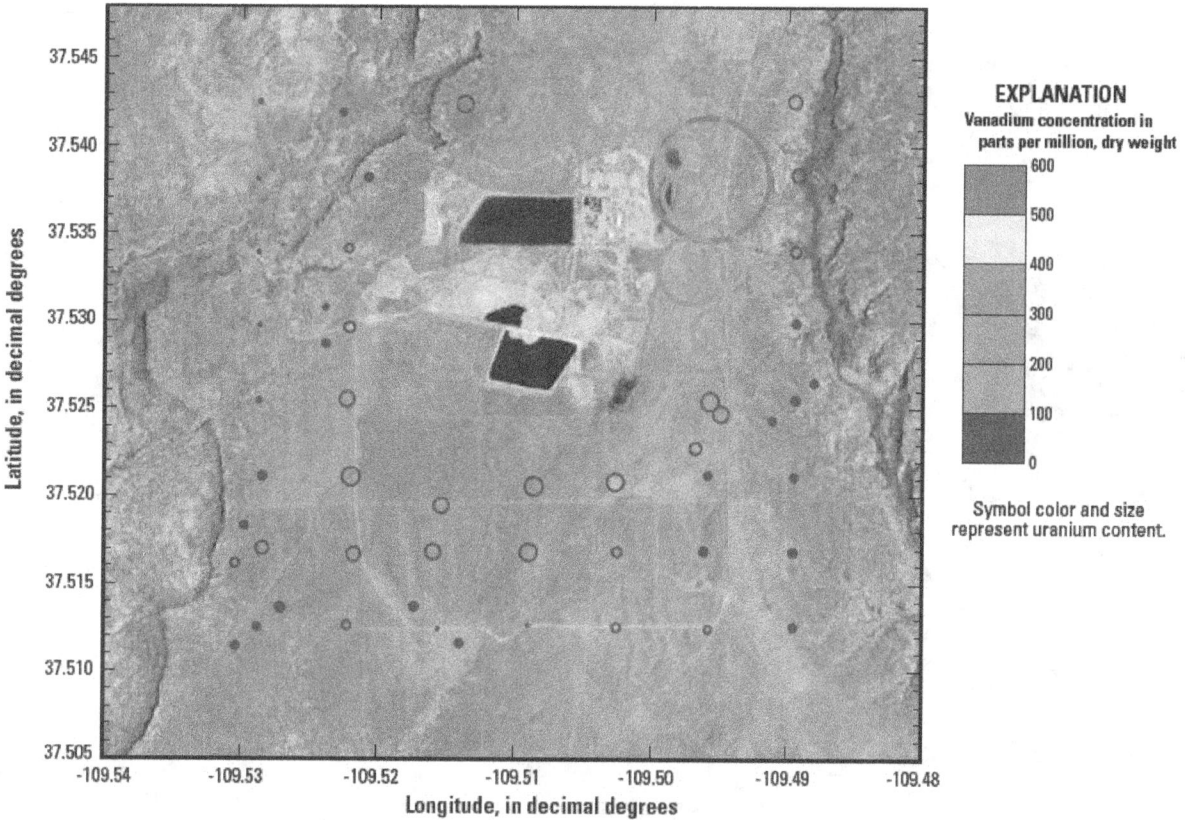

Figure 42. Vanadium concentration in plant-tissue samples collected from big sagebrush (*Artemisia tridentata*) in areas surrounding and within the White Mesa uranium mill, San Juan County, Utah, during September 2009.

sagebrush ranged from 7.4 to 11.4 percent (dry weight). In contrast to the spatial distribution of U and V concentrations in plant tissue samples, calcium concentrations did not display any spatial pattern related to eolian transport (fig. 43). The observed distribution of calcium is consistent with a chemical element uniformly distributed in the soil and rock material of the study site and inconsistent with a chemical element that would be enriched from material transported and stockpiled at the mill site, such as U and V.

Cottonwood Tree Coring

Cottonwood trees growing adjacent to five of the springs that were routinely sampled during the study (Oasis, Mill, Entrance, Cow Camp, and Ruin Springs; fig. 19) were cored using standard tree coring methods (Yanosky and Vroblesky, 1992). Previous work has indicated that chemical analyses of tree cores can provide insight into the historical concentration of selected contaminants in shallow groundwater systems (Yanosky and Vroblesky, 1989a; Yanosky and Vroblesky, 1989b; Yanosky and Vroblesky, 1992); therefore, chemical analyses of cores from cottonwood trees growing adjacent to springs surrounding the White Mesa mill site could provide a

good proxy for the historical reconstruction of U concentrations in groundwater before and after mill operation.

The outer 2 cm of each tree core were analyzed for U content at the USGS National Water Quality Laboratory. Dating of two of the five tree cores indicated that (fig. 44) it is likely that the outer 2 cm of core material grew during mill operation. Chemical analysis of the outer tree-core material did not detect a U concentration above the lower reporting limit of 0.1 micrograms per gram (µg/g), dry weight (table 16). Because U could not be detected in the five outer tree-core samples, additional U analyses of older core material would not be useful for reconstructing historical trends in spring-water U concentration; therefore, additional samples were not analyzed.

Environmental Implications

The mill site has been in operation since 1980 and is currently (2010) the only conventional uranium mill operating in the United States. In 2007, the Ute Mountain Ute Tribe requested that the EPA and USGS conduct an independent evaluation of potential offsite migration of radionuclides and selected trace elements associated with ore storage and the

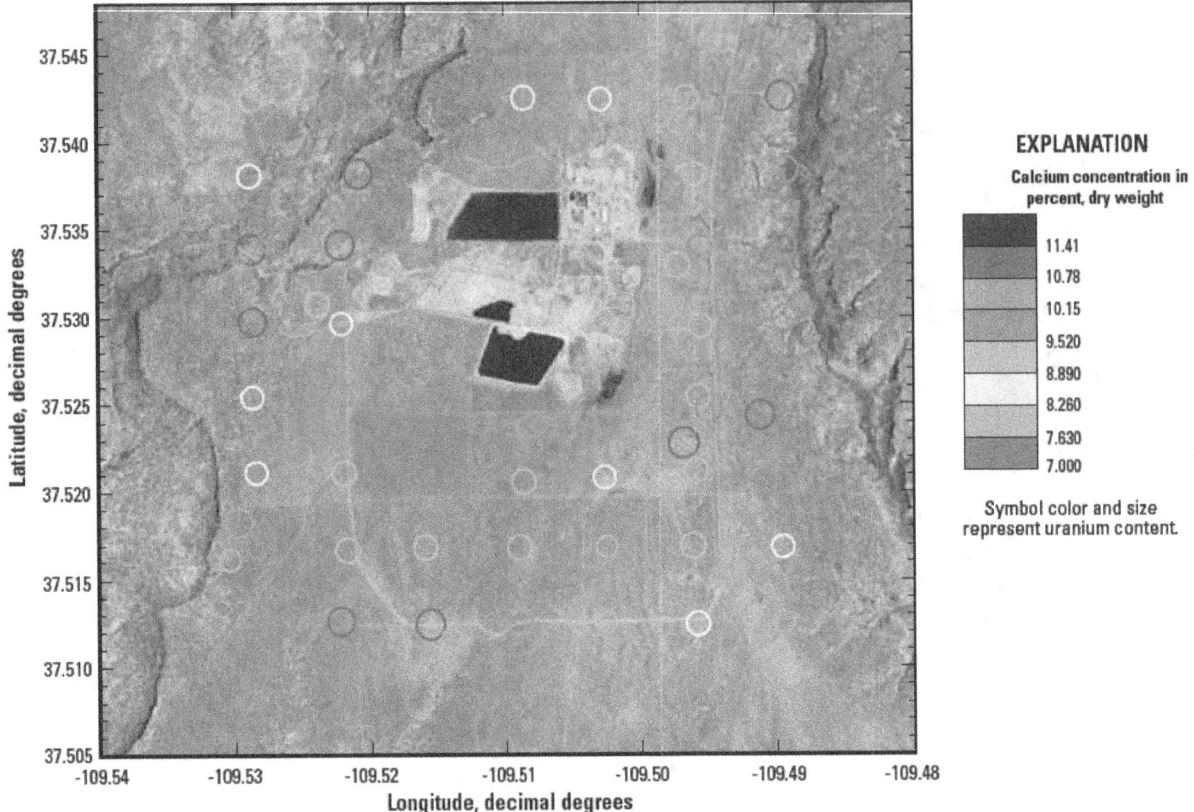

Figure 43. Calcium concentration in plant-tissue samples collected from big sagebrush (*Artemisia tridentata*) in areas surrounding and within the White Mesa uranium mill, San Juan County, Utah, during September 2009.

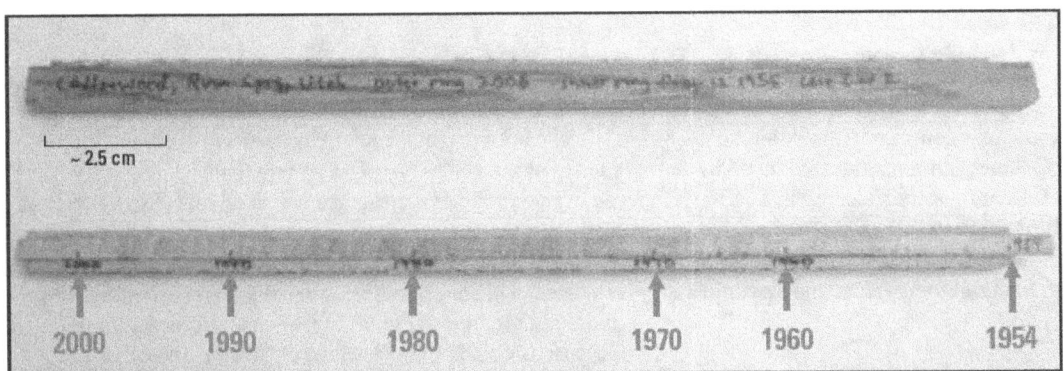

Figure 44. Photograph of dated tree core collected from near Ruin Spring, Utah. Core prepared and dated by T. Yanosky, U.S. Geological Survey (retired).

Table 16. Analytical results from tree cores collected at spring sites surrounding the White Mesa mill site near Blanding, Utah.

[**Abbreviations**: mm/dd/yyyy, month/day/year; µg/g, micrograms per gram; <, less than]

Tree–coring site	Sample date (mm/dd/yyyy)	Uranium concentration, (µg/g dry weight)	Water present in biota tissue, (percent of dry weight)
Oasis Spring	11/12/2008	<0.1	69
Mill Spring	11/12/2008	<0.1	58
Entrance Spring	11/11/2008	<0.1	86
Cow Camp Spring	11/13/2008	<0.1	69
Ruin Spring	11/12/2008	<0.1	69

milling process to tribal lands and Bureau of Land Management (BLM) managed properties adjacent to the mill site. Specific objectives of this study were (1) to better understand recharge sources and residence times of groundwater surrounding the mill site, (2) to determine the current concentrations of U and associated trace elements in groundwater surrounding the mill site, (3) to differentiate natural from anthropogenic contaminant sources to groundwater resources surrounding the mill site, (4) to assess the solubility and potential for offsite transport of U-bearing minerals in groundwater surrounding the mill site, and (5) to use stream-sediment and plant-material samples from areas surrounding the mill site to identify potential areas of offsite contamination and contaminant sources. The study results are summarized in terms of implications for offsite migration of contaminants from the mill site (fig. 45).

Age-dating methods and an evaluation of groundwater recharge temperatures using dissolved-gas samples were used to assess the recharge source and the residence time of groundwater at various sampling sites surrounding the mill site. The apparent age and probable recharge temperatures estimated from these methods for water derived from wells completed in the surficial aquifer indicate that the aquifer is recharged locally by precipitation. Tritium/helium age-dating of water samples collected from Cow Camp Spring, Oasis Spring, and Entrance Spring yielded apparent ages of recent to 18 years. This apparent age indicates a localized and potentially induced flow path from artificial recharge to the surficial aquifer. Potential sources of artificial recharge include infiltrating water from the unlined wildlife refuge ponds located to the northeast of the mill site and irrigated agriculture in the fields surrounding Blanding, Utah. Water samples with apparent ages greater than 50 years, including wells completed in the Dakota Sandstone/Burro Canyon and Navajo Sandstone aquifers, indicate little to no current risk of contamination from mill operations because the mill only has been in operation since 1980.

Water samples from Entrance Spring were found to be the most isotopically enriched relative to all the water samples that were collected during the study. The $\delta^{18}O$ and δD data indicate that water discharging from Entrance Spring contains the isotopic fingerprint of water from Recapture Reservoir, which is used as facilities water on the mill site and as an irrigation source for fields surrounding the town of Blanding. Infiltration of the facilities water or excess irrigation water could contribute to the enriched isotopic fingerprint observed for Entrance Spring.

Stable isotopes of sulfur and oxygen in sulfate were used to identify potential leakage from the tailings cells to areas outside the mill site. Hurst and Solomon (2008) found that water samples from the tailings cells were enriched in $\delta^{18}O_{sulfate}$ relative to other water samples on the mill property. In addition, Hurst and Solomon found that the sulfuric acid used during ore processing resulted in relatively consistent values of $\delta^{34}S_{sulfate}$ in water samples from the tailings cells. None of the spring or monitoring-well samples collected from areas

surrounding the mill site contain $\delta^{18}O_{sulfate}$ and $\delta^{34}S_{sulfate}$ isotopic signatures indicative of recharge from tailings cells within the mill boundary. Similarities in the $\delta^{34}S_{sulfate}$ values in water samples from the wildlife ponds and tailings cells indicate a possible contaminant linkage originating from the tailings cells (Hurst and Solomon, 2008) that could be related to eolian transport of aerosols from the cells. To date (2010), the $\delta^{34}S_{sulfate}$ or $\delta^{18}O_{sulfate}$ values measured in wells and springs surrounding the White Mesa mill site do not have an isotopic signature characteristic of the tailings cells. Because the wildlife ponds are actively leaking, it is likely that future groundwater samples from the surficial aquifer at sites within and adjacent to the mill site could exhibit decreasing $\delta^{34}S_{sulfate}$ values .

All dissolved uranium concentrations in groundwater at downgradient sites sampled during this study, except for Entrance Spring and the September 2008 and September 2009 samples collected at Mill Spring, had dissolved-U concentrations in the range expected for naturally occurring U and that of upgradient sites. Uranium isotopes were used to help distinguish the source of U in the groundwater samples collected from all sites during the study. The uranium isotope data indicate that the mill is not a source of uranium in the groundwater in the unconfined aquifer at any sites monitored during the study, with the possible exception of Entrance Spring. The $^{234}U/^{238}U$ activity ratio values for water-quality samples collected at Entrance Spring, and the decrease in this ratio concomitant with an increase in the concentration of dissolved U, indicate that there could be some mixing of uranium ore with groundwater at the spring. A possible mechanism for this mixing is the eolian transport of small sized particles blown off the ore storage pads, deposited in the Entrance Spring drainage, and then dissolved in surface runoff.

Water-quality data collected during the study from 2007 through 2009 were summarized. With the exception of arsenic, thallium, and uranium, the concentration of most trace elements in water samples collected during the study were below both the MCLs and MCLGs established by the U.S. Environmental Protection Agency. Water samples from Entrance Spring had the highest median U concentration compared to other water samples collected from wells and springs monitored during the study. If the elevated uranium concentrations observed in Entrance Spring are not the result of natural sources, a possible pathway from the mill site to the spring could be airborne transport of ore with subsequent dissolution of the wind deposited material in the Entrance Spring drainage. This pathway is feasible for several reasons: (1) the ore to be processed in the mill is stored uncovered on ore storage pads directly across from Entrance Spring, and much of this material is fine grained, which easily can be transported by the wind; (2) starting approximately three years ago trucks delivering ore were covered, prior to that time trucks delivering ore were possibly uncovered and turned onto the mill from Highway 191, directly across from Entrance Spring; and (3) as discussed in the "Uranium Mobility" section, any solid-phase U in contact with infiltrating water would dissolve readily, and any aqueous-phase U likely would remain in solution. The

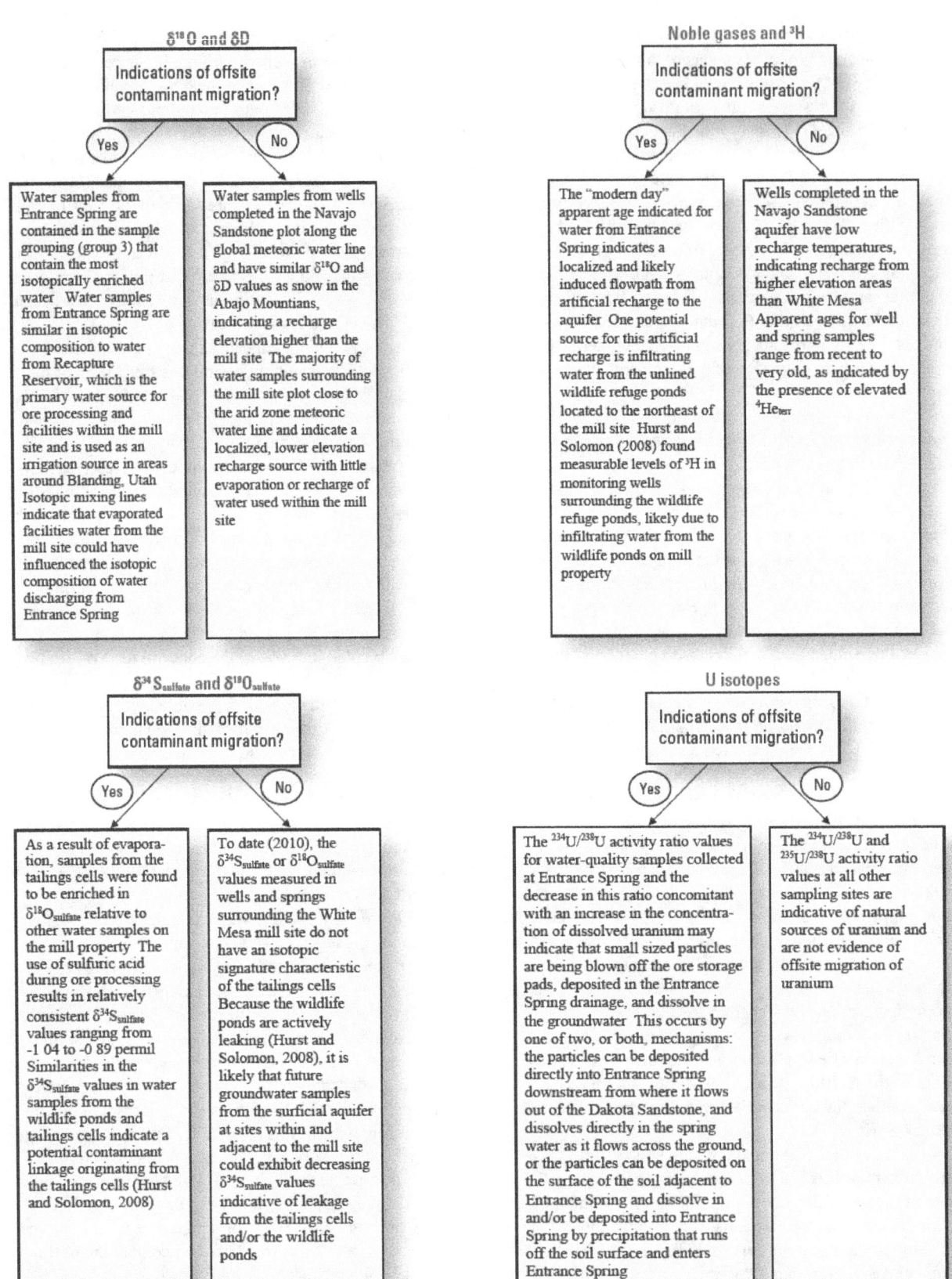

Figure 45. Diagram summarizing study results with respect to offsite contaminant migration from the White Mesa mill site, San Juan County, Utah.

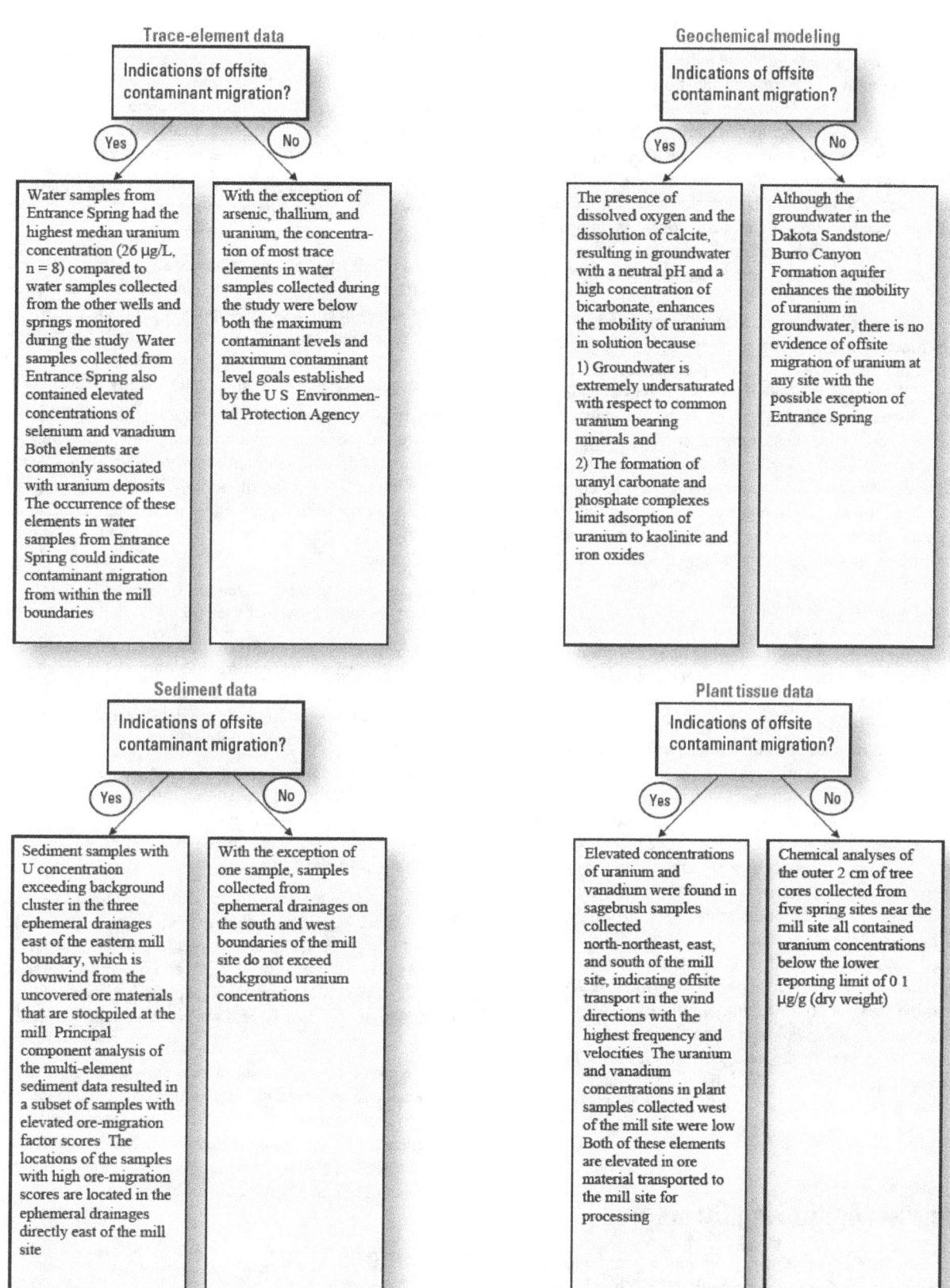

Figure 45. Diagram summarizing study results with respect to offsite contaminant migration from the White Mesa mill site, San Juan County, Utah.—Continued

tailings cells are not a likely source of U at Entrance Spring. An analysis of the groundwater flow paths on the White Mesa indicate that the prevailing groundwater-flow direction is toward the south, and any leakage from tailings is unlikely to flow east toward Entrance Spring. Water samples collected from Entrance Spring also contained elevated concentrations of selenium and vanadium. All three of these constituents commonly are associated with U deposits, and their elevated levels at Entrance Spring could indicate contaminant migration from within the mill boundaries or contact with undiscovered and naturally occurring U ore bodies in the vicinity of the mill site.

The mobility of U in groundwater is determined by U solution-mineral equilibria and sorption reactions that are a function of pH, redox conditions, the presence of complexing agents, and the presence of other metals, such as vanadium, that can induce coprecipitation. Much of the groundwater in the study area contained measurable dissolved oxygen, and the dissolution of calcite along potential groundwater flow paths resulted in groundwater with neutral pH and a high concentration of bicarbonate, which enhances the mobility of U. Although the groundwater in the surficial aquifer enhances the mobility of U in groundwater, there is no evidence of offsite migration of U at any of the monitoring sites with the possible exception of Entrance Spring.

Sediment samples were collected from ephemeral drainages surrounding the mill site and were analyzed for major and trace constitutes to identify potential offsite transport of contaminants from within mill boundaries. Sediment samples from three ephemeral drainages east of the eastern mill boundary, which are downwind from the uncovered ore materials that are stockpiled at the mill, had U concentrations exceeding background. One of these three ephemeral drainages houses Entrance Spring, which contains anomalous isotopic values and trace-element concentration data relative to water samples collected from other parts of the study area. With the exception of one sample, samples collected from ephemeral drainages on the south and west boundaries of the mill site did not exceed background U concentrations.

Tissue samples were collected from big sagebrush (*Artemisia tridentata*) in areas surrounding the White Mesa mill site to determine areas of offsite migration of ore and associated material, primarily from eolian transport. Elevated concentrations of U and V were found in sagebrush samples collected north-northeast, east, and south of the mill site, indicating offsite transport in predominant wind directions. The U and V concentrations in plant samples collected west of the mill site were low.

Potential Monitoring Strategies

If environmental monitoring programs are continued or newly implemented in areas surrounding the White Mesa mill site, the following suggestions with respect to sampling media, sampling intervals, and monitoring constituents should be considered:

- Because of the continued operation of the White Mesa mill, quarterly monitoring of field parameters and major- and trace-element concentrations in selected springs and wells sampled during this study should continue. The sampling sites should include Mill Spring, Entrance Spring, Cow Camp Spring, Ruin Spring, East well, and West well.

- Because of the elevated uranium concentrations measured at Entrance Spring, annual monitoring for U isotopes, $\delta^{34}S_{sulfate}$, $\delta^{18}O$, and δD is suggested.

- Annual monitoring of background water quality (field parameters and major- and trace-element concentrations) at Oasis Spring is needed to supplement geochemical background data collected during the study. If funds allow, Millview well should be re-drilled and annually sampled to provide additional background water-quality data.

- Study results indicate that plant sampling is a useful tool to detect offsite contaminant migration; therefore, big sagebrush should be sampled every three years in areas east of the mill site using the same grid sampling program used during the initial study. Plant tissue samples should be analyzed for the same constituents determined for the current study.

- Consideration should be given to off-site fugitive dust monitoring in areas east of the mill site.

- Consideration should be given to drilling a new monitoring well upgradient from the current locations of the East and West wells because it would be better positioned to act as an "early warning" system for the detection of groundwater contamination from mill activities.

- Because of the elevated uranium concentrations detected in ephemeral drainages east of the mill site, consideration should be given to collection of sediment samples from the two remaining unsampled, ephemeral watersheds directly east of the mill site

- Monitoring programs within the mill site should consider adding other key constituents that can provide additional insight into potential contaminant sources and processes, such as U isotopes, $\delta^{34}S_{sulfate}$, $\delta^{18}O$, and δD. Furthermore, additional isotopic data from Recapture Reservoir are needed to better identify seasonal variations in $\delta^{18}O$ and δD.

- Future monitoring data should be archived in a maintained and easily accessible database, similar to the database used to archive the data collected during the current study. Open access to all data collected during the current and any future studies will be critical in identifying long-term trends in potential off-site contamination.

References Cited

Aeshbach-Hertig, W., Peeters, F., Beyerle, U., and Kipfer, R., 1999, Interpretation of dissolved atmospheric noble gases in natural waters: Water Resources Research, v. 35, p. 2779–2792.

Appelo, C.A.J., and Postma, D., 2005, Geochemistry, ground-water and pollution (2d ed.): New York, A.A. Balkema Publishers, 649 p.

Ballentine, C.J., and Hall, C.M., 1999, Determining paleo-temperature and other variables by using an error-weighted, nonlinear inversion of noble gas concentrations in water: Geochemica Cosmochimica Acta, v. 63, p. 2315–2336.

Bureau of Land Management, 2011, Transportation policy for shipments of Colorado Plateau uranium ores to the White Mesa Uranium Mill, accessed December 16, 2011, at URL https://www.blm.gov/ut/enbb/files/Appendix_D_Transportation_Policy.pdf.

Busche, F.D., 1989, Using plants as an exploration tool for gold: Journal of Geochemical Exploration, v. 32, p. 199–209.

Christie, O.H.J., Esbensen, K., Meyer, T., and Wold, S., 1984, Aspects of pattern recognition in organic geochemistry: Organic Geochemistry, v. 6, p. 885–891.

Clarke, W.B., Jenkins, W.J., and Top, Z., 1976, Determination of tritium by mass spectrometric measurements of ^{3}He: International Journal of Applied Radiation Isotopes, v. 27, p. 515–522.

Craig, H., 1961, Standard for reporting concentrations of deuterium and oxygen-18 in natural waters: Science, v. 133, no. 3467, p. 1833–1834.

Cutter, B.E., and Guyrtte, R.P., 1993, Anatomical, chemical and ecological factors affecting tree species choice in dendrochemistry studies: Journal of Environmental Quality, v. 22, p. 611–619.

Davis, J.A., Curtis, G.P., Wilkins, M.J., Kohler, M., Fox, P., Naftz, D.L., and Lloyd, J.R., 2006, Processes affecting transport of uranium in a suboxic aquifer: Physics and Chemistry of the Earth, v. 31, p. 548–555.

Denison Mines, 2008, Environmental report in support of construction, Tailings Cell 4b, White Mesa uranium mill, Blanding, Utah: Denver, CO, Denison Mines (USA) Corp., 5 p.

Denison Mines, 2010, White Mesa mill site, accessed July 30, 2010, at URL http://www.denisonmines.com/SiteResources/ViewContent.asp?DocID=96&v1ID=&RevID=654&lang=1.

Drever, J.I., 1997, The geochemistry of natural waters (3d ed.): Upper Saddle River, NJ, Prentice Hall, 436 p.

Duewer, D.L., Kowalski, B.R., and Schatzki, T.F., 1975, Source identification of oil spills by pattern recognition analysis of natural elemental composition: Analytical Chemistry, v. 47, p. 1573–1578.

Duff. M.C., and Amrhein, C., 1996, Uranium (VI) adsorption on goethite and soil in carbonate solutions: Soil Science Society of America Journal, v. 60, p. 1393–1400.

Echevarria, G., Sheppard, M.I., and Morel, J.L., 2001, Effect of pH on the sorption of uranium in soils: Journal of Environmental Radioactivity, v. 53, p. 257–264

Fishman, M.J., and Friedman, L.C., 1989, Methods for determination of inorganic substances in water and fluvial sediments: U.S. Geological Survey Techniques of Water Resources Investigations, book 5, chap. A1, 545 p.

Fitts, R. C., 2002, Groundwater Science: San Diego, CA, Academic Press, 129 p.

Freethey, G.W. and Cordy, G.E., 1991, Geohydrology of Mesozoic Rocks in the Upper Colorado River Basin in Arizona, Colorado, New Mexico, Utah, and Wyoming, excluding the San Juan Basin: U.S. Geological Survey Professional Paper 1411–C, 118 p.

Freeze, R.A., and Cherry, J.A., 1979, Groundwater: Englewood Cliffs, Prentice Hall, 606 p.

Gough, L.P., and Erdman, J.A., 1980, Seasonal differences in the element content of Wyoming Big Sagebrush: Journal of Range Management, v. 33, p. 374–378.

Gough, L.P., and Erdman, J.A., 1983, Baseline elemental concentrations for Big Sagebrush from Western U.S.A.: Journal of Range Management, v. 36, p. 718–722.

Hageman, P.L., Brown, Z.A., and Welsch, E., 2002, Arsenic and selenium by flow injection or continuous flow-hydride generation-atomic absorption spectrophotometry, chap. L in Taggart, J. E., ed., Analytical methods for chemical analysis of geologic and other materials, U.S. Geological Survey, Open-File Report 02–223, 9 p.

Hansen D.T. and Fish, R.H., 1993, Soil Survey of San Juan County, Utah, Central Part: U.S. Department of Agriculture, Soil Conservation Service, 208 p.

Haynes, D.D., Vogel, J.D., and Wyant, D.G., 1972, Geology, structure, and uranium deposits of the Cortez quadrangle, Colorado and Utah: U.S. Geologic Survey Miscellaneous Geologic Investigations Map I–629.

Heilweil, V.M., and Susong, D.D., 2007, Assessment of artificial recharge at Sand Hollow Reservoir, Washington County, Utah, updated to Conditions through 2006: U.S. Geological Survey Scientific Investigations Report 2007–5023, 14 p.

Hem, J.D., 1989, Study and interpretation of the chemical characteristics of natural water: U.S. Geological Survey Water-Supply Paper 2254, 263 p.

Heydorn, K., and Thuesen, I., 1989, Classification of ancient Mesopotamian ceramics and clay using SIMCA for supervised pattern recognition, Chemometrics and Intelligent Laboratory Systems. v. 7, p. 181–188.

Hurst, T.G., and Solomon, D.K., 2008, Summary of work completed, data results, interpretations and recommendations for the July 2007 sampling event at the Denison Mines, USA, White Mesa Uranium Mill near Blanding, Utah, accessed August 2, 2010, at URL http://www.radiationcontrol.utah.gov/Uranium_Mills/IUC/uofu_gwifstudy/finalreport.pdf.

Hsi, C-K.D., and Langmuir, D., 1985, Adsorption of uranyl onto ferric oxy-hydroxides: Applications of the surface complexation site-binding model: Geochemica Cosmochimica Acta, v. 49, p. 1931–1941.

Infometrix, 2010, Comprehensive chemometrics modeling software, accessed June 27, 2010, at URL http://infometrix.com/software/pirouette.html.

Intera, Incorporated, 2006, Background groundwater quality report: existing wells for Denison Mines (USA) Corp.'s White Mesa mill site, San Juan County, Utah, 62 p.

International Atomic Energy Agency, 2010, Global network of isotopes in precipitation: The GNIP database. Accessed August 16, 2010 at: http://www-naweb.iaea.org/napc/ih/IHS_resources_gnip.html.

International Uranium Corp., 2010, White Mesa mill, accessed October 7, 2010, at URL http://www.wma-minelife.com/uranium/mill/ef.htm.

Johnson, H.S., and Thordarson, W., 1966, Uranium Deposits of the Moab, Monticello, White Canyon, and Monument Valley Districts Utah and Arizona: U.S. Geological Survey Bulletin 1222–H, 53 p.

Jordan, F., Waugh, W.J., Glenn, E.P., Sam, L., Thompson, T., and Thompson, T.L., 2008, Natural bioremediation of a nitrate-contaminated soil-and-aquifer system in a desert environment: Journal of Arid Environments, v. 72, p. 748–763.

Ketterer, M.E., Wetzel, W.C., Layman, R.R., Matisoff, G., Bonniwell, E.C., 2000, Isotopic studies of sources of uranium in sediments of the Ashtabula River, Ohio, USA: Environmental Science and Technology, v. 34, p. 966–972.

Ketterer, M.E., Hafer, K.M., Link, C.L., Royden, C.S., Hartsock, W.J., 2003, Anthropogenic [236]U at Rocky Flats, Ashtabula River Harbor, and Mersey estuary: three case studies by sector inductively coupled plasma mass spectrometry: Journal of Environmental Radioactivity, v. 67, p. 191–206.

Kimball, B.A., 1981, Geochemistry of spring water, southeastern Uinta Basin, Utah and Colorado: U.S. Geological Survey Water-Supply Paper 2074, 30 p.

Kipfer, R., Aeschbach-Hertig, W., Peeters, F., and Stute, M., 2002, Noble gases in lakes and ground waters, in Porcelli, D., and others, eds., Noble gases in geochemistry and cosmochemistry: Reviews in Mineralogy and Geochemistry, v. 47, p. 615–700.

Kirby, S., 2008, Geologic and hydrologic characterization of the Dakota-Burro Canyon aquifer near Blanding, San Juan County, Utah: Special Study 123 Utah Geological Survey, 52 p.

Kraemer, T.F., and Genereux, D.P., 1998, Applications of uranium- and thorium- series radionuclides in catchment hydrology studies, in Kendall, C., and McDonnell, J.J., eds., Isotope Tracers in Catchment Hydrology: New York, Elsevier, p. 679–722.

MacCarthy, P., DeLuca, S.J., Voorhees, K.J., Malcolm, R.L., and Thurman, E.M., 1985, Pyrolysis-mass spectrometry/pattern recognition on a well-characterized suite of humic samples: Geochimica et Cosmochimica Acta, v.49, p. 2091–2096.

Meglen, R.R., and Erickson, G.A., 1983, Application of pattern recognition to the evaluation of contamination from oil shale retorting, in Francis, C.W., and Auerbach, S.I., eds., Environment and Solid Wastes: Characterization, Treatment, and Disposal: Boston, MA, Butterworths, p. 369–381.

Mello, E., Monna, D., and Oddone, M., 1988, Discriminating sources of Mediterranean marbles: A pattern recognition approach: Archaeometry, v. 30, p. 102–108.

Miesch, A.T., 1961, Classification of elements in Colorado Plateau uranium deposits and multiple stages of mineralization: U.S. Geological Survey Professional Paper P 0424–B, p. B289–B291.

Miesch, A.T., 1962, Compositions of sandstone host rocks of uranium deposits: Transactions of the Society of Mining Engineers of American Institute of Mining, Metallurgical and Petroleum Engineers, Inc., v. 223, p. 178–184.

Miesch, A.T., 1963, Distributions of elements in Colorado Plateau uranium deposits—a preliminary report: U.S. Geological Survey Bulletin 1147–E, 57 p.

Miller, W.R., Wanty, R.B., and McHugh, J.B., 1984, Application of mineral-solution equilibria to geochemical exploration for sandstone-hosted uranium deposits in two basins in west central Utah: Economic Geology, v. 79, p. 266–283.

Morrison, S.J., Metzler, D.R., and Dwyer, B.P., 2002, Removal of As, Mn, Mo, Se, U, V and Zn from groundwater by zero-valent iron in a passive treatment cell: Reaction progress modeling: Journal of Contaminant Hydrology, v. 56, p. 99–116.

Mueller, D. K., and Titus, C. J., 2005, Quality of nutrient data from streams and ground water sampled during water years 1992–2001: U.S. Geological Survey Scientific Investigations Report 2005–5106, 27 p.

Murphy, Christine M., Briggs, Paul H., Adrian, Betty M., Wilson, Steve A., Hageman, Phil L., Theodorakos, Pete M., 1997, Chain of custody; recommendations for acceptance and analysis of evidentiary geochemical samples: U.S. Geological Survey Circular 1138, 26 p.

Naftz, D.L., 1996a, Pattern-recognition analysis and classification modeling of selenium-producing areas: Journal of Chemometrics, v. 10, p. 309–324.

Naftz, D.L., 1996b, Using geochemical and statistical tools to identify irrigated areas that might contain high selenium concentrations in surface water, U.S. Geological Survey Fact Sheet FS–077–96.

Naftz, D.L., Peterman, Z.E., and Spangler, L.E., 1997, Using 87Sr values to identify sources of salinity to a freshwater aquifer, Greater Aneth Oil Field, Utah, U.S.A.: Chemical Geology, v. 141, p. 195–209.

National Oceanic and Atmospheric Administration, 2010, National Climatic Data Center: Climate data online, accessed August 20, 2010, at URL http://www7 ncdc.noaa.gov/CDO/cdo.

Northrop, H.R., Goldhaber, M.B., Whitney, G., Landis, G.P., and Rye, R.O., 1990, Genesis of the tabular-type vanadium-uranium deposits of the Henry Basin, Utah: Economic Geology, v. 85, p. 215–269.

Parkhurst, D.A., and Appelo, C.A.J., 2010, User's guide to PHREEQC (Ver. 2)—A computer program for speciation, batch-reaction, one-dimensional transport, and inverse geochemical calculations, accessed September 22, 2010, at URL http://wwwbrr.cr.usgs.gov/projects/GWC_coupled/phreeqc/html/final html.

Peacock, T.R., and Crock, J.G., 2002, Plant material preparation and determination of weight percent ash, chap. B *in* Taggart, Joseph E., ed., Analytical methods for chemical analysis of geologic and other materials: U.S. Geological Survey, Open-File Report 02–223, 6 p.

Peacock, T.R., Taylor, C.D., and Theodorakos, P.M., 2002, Stream-sediment sample preparation, chap. A2 *in* Taggart, Joseph E., ed., Analytical methods for chemical analysis of geologic and other materials: U.S. Geological Survey, Open-File Report 02–223, 6 p.

Peterson, D.M., Cummins, L.E., and Miller, J.D., 2008, DOE remediation of uranium mills: A progress report: Southwest Hydrology, November/December issue, p. 26–28.

Révész, Kinga, and Coplen, T.B., 2008a, Determination of the $\delta(^2H/^1H)$ of water: RSIL lab code 1574, chap. C1 *in* Révész, Kinga, and Coplen, T.B., eds., Methods of the Reston Stable Isotope Laboratory: Reston, VA, U.S. Geological Survey Techniques and Methods v. 10, 27 p.

Révész, Kinga, and Coplen, T.B., 2008b, Determination of the $\delta(180/160)$, of water: RSIL lab code 489, chap. C2 *in* Révész, Kinga, and Coplen, T.B., eds., Methods of the Reston Stable Isotope Laboratory: Reston, VA, U.S. Geological Survey Techniques and Methods v. 10, 28 p.

Révész, Kinga, and Qi, Haiping., 2006, Determination of the $\delta(^{34}S/^{32}S)$ of sulfate in water: RSIL lab code 1951, chap. C10 in Révész, Kinga, and Coplen, T.B., eds., Methods of the Reston Stable Isotope Laboratory: Reston, VA, U.S. Geological Survey Techniques and Methods v. 10, 33 p.

Ries, K.G., Guthrie, J.D., Rea, A.H., Steeves, P.A., and Stewart, D.W., 2008, StreamStats: A water resources web application: U.S. Geological Survey Fact Sheet FS 2008–3067, 6 p.

Rose, A.W., Hawkes, H.E., and Webb, J.S., 1979, Geochemistry in Mineral Exploration: New York, Academic Press, 657 p.

Sheldon, A., 2002, Diffusion of radiogenic helium in shallow ground water—Implications for crustal degassing: Salt Lake City, UT, University of Utah, Ph.D. dissertation, 185 p.

Sherman, H.M., Gierke, J.S., and Anderson, C.P., 2007, Controls on spatial variability of uranium in sandstone aquifers: Ground Water Monitoring and Remediation, v. 27, no. 2, p. 106–118.

Smith, S.C., and Kretschmer, E.L., 1992, Gold patterns in big sagebrush over the CX and Mag deposits, Pinson Mine, Humboldt County, Nevada: Journal of Geochemical Exploration, v. 46, p. 147–161.

Solomon, D.K., and Cook, P.G., 2000, ^3H and ^3He, *in* Cook, P.G., and Herczeg, A.L., eds., Environmental tracers in subsurface hydrology: Boston, MA, Kluwer Academic Publishers, p. 397–424.

Spangler, L.E., Naftz, D.L., and Peterman, Z.E., 1996, Hydrology, chemical quality, and characterization of salinity in the Navajo Aquifer in and near the Greater Aneth Oil Field: San Juan County, Utah, U.S. Geological Survey Water-Resources Investigations Report 96–4155, 90 p.

Stewart, K.C., and McKown, D.M., 1995, Sagebrush as a sampling medium for gold exploration in the Great Basin—evaluation from a greenhouse study: Journal of Geochemical Exploration, v. 54, p. 19–26.

Stute, M., and Schlosser, P., 2001, Atmospheric noble gases, *in* Cook, P., and Herczeg, A.L., eds., Environmental tracers in subsurface hydrology: Boston, Massachusetts, Kluwer Academic Publishers, p. 349–377.

Titan Environmental Corporation, 1994, Hydrogeologic evaluation of White Mesa Uranium Mill: Englewood, Colorado, Titan Environmental Corporation, variously paged.

U.S. Department of Energy, 2005, Remediation of the Moab Uranium Mill Tailings, Grand and San Juan Counties, Utah, Final Environmental Impact Statement, U.S. Department of Energy data available online, accessed October 12, 2010, at URL http://www.gjem.energy.gov/moab/documents/eis/final_eis/Volume_II/_Contents.pdf.

U.S. Environmental Protection Agency, 1994a, Method 200.8: Determination of trace elements in waters and wastes by inductively coupled plasma-mass spectrometry, U.S. Environmental Protection Agency data available online, accessed June 20, 2011, at URL http://www.epa.gov/water-science/methods/method/files/200_8.pdf.

U.S. Environmental Protection Agency, 1994b, Method 200.7: Determination of metals and trace elements in water and wastes by inductively coupled plasma-atomic emission spectrometry U.S. Environmental Protection Agency data available online at URL http://www.epa.gov/waterscience/methods/method/files/200_7.pdf.

U.S. Environmental Protection Agency, 2009, Drinking-water contaminants, U.S. Environmental Protection Agency data available online, accessed April 13, 2010, at URL http://www.epa.gov/safewater/contaminants/index.html#7.

U.S. Geological Survey, 2006, Collection of water samples (ver. 2.0): U.S. Geological Survey Techniques of Water-Resources Investigations, book 9, chap. A4, September, accessed August 1, 2010 at URL http://pubs.water.usgs.gov/twri9A4/.

U.S. Geological Survey, 2010a, Geochemistry of stream sediments from NURE-HSSR, USGS data available online, accessed June 28, 2010, at URL http://mrdata.usgs.gov/geochemistry/nuresed html.

U.S. Geological Survey, 2010b, National Water Quality Laboratory catalog, USGS data available online, accessed August 27, 2010, at URL http://wwwnwql.cr.usgs.gov/USGS/catalog/index.cfm?a=bs&sa=l&sap=2114.

U.S. Geological Survey, 2010c, National Water Quality Laboratory, Chain of custody, USGS data available online, accessed September 13, 2010, at URL http://wwwnwql.cr.usgs.gov/htmls/QUAX0030.4controlled.pdf.

U.S. Geological Survey, variously dated, National field manual for the collection of water-quality data: U.S. Geological Survey Techniques of Water-Resources Investigations, book 9, chaps. A1–A9, available online at URL http://pubs.water.usgs.gov/twri9A.

Utah Division of Water Quality, 2006, Recapture reservoir, accessed August 11, 2010, at URL http://www.waterquality.utah.gov/watersheds/lakes/RECAPTUR.pdf.

Voorhees, K.J., and Tsao, R., 1985, Smoke aerosol analysis by pyrolysis-mass spectrometry/pattern recognition for assessment of fuels involved in flaming combustion: Analytical Chemistry, v. 57, p. 1630–1636.

Vroblesky, D.A., Joshi, M., Morrell, J., and Peterson, J.E., 2003, Evaluation of passive diffusion bag samplers, dialysis, samplers, and nylon-screen samplers in selected wells at Andersen Air Force Base, Guam, March-April 2002: U.S. Geological Survey Water-Resources Investigation Report 03–4157, 29 p.

Welch, A.H., and Preissler, A.M., 1986, Aqueous geochemistry of the Bradys Hot Springs Geothermal Area, Churchill County, Nevada, in Subitzky, Seymour and others, eds., Selected Papers in the Hydrologic Sciences: U.S. Geological Survey Water-Supply Paper 2290, p. 17–36.

Whitfield, M.S., Thordarson, W., Oatfield, W.J., Zimmerman, E.A., and Rueger, B.F., 1983, Regional hydrology of the Blanding-Durango area, southern paradox Basin, Utah and Colorado: U.S. Geological Survey Water-Resources Investigations Report 83–4218, 88 p.

Wilde, F.D., ed., 2004, Cleaning of Equipment for water sampling (ver. 2.0): U.S. Geological Survey Techniques of Water-Resources Investigations, book 9, chap. A3, April, accessed June 20, 2011, at URL http://pubs.water.usgs.gov/twri9A3/.

Wilde, F.D., 2008, Guidelines for field-measured water-quality properties (ver. 2.0): U.S. Geological Survey Techniques of Water-Resources Investigations, book 9, chap. A6, section 6.0, October, available online at URL http://pubs.water.usgs.gov/twri9A/.

Wilkowske, C.D., Rowland, R.C., and Naftz, D.L., 2002, Selected hydrologic data for the field demonstration of three permeable reactive barriers near Fry Canyon, Utah, 1996–2000: U.S. Geological Survey Open-File Report 01–361, 102 p.

Wilt, M.F., Geddes, J.D., Tamma, R.V., Miller, G.C., and Everett, R.L., 1992, Interspecific variation of phenolic concentrations in persistent leaves among six taxa from subgenus Tridentatae of Artemista: Biochemical Systematics and Ecology, v. 20, p. 41–52.

Witkind, I.J., 1964, Geology of the Abajo Mountains area, San Juan County, Utah: U.S. Geological Survey Professional Paper 453, 110 p.

Western Regional Climate Center, 2010, Monthly climate summary for Blanding, Utah, accessed August 5, 2010, at URL http://www.wrcc.dri.edu/cgi-bin/cliMAIN.pl?utblan.

Yankosky, T.M., and Vroblesky, D.A., 1989a, Botanical comments on trees as chemical recorders of ground-water contamination: U.S. Geological Survey Open-File Report 89–0409, 110 p.

Yankosky, T.M., and Vroblesky, D.A., 1989b, Use of tree-ring chemistry to document historical ground-water contamination and to estimate aquifer properties, Aberdeen Proving Ground, Maryland: Eos, Transactions American Geophysical Union, v. 70, 326 p.

Yanosky, T.M., and Vroblesky, D.A., 1992, Relation of nickel contamination in tree rings to groundwater contamination: Water Resources Research, v. 28, p. 2077–2083.

Zhu, C., and Burden, D.S., 2001, Mineralogical compositions of aquifer matrix as necessary initial conditions in reactive contaminant transport models: Journal of Contaminant Hydrology, v. 51, p. 145–161.

Zielinski, R.A., Chafin, D.T., Banta, E.R., and Szabo, B.J., 1997, Use of ^{234}U and ^{238}U isotopes to evaluate contamination of near-surface groundwater with uranium-mill effluent: a case study in south-central Colorado, U.S.A.: Environmental Geology, v. 32, no. 2, p. 124–136.

Appendix 1

Appendix 1.   Field and laboratory data for water samples collected near the White Mesa uranium mill, San Juan County, Utah, June 2007–October 2009.

[**Abbreviations**: ANC, acid neutralizing capacity; $CaCO_3$, calcium carbonate; cm^3/g at STP, cubic centimeters of gas per gram at standard temperature (25°C) and pressure (1 bar); E, estimated; ft, feet; gal/min, gallons per minute; hh:mm, hour:minute; LSD, land surface datum; M, presence verified but not quantified; mg/L, milligrams per liter; mm/dd/yyyy, month/day/year; mm Hg, millimeters of mercury; mV, millivolts; permil, parts per thousand; pCi/L, picocuries per liter; SHE, standard hydrogen electrode; SiO_2, silicon dioxide; U, analyzed for but not detected; µg/L, micrograms per liter; µS/cm, microsiemens per centimeter; °C, degrees Celcius; <, less than; —, no data]

Station name	Station number	Date (mm/dd/yyyy)	Time (hh:mm)	Depth to water level, (ft below LSD)	Dissolved oxygen, field, (mg/L)	Flow rate, instantaneous, (gal/min)	pH, field, (standard units)	pH, lab, (standard units)	Specific conductance, lab, (µS/cm)	Specific conductance, field (µS/cm)	Temperature, field, (°C)	Total partial pressure dissolved gasses, field, (mm Hg)
Reference Spring North	373550109341701	06/21/2007		—	—	—	—	—	—	—	—	—
(D–38–22)23cda–1 South well	372756109280901	09/11/2007		—	1 8	—	8 1	—	—	530	20 7	—
		12/11/2007	15:00	0 00	<0 1	—	7 6	8 1	446	401	21 0	—
		03/11/2008	15:05	—	<0 1	—	7 9	8 0	451	480	13 3	—
		11/12/2008	10:05	—	<0 1	—	8 0	8 3	450	456	21 0	—
(D–38–22)23acb–1 North well	372817109275701	09/11/2007	11:00	—	1	—	7 9	—	—	393	23 6	—
		12/11/2007	12:00	—	<0 1	—	7 6	8 1	396	390	22 1	—
		03/11/2008	11:20	—	<0 1	—	8 0	8 0	405	420	12 8	—
		11/11/2008	09:41	—	0 6	—	8 1	8 2	424	432	19 6	—
(D–38–22)23bba–S1 Right Hand Fork Seep	372832109282001	03/12/2008	17:00	—	—	—	—	—	—	—	—	—
(D–38–22) 8dcd–1 West well	372930109310701	06/21/2007										
		09/11/2007	15:30	—	3 4	—	6 7	—	—	4,620	17 2	—
		12/13/2007	13:00	84 70	1 1	—	6 8	6 9	5,140	4,960	15 1	—
		03/13/2008	11:40	84 60	1 7	—	6 7	6 8	5,250	5,100	15 6	—
		09/16/2008	14:15	84 70	<0 1	—	6 4	6 8	5,210	5,220	24 1	—
		11/13/2008	10:14	84 56	3 1	—	6 5	7 1	5,130	5,120	17 2	—
		12/08/2008	14:50	84 36	—	—	—	—	—	—	—	—
		12/08/2008	14:55	84 36	—	—	—	—	—	—	—	—
		12/08/2008	15:00	84 36	—	—	—	—	—	—	—	—
		04/21/2009	11:00	84 77	3 9	—	6 4	6 9	5,030	5,230	21 9	—
		09/22/2009	10:30	84 69	<0 1	—	6 7	—	—	5,200	19 3	—
		10/19/2009	15:00	84 62	—	—	—	—	—	—	—	—
		10/19/2009	15:05	84 62	—	—	—	—	—	—	—	—
		10/19/2009	15:10	84 62	—	—	—	—	—	—	—	—
(D–38–22)10cbc Anasazi Pond near spillway	372943109293201	09/18/2008	10:50	—	4 0	—	7 4	7 5	220	225	18 6	—
(D–38–22)10bcc–1 East well	372954109293601	06/21/2007		—	—	—	—	—	—	—	—	—
		09/11/2007	13:30	—	5 5	—	8 0	—	—	558	15 9	—
		12/14/2007	10:45	55 28	3 4	—	8 0	8 0	565	518	15 1	—
		03/13/2008	17:00	55 30	1 3	—	8 3	8 2	614	663	19 5	—
(D–38–22)10bcc–1 East well		09/16/2008	10:45	55 77	1 3	—	7 6	7 8	615	635	19 1	—
		11/13/2008	14:00	55 30	3 0	—	7 8	8 2	615	641	18 2	—
		04/21/2009	15:50	55 45	0 4	—	7 8	7 9	615	642	22 9	—
		09/22/2009	15:40	55 50	0 5	—	8 0	8 1	646	712	—	—

Appendix 1. Field and laboratory data for water samples collected near the White Mesa uranium mill, San Juan County, Utah, June 2007–October 2009.—Continued

[**Abbreviations**: ANC, acid neutralizing capacity; CaCO₃, calcium carbonate; cm³/g at STP, cubic centimeters of gas per gram at standard temperature (25°C) and pressure (1 bar); E, estimated; ft, feet; gal/min, gallons per minute; hh:mm, hour:minute; LSD, land surface datum; M, presence verified but not quantified; mg/L, milligrams per liter; mm/dd/yyyy, month/day/year; mm Hg, millimeters of mercury; mV, millivolts; permil, parts per thousand; pCi/L, picocuries per liter; SHE, standard hydrogen electrode; SiO₂, silicon dioxide; U, analyzed for but not detected; µg/L, micrograms per liter; µS/cm, microsiemens per centimeter; °C, degrees Celcius; <, less than; —, no data]

Station name	Station number	Date (mm/dd/yyyy)	Time (hh:mm)	Depth to water level, (ft below LSD)	Dissolved oxygen, field, (mg/L)	Flow rate, instantaneous, (gal/min)	pH, field, (standard units)	pH, lab, (standard units)	Specific conductance, lab, (µS/cm)	Specific conductance, field (µS/cm)	Temperature, field, (°C)	Total partial pressure dissolved gasses, field, (mm Hg)
(D–38–22) 8bad–S1 Ruin Spring	373006109312301	06/01/2007		—	—	—	—	—	—	—	—	—
		09/11/2007	16:00	—	4 6	—	7 2	—	—	1,350	18 5	—
		12/13/2007	09:30	—	—	0 09	7 3	7 6	1,380	1,580	15 0	—
		03/13/2008	12:20	—	—	—	7 5	7 6	1,160	1,170	10 5	—
		06/18/2008	15:20	—	—	0 89	7 2	7 8	1,240	1,240	13 3	—
		09/17/2008	12:20	—	10 6	0 86	7 2	7 7	1,250	1,250	16 7	—
		11/11/2008	13:45	—	8 9	0 75	7 5	8	1,280	1,290	15 9	—
		04/22/2009	10:15	—	9 5	0 74	7 4	7 3	1,330	1,340	12 3	—
		09/23/2009	13:30	—	—	0 68	7 3	7 8	1,380	1,350	16 8	—
(D–38–22) 4adb South Mill Pond	373052109294901	03/12/2008	14:40	—	11 8	—	9 6	7 3	146	193	10 7	—
(D–37–22)32ddc–1 MW3A	373116109305601	12/09/2008	09:15	80 72	—	—	—	—	—	—	—	—
		12/09/2008	09:20	80 72	—	—	—	—	—	—	—	—
		10/20/2009	08:20	80 47	—	—	—	—	—	—	—	—
		10/20/2009	08:25	80 47	—	—	—	—	—	—	—	—
(D–37–22)31dcb–S1 Cow Camp Spring	373122109321501	09/18/2007	17:00	—	0 3	—	7 1	—	—	1,600	15 7	601
		09/19/2007	19:00	—	0 3	—	7 1	—	—	1,600	15 7	—
		03/12/2008	13:25	—	4 6	15 7	7 5	7 7	1,540	1,490	15 1	—
		06/18/2008	14:20	—	—	0 27	7 7	8 1	1,530	1,530	23 0	—
		09/17/2008	13:35	—	5 3	2 2	7 1	7 6	1,510	1,480	15 4	—
		11/13/2008	15:55	—	8 5	1 8	8 0	8 3	1,530	1,540	8 1	—
		04/22/2009	11:15	—	7 4	2 2	8 0	8 0	1,500	1,540	17 0	—
		09/23/2009	10:15	—	—	2 0	8 1	8 2	1,550	1,560	13 4	—
(D–37–22)32bab–S1 Mill Spring	373158109312601	03/12/2008	11:55	—	7 0	0 17	7 3	7 6	1,330	1,310	6 2	—
		09/18/2008	09:20	—	9 0	—	7 4	7 7	2,400	2,350	13 4	—
		11/12/2008	15:00	—	11 1	0 12	7 5	8 0	1,820	1,830	4 8	—
		04/23/2009	08:55	—	8 8	0 12	7 3	7 8	1,090	1,130	8 9	—
		09/24/2009	11:10	—	—	—	7 4	7 5	3,660	3,710	13 7	—
(D–37–22)32bab–S1 Mill Spring		10/19/2009	17:00	—	—	—	—	—	—	—	12 0	—
(D–37–22)27ccc–S1 Entrance Spring	373202109293401	06/21/2007										
		09/20/2007	14:00	—	4 4	—	7 5	—	—	731	20 8	601
		12/13/2007	15:30	—	9 4	0 05	7 9	8 0	1,070	994	4 5	—
		03/13/2008	16:10	—	—	—	8 1	8 0	959	975	8 5	—
		06/19/2008	09:15	—	—	20 2	7 6	7 9	984	939	18 0	—
		07/22/2008	12:00	—	—	—	—	—	—	256	—	—
		09/17/2008	10:00	—	8 7	22 9	7 5	7 8	920	910	15 4	—
		11/11/2008	12:45	—	14 6	15 7	7 9	7 7	919	944	8 5	—
		04/22/2009	09:30	—	6 3	22 9	7 3	7 5	884	915	10 3	—
		09/23/2009	09:00	—	3 4	18 0	7 3	7 7	971	960	10 8	—

Appendix 1. Field and laboratory data for water samples collected near the White Mesa uranium mill, San Juan County, Utah, June 2007–October 2009.—Continued

[**Abbreviations**: ANC, acid neutralizing capacity; CaCO$_3$, calcium carbonate; cm^3/g at STP, cubic centimeters of gas per gram at standard temperature (25°C) and pressure (1 bar); E, estimated; ft, feet; gal/min, gallons per minute; hh:mm, hour:minute; LSD, land surface datum; M, presence verified but not quantified; mg/L, milligrams per liter; mm/dd/yyyy, month/day/year; mm Hg, millimeters of mercury; mV, millivolts; permil, parts per thousand; pCi/L, picocuries per liter; SHE, standard hydrogen electrode; SiO$_2$, silicon dioxide; U, analyzed for but not detected; µg/L, micrograms per liter; µS/cm, microsiemens per centimeter; °C, degrees Celcius; <, less than; —, no data]

Station name	Station number	Date (mm/dd/yyyy)	Time (hh:mm)	Depth to water level, (ft below LSD)	Dissolved oxygen, field, (mg/L)	Flow rate, instantaneous, (gal/min)	pH, field, (standard units)	pH, lab, (standard units)	Specific conductance, lab, (µS/cm)	Specific conductance, field (µS/cm)	Temperature, field, (°C)	Total partial pressure dissolved gasses, field, (mm Hg)
(D–37–22)28acc–1 MW18	373233109301001	12/09/2008	08:45	70 66	—	—	—	—	—	—	—	—
		12/09/2008	08:50	70 66	—	—	—	—	—	—	—	—
		12/09/2008	08:55	70 66	—	—	—	—	—	—	—	—
		10/20/2009	08:55	69 80	—	—	—	—	—	—	—	—
		10/20/2009	09:00	69 80	—	—	—	—	—	—	—	—
		10/20/2009	09:05	69 80	—	—	—	—	—	—	—	—
(D–37–22)10cdc–1 Lyman well	373442109291501	12/12/2007	11:00	—	0 1	—	7 5	7 1	819	824	14 0	—
(D–37–22) 8dba–1 Millview well	373501109310801	09/18/2007	12:00	—	3 5	—	7 1	—	—	636	14 0	619
(D–37–22) 2aad–1 Bayless well	373612109273201	12/12/2007	13:30	27 78	7 6	—	7 5	7 4	956	963	13 3	—
(D–36–22)19aad–S1 Oasis Spring	373850109315301	09/19/2007	15:55	—	3 7	—	7 0	—	—	627	13 9	563
		09/18/2008	14:00	—	4 6	—	7 5	7 5	679	610	13 5	—
		11/12/2008	12:15	—	8 0	—	7 4	8 1	604	613	4 3	—
		04/23/2009	10:45	—	9 2	—	7 7	7 5	578	582	6 9	—
		09/24/2009	09:10	—	—	—	7 6	7 5	655	663	9 6	—
(D–36–22)12dbc Recapture Reservoir	374002109263501	04/23/2009	12:40	—	11 2	—	8 4	8 3	263	267	12 5	—

Appendix 1. Field and laboratory data for water samples collected near the White Mesa uranium mill, San Juan County, Utah, June 2007–October 2009.—Continued

[**Abbreviations**: ANC, acid neutralizing capacity; CaCO$_3$, calcium carbonate; cm^3/g at STP, cubic centimeters of gas per gram at standard temperature (25°C) and pressure (1 bar); E, estimated; ft, feet; gal/min, gallons per minute; hh:mm, hour:minute; LSD, land surface datum; M, presence verified but not quantified; mg/L, milligrams per liter; mm/dd/yyyy, month/day/year; mm Hg, millimeters of mercury; mV, millivolts; permil, parts per thousand; pCi/L, picocuries per liter; SHE, standard hydrogen electrode; SiO$_2$, silicon dioxide; U, analyzed for but not detected; µg/L, micrograms per liter; µS/cm, microsiemens per centimeter; °C, degrees Celcius; <, less than; —, no data]

Station name	Station number	Date (mm/dd/yyyy)	Time (hh:mm)	Redox potential, relative to SHE, (mV)	Sampling depth, (ft)	Residue on evaporation at 180°C, dissolved, (mg/L)	Calcium, dissolved, (mg/L)	Magnesium, dissolved, (mg/L)	Potassium, dissolved, (mg/L)	Sodium, dissolved, (mg/L)	ANC, lab, (mg/L as CaCO$_3$)	Alkalinity, field, (mg/L as CaCO$_3$)
Reference Spring North	373550109341701	06/21/2007		—	—	—	—	—	—	—	—	—
(D–38–22)23cda–1 South well	372756109280901	09/11/2007		—	—	—	—	—	—	—	—	—
		12/11/2007	15:00	—	—	281	24 8	19 3	3 15	43 7	191	182
		03/11/2008	15:05	–230	—	271	24 9	19 8	3 07	44 3	190	181
		11/12/2008	10:05	–286	—	274	24 6	20	3 16	47 5	191	185
(D–38–22)23acb–1 North well	372817109275701	09/11/2007	11:00	—	—	—	—	—	—	—	—	—
		12/11/2007	12:00	—	—	241	25 1	21 7	3 42	29 8	186	184
		03/11/2008	11:20	–420	—	237	25 3	21 6	3 43	29 5	186	177
		11/11/2008	09:41	–328	—	262	24 6	20 4	4 38	44 1	201	195
(D–38–22)23bba–S1 Right Hand Fork Seep	372832109282001	03/12/2008	17:00	—	—	—	—	—	—	—	—	—
(D–38–22) 8dcd–1 West well	372930109310701	06/21/2007		—	—	—	—	—	—	—	—	—
		09/11/2007	15:30	—	—	—	—	—	—	—	—	—
		12/13/2007	13:00	65	100	5,000	456	220	18 5	622	348	332
		03/13/2008	11:40	–37	100	5,060	477	229	18 7	663	379	382
		09/16/2008	14:15	–100	96	5,040	441	220	18 4	648	386	376
		11/13/2008	10:14	49	95 5	4,980	461	239	75 8	618	380	370
		12/08/2008	14:50		87	—	—	—	—	—	—	—
		12/08/2008	14:55	—	94	—	—	—	—	—	—	—
		12/08/2008	15:00	—	107	—	—	—	—	—	—	—
		04/21/2009	11:00	76	96	4,990	465	234	19 6	602	374	354
		09/22/2009	10:30	29	96	4,980	526	275	22 2	726	—	344
		10/19/2009	15:00	—	90	—	—	—	—	—	—	—
		10/19/2009	15:05	—	99	—	—	—	—	—	—	—
		10/19/2009	15:10	—	108	—	—	—	—	—	—	—
(D–38–22)10cbc Anasazi Pond near spillway	372943109293201	09/18/2008	10:50	—	—	149	35 1	3 02	6 49	0 27	113	108
(D–38–22)10bcc–1 East well	372954109293601	06/21/2007		—	—	—	—	—	—	—	—	—
		09/11/2007	13:30	—	—	—	—	—	—	—	—	—
		12/14/2007	10:45	21	66	363	9 01	2 81	1 39	127	203	202
		03/13/2008	17:00	–99	65 5	380	13 3	4 09	1 58	122	221	218
(D–38–22)10bcc–1 East well		09/16/2008	10:45	–64	66	374	6 74	1 41	1 23	135	224	213
		11/13/2008	14:00	–28	66 2	390	8 65	2 4	1 44	130	228	222
		04/21/2009	15:50	–203	76	375	10 9	3 13	1 45	122	230	219
		09/22/2009	15:40	7	77	384	7 61	2 42	1 4	124	240	209
(D–38–22) 8bad–S1 Ruin Spring	373006109312301	06/01/2007		—	—	—	—	—	—	—	—	—
		09/11/2007	16:00	—	—	—	—	—	—	—	—	—
		12/13/2007	09:30	—	—	1,070	154	34	3 52	114	194	186
		03/13/2008	12:20	—	—	828	129	29 7	2 53	89	190	184
		06/18/2008	15:20	—	—	942	142	30 7	3 01	99 8	196	191
		09/17/2008	12:20	—	—	952	142	31	3 34	106	196	193
		11/11/2008	13:45	—	—	965	152	32 4	3 53	110	196	179

Appendix 1. Field and laboratory data for water samples collected near the White Mesa uranium mill, San Juan County, Utah, June 2007–October 2009.—Continued

[**Abbreviations**: ANC, acid neutralizing capacity; CaCO$_3$, calcium carbonate; cm^3/g at STP, cubic centimeters of gas per gram at standard temperature (25°C) and pressure (1 bar); E, estimated; ft, feet; gal/min, gallons per minute; hh:mm, hour:minute; LSD, land surface datum; M, presence verified but not quantified; mg/L, milligrams per liter; mm/dd/yyyy, month/day/year; mm Hg, millimeters of mercury; mV, millivolts; permil, parts per thousand; pCi/L, picocuries per liter; SHE, standard hydrogen electrode; SiO$_2$, silicon dioxide; U, analyzed for but not detected; µg/L, micrograms per liter; µS/cm, microsiemens per centimeter; °C, degrees Celcius; <, less than; —, no data]

Station name	Station number	Date (mm/dd/yyyy)	Time (hh:mm)	Redox potential, relative to SHE, (mV)	Sampling depth, (ft)	Residue on evaporation at 180°C, dissolved, (mg/L)	Calcium, dissolved, (mg/L)	Magnesium, dissolved, (mg/L)	Potassium, dissolved, (mg/L)	Sodium, dissolved, (mg/L)	ANC, lab, (mg/L as CaCO$_3$)	Alkalinity, field, (mg/L as CaCO$_3$)
		04/22/2009	10:15	—	—	963	145	30 9	2 99	108	193	194
		09/23/2009	13:30	—	—	1,040	148	32 5	3 41	114	197	189
(D–38–22) 4adb South Mill Pond	373052109294901	03/12/2008	14:40	—	—	98	22 7	1 95	5 9	0 91	70	60 9
(D–37–22)32ddc–1 MW3A	373116109305601	12/09/2008	09:15	—	82	—	—	—	—	—	—	—
		12/09/2008	09:20	—	86	—	—	—	—	—	—	—
		10/20/2009	08:20	—	81	—	—	—	—	—	—	—
		10/20/2009	08:25	—	90	—	—	—	—	—	—	—
(D–37–22)31dcb–S1 Cow Camp Spring	373122109321501	09/18/2007	17:00	—	—	—	—	—	—	—	—	—
		09/19/2007	19:00	—	—	—	—	—	—	—	—	—
		03/12/2008	13:25	—	—	1,010	90 9	24 5	5 65	208	261	254
		06/18/2008	14:20	—	—	1,020	89 8	24 5	6 24	210	267	251
		09/17/2008	13:35	—	—	1,020	88	24 6	5 69	213	264	259
		11/13/2008	15:55	—	—	1,020	92 5	25	6 02	215	267	259
		04/22/2009	11:15	—	—	1,020	89 4	23 7	6 02	209	265	261
		09/23/2009	10:15	—	—	1,020	92 7	25 2	5 78	216	268	259
(D–37–22)32bab–S1 Mill Spring	373158109312601	03/12/2008	11:55	—	—	950	142	33 5	1 98	124	295	284
		09/18/2008	09:20	—	—	1,870	219	57 2	3 35	276	472	512
		11/12/2008	15:00	—	—	1,420	200	52 4	2 66	194	365	350
		04/23/2009	08:55	—	—	752	108	26 5	1 47	101	298	284
		09/24/2009	11:10	—	—	2,900	384	99 9	2 89	440	624	609
(D–37–22)32bab–S1 Mill Spring		10/19/2009	17:00	—	—	—	—	—	—	—	—	—
(D–37–22)27ccc–S1 Entrance Spring	373202109293401	06/21/2007		—	—	—	—	—	—	—	—	—
		09/20/2007	14:00	—	—	—	—	—	—	—	—	—
		12/13/2007	15:30	—	—	728	116	34 4	3 46	75 8	245	239
		03/13/2008	16:10	—	—	620	101	29 2	4 25	66 1	235	240
		06/19/2008	09:15	—	—	647	100	30 5	2 06	70 5	252	240
		07/22/2008	12:00	—	—	—	—	—	—	—	—	—
		09/17/2008	10:00	—	—	613	94 2	29 1	2 88	62 9	246	234
		11/11/2008	12:45	—	—	606	100	34 2	3 12	72	229	216
		04/22/2009	09:30	—	—	476	89	26 4	3 97	64 2	244	230
		09/23/2009	09:00	—	—	630	87 4	33 7	1 69	68 6	237	223
(D–37–22)28acc–1 MW18	373233109301001	12/09/2008	08:45	—	79	—	—	—	—	—	—	—
		12/09/2008	08:50	—	99	—	—	—	—	—	—	—
		12/09/2008	08:55	—	129	—	—	—	—	—	—	—
		10/20/2009	08:55	—	89	—	—	—	—	—	—	—
		10/20/2009	09:00	—	114	—	—	—	—	—	—	—
		10/20/2009	09:05	—	139	—	—	—	—	—	—	—
(D–37–22)10cdc–1 Lyman well	373442109291501	12/12/2007	11:00	—	—	566	106	31 6	1 95	27 3	218	217

Appendix 1.   Field and laboratory data for water samples collected near the White Mesa uranium mill, San Juan County, Utah, June 2007–October 2009.—Continued

[**Abbreviations**: ANC, acid neutralizing capacity; CaCO₃, calcium carbonate; cm³/g at STP, cubic centimeters of gas per gram at standard temperature (25°C) and pressure (1 bar); E, estimated; ft, feet; gal/min, gallons per minute; hh:mm, hour:minute; LSD, land surface datum; M, presence verified but not quantified; mg/L, milligrams per liter; mm/dd/yyyy, month/day/year; mm Hg, millimeters of mercury; mV, millivolts; permil, parts per thousand; pCi/L, picocuries per liter; SHE, standard hydrogen electrode; SiO₂, silicon dioxide; U, analyzed for but not detected; µg/L, micrograms per liter; µS/cm, microsiemens per centimeter; °C, degrees Celcius; <, less than; —, no data]

Station name	Station number	Date (mm/dd/yyyy)	Time (hh:mm)	Redox potential, relative to SHE, (mV)	Sampling depth, (ft)	Residue on evaporation at 180°C, dissolved, (mg/L)	Calcium, dissolved, (mg/L)	Magnesium, dissolved, (mg/L)	Potassium, dissolved, (mg/L)	Sodium, dissolved, (mg/L)	ANC, lab, (mg/L as CaCO₃)	Alkalinity, field, (mg/L as CaCO₃)
(D–37–22) 8dba–1 Millview well	373501109310801	09/18/2007	12:00	—	—	—	—	—	—	—	—	—
(D–37–22) 2aad–1 Bayless well	373612109273201	12/12/2007	13:30	—	—	625	113	28 1	2 72	52 8	281	277
(D–36–22)19aad–S1 Oasis Spring	373850109315301	09/19/2007	15:55	—	—	—	—	—	—	—	—	—
		09/18/2008	14:00	−37	—	436	82 6	15 9	1 7	38 2	237	246
		11/12/2008	12:15	20	—	386	74 3	16	1 27	35 9	187	170
		04/23/2009	10:45	−122	—	376	67	14 2	1 3	33 7	176	169
		09/24/2009	09:10	—	—	410	84 2	13 9	3 11	33 5	187	183
(D–36–22)12dbc Recapture Reservoir	374002109263501	04/23/2009	12:40	—	1	158	34 2	4 92	1 66	12 6	109	105

Appendix 1. Field and laboratory data for water samples collected near the White Mesa uranium mill, San Juan County, Utah, June 2007–October 2009.—Continued

[**Abbreviations**: ANC, acid neutralizing capacity; CaCO$_3$, calcium carbonate; cm^3/g at STP, cubic centimeters of gas per gram at standard temperature (25°C) and pressure (1 bar); E, estimated; ft, feet; gal/min, gallons per minute; hh:mm, hour:minute; LSD, land surface datum; M, presence verified but not quantified; mg/L, milligrams per liter; mm/dd/yyyy, month/day/year; mm Hg, millimeters of mercury; mV, millivolts; permil, parts per thousand; pCi/L, picocuries per liter; SHE, standard hydrogen electrode; SiO$_2$, silicon dioxide; U, analyzed for but not detected; µg/L, micrograms per liter; µS/cm, microsiemens per centimeter; °C, degrees Celcius; <, less than; —, no data]

Station name	Station number	Date (mm/dd/yyyy)	Time (hh:mm)	Bicarbonate, field, dissolved, (mg/L)	Carbonate, field, dissolved, (mg/L)	Chloride, dissolved, (mg/L)	Fluoride, dissolved, (mg/L)	Hydrogen sulfide, dissolved, (mg/L)	Silica, dissolved, (mg/L as SiO$_2$)	Sulfate, dissolved, (mg/L as SO$_4$)	Sulfide, dissolved, field, (mg/L)	Nitrate + nitrite, dissolved, (mg/L as N)
Reference Spring North	373550109341701	06/21/2007		—	—	—	—	—	—	—	—	—
(D–38–22)23cda–1 South well	372756109280901	09/11/2007		—	—	—	—	—	—	—	—	—
		12/11/2007	15:00	221	—	1 58	0 18	M	18 2	46 6	—	<0 04
		03/11/2008	15:05	221	—	1 54	0 17	U	18 2	48 2	<0 20	<0 04
		11/12/2008	10:05	226	—	1 53	0 17	M	16 7	49	<0 20	<0 04
(D–38–22)23acb–1 North well	372817109275701	09/11/2007	11:00	—	—	—	—	—	—	—	—	—
		12/11/2007	12:00	225	—	0 92	0 19	M	18 3	30 8	—	<0 04
		03/11/2008	11:20	216	—	0 86	0 16	U	19	30 8	<0 20	<0 04
		11/11/2008	09:41	237	—	1 09	0 2	M	16 4	32 1	<0 20	<0 04
(D–38–22)23bba–S1 Right Hand Fork Seep	372832109282001	03/12/2008	17:00	—	—	—	—	—	—	—	—	—
(D–38–22) 8dcd–1 West well	372930109310701	06/21/2007		—	—	—	—	—	—	—	—	—
		09/11/2007	15:30	—	—	—	—	—	—	—	—	—
		12/13/2007	13:00	404	—	11 7	0 19	U	15 6	3,000	—	0 24
		03/13/2008	11:40	466	—	10 9	0 15	—	16 4	3,050	<0 20	E0 02
		09/16/2008	14:15	458	—	10 2	0 12	U	14 2	3,030	<0 20	E0 02
		11/13/2008	10:14	451	—	11	0 13	U	44 9	3,050	<0 20	0 33
		12/08/2008	14:50	—	—	—	—	—	—	—	—	—
		12/08/2008	14:55	—	—	—	—	—	—	—	—	—
		12/08/2008	15:00	—	—	—	—	—	—	—	—	—
		04/21/2009	11:00	431	—	11 2	0 16	—	16 8	3,090	<0 20	0 07
		09/22/2009	10:30	419	—	11	0 1	U	21	3,070	<0 20	0 16
		10/19/2009	15:00	—	—	—	—	—	—	—	—	—
		10/19/2009	15:05	—	—	—	—	U	—	—	—	—
		10/19/2009	15:10	—	—	—	—	—	—	—	—	—
				132	—							
(D–38–22)10cbc Anasazi Pond near spillway	372943109293201	09/18/2008	10:50	—	—	0 8	E0 08	—	5 82	0 52	—	<0 04
(D–38–22)10bcc–1 East well	372954109293601	06/21/2007		—	—	—	—	—	—	—	—	—
		09/11/2007	13:30	246	—	—	—	—	—	—	—	—
		12/14/2007	10:45	266	—	14	0 78	U	8 99	60 7	—	2 11
		03/13/2008	17:00	260	—	14 3	0 73	—	8 4	64 4	<0 20	1 51
		09/16/2008	10:45	271	—	14 3	0 89	U	6 94	64 5	<0 20	0 85
(D–38–22)10bcc–1 East well		11/13/2008	14:00	268	—	14 6	0 87	U	7 54	67 2	<0 20	0 8
		04/21/2009	15:50	255	—	14 3	0 83	—	7 95	68 6	<0 20	0 55
		09/22/2009	15:40	—	—	14 7	0 97	U	7 73	68 3	<0 20	0 54
(D–38–22) 8bad–S1 Ruin Spring	373006109312301	06/01/2007		—	—	—	—	—	—	—	—	—
		09/11/2007	16:00	227	—	—	—	—	—	—	—	—
		12/13/2007	09:30	224	—	26 6	0 58	—	12 1	516	—	1 45
		03/13/2008	12:20	233	—	20 3	0 6	—	11 5	382	—	1 56
		06/18/2008	15:20	235	—	23 9	0 54	—	11	442	—	1 59
		09/17/2008	12:20	218	—	24 1	0 56	—	11 4	453	—	1 63

Appendix 1.   Field and laboratory data for water samples collected near the White Mesa uranium mill, San Juan County, Utah, June 2007–October 2009.—Continued

[**Abbreviations**: ANC, acid neutralizing capacity; CaCO₃, calcium carbonate; cm³/g at STP, cubic centimeters of gas per gram at standard temperature (25°C) and pressure (1 bar); E, estimated; ft, feet; gal/min, gallons per minute; hh:mm, hour:minute; LSD, land surface datum; M, presence verified but not quantified; mg/L, milligrams per liter; mm/dd/yyyy, month/day/year; mm Hg, millimeters of mercury; mV, millivolts; permil, parts per thousand; pCi/L, picocuries per liter; SHE, standard hydrogen electrode; SiO₂, silicon dioxide; U, analyzed for but not detected; μg/L, micrograms per liter; μS/cm, microsiemens per centimeter; °C, degrees Celcius; <, less than; —, no data]

Station name	Station number	Date (mm/dd/yyyy)	Time (hh:mm)	Bicarbonate, field, dissolved, (mg/L)	Carbonate, field, dissolved, (mg/L)	Chloride, dissolved, (mg/L)	Fluoride, dissolved, (mg/L)	Hydrogen sulfide, dissolved, (mg/L)	Silica, dissolved, (mg/L as SiO₂)	Sulfate, dissolved, (mg/L as SO₄)	Sulfide, dissolved, field, (mg/L)	Nitrate + nitrite, dissolved, (mg/L as N)
		11/11/2008	13:45	237	—	25 3	0 57	—	12	460	—	1 61
		04/22/2009	10:15	230	—	24 6	0 51	—	10 8	478	—	1 53
		09/23/2009	13:30	62 3	5 6	25 6	0 57	—	12 6	507	—	1 43
(D–38–22) 4adb South Mill Pond	373052109294901	03/12/2008	14:40	—	—	1 76	<0 12	—	3 86	0 9	—	<0 04
(D–37–22)32ddc–1 MW3A	373116109305601	12/09/2008	09:15	—	—	—	—	—	—	—	—	—
		12/09/2008	09:20	—	—	—	—	—	—	—	—	—
		10/20/2009	08:20	—	—	—	—	—	—	—	—	—
		10/20/2009	08:25	—	—	—	—	—	—	—	—	—
(D–37–22)31dcb–S1 Cow Camp Spring	373122109321501	09/18/2007	17:00	—	—	—	—	—	—	—	—	—
		09/19/2007	19:00	309	—	—	—	—	—	—	—	—
		03/12/2008	13:25	306	—	112	0 42	—	19 2	356	—	0 11
		06/18/2008	14:20	316	—	116	0 42	—	18 5	362	—	<0 04
		09/17/2008	13:35	316	—	111	0 43	—	17 4	357	—	0 07
		11/13/2008	15:55	318	—	116	0 47	—	18 4	360	—	0 04
		04/22/2009	11:15	—	—	113	0 46	—	18 4	359	—	<0 04
		09/23/2009	10:15	346	—	117	0 44	—	18 8	366	—	E0 03
(D–37–22)32bab–S1 Mill Spring	373158109312601	03/12/2008	11:55	624	—	34 2	0 58	—	14 2	378	—	<0 04
		09/18/2008	09:20	426	—	63 1	0 64	—	16 9	815	—	E0 02
		11/12/2008	15:00	346	—	39 1	0 64	—	15 8	636	—	<0 04
		04/23/2009	08:55	—	—	22 7	0 64	—	15 9	282	—	<0 04
		09/24/2009	11:10	—	—	84 8	0 73	—	19 9	1,570	—	<0 04
		10/19/2009	17:00	—	—	—	—	—	—	—	—	—
(D–37–22)27ccc–S1 Entrance Spring	373202109293401	06/21/2007		—	—	—	—	—	—	—	—	—
		09/20/2007	14:00	291	—	—	—	—	—	—	—	—
		12/13/2007	15:30	292	—	79 8	0 63	—	15 4	206	—	1 38
		03/13/2008	16:10	292	—	59 4	0 64	—	15 6	172	—	1 85
		06/19/2008	09:15	—	—	62 8	0 71	—	15 8	177	—	2 48
		07/22/2008	12:00	286	—	—	—	—	—	—	—	—
		09/17/2008	10:00	264	—	54 1	0 68	—	12 3	162	—	1 45
		11/11/2008	12:45	280	—	59 8	0 64	—	9 81	174	—	1 1
		04/22/2009	09:30	272	—	40 5	0 56	—	15 4	129	—	1 95
		09/23/2009	09:00	—	—	60 7	0 64	—	14	188	—	0 43
(D–37–22)28acc–1 MW18	373233109301001	12/09/2008	08:45	—	—	—	—	—	—	—	—	—
		12/09/2008	08:50	—	—	—	—	—	—	—	—	—
		12/09/2008	08:55	—	—	—	—	—	—	—	—	—
		10/20/2009	08:55	—	—	—	—	M	—	—	—	—
		10/20/2009	09:00	—	—	—	—	M	—	—	—	—
		10/20/2009	09:05	264	—	—	—	—	—	—	—	—

Appendix 1. Field and laboratory data for water samples collected near the White Mesa uranium mill, San Juan County, Utah, June 2007–October 2009.—Continued

[**Abbreviations**: ANC, acid neutralizing capacity; CaCO₃, calcium carbonate; cm³/g at STP, cubic centimeters of gas per gram at standard temperature (25°C) and pressure (1 bar); E, estimated; ft, feet; gal/min, gallons per minute; hh:mm, hour:minute; LSD, land surface datum; M, presence verified but not quantified; mg/L, milligrams per liter; mm/dd/yyyy, month/day/year; mm Hg, millimeters of mercury; mV, millivolts; permil, parts per thousand; pCi/L, picocuries per liter; SHE, standard hydrogen electrode; SiO₂, silicon dioxide; U, analyzed for but not detected; µg/L, micrograms per liter; µS/cm, microsiemens per centimeter; °C, degrees Celcius; <, less than; —, no data]

Station name	Station number	Date (mm/dd/ yyyy)	Time (hh:mm)	Bicarbonate, field, dissolved, (mg/L)	Carbonate, field, dissolved, (mg/L)	Chloride, disolved, (mg/L)	Fluoride, dissolved, (mg/L)	Hydrogen sulfide, dissolved, (mg/L)	Silica, dissolved, (mg/L as SiO₂)	Sulfate, dissolved, (mg/L as SO₄)	Sulfide, dissolved, field, (mg/L)	Nitrate + nitrite, dissolved, (mg/L as N)
(D–37–22)10cdc–1 Lyman well	373442109291501	12/12/2007	11:00	—	—	23 4	0 53	U	19 1	173	—	0 98
(D–37–22) 8dba–1 Millview well	373501109310801	09/18/2007	12:00	338	—	—	—	—	—	—	—	—
(D–37–22) 2aad–1 Bayless well	373612109273201	12/12/2007	13:30	—	—	55 1	0 51	U	14 5	153	—	0 12
(D–36–22)19aad–S1 Oasis Spring	373850109315301	09/19/2007	15:55	300	—	—	—	—	—	—	—	—
		09/18/2008	14:00	207	—	29 7	0 4	—	17 3	70 2	—	E0 04
		11/12/2008	12:15	206	—	30 6	0 38	—	12 3	83 9	—	0 54
		04/23/2009	10:45	223	—	28 6	0 38	—	12 1	82 6	—	0 8
		09/24/2009	09:10	110	9 1	30	0 26	—	13 7	106	—	0 07
(D–36–22)12dbc Recapture Reservoir	374002109263501	04/23/2009	12:40			2 84	0 16	—	4 7	25 5	—	<0 04

Appendix 1.   Field and laboratory data for water samples collected near the White Mesa uranium mill, San Juan County, Utah, June 2007–October 2009.—Continued

[**Abbreviations**: ANC, acid neutralizing capacity; CaCO₃, calcium carbonate; cm³/g at STP, cubic centimeters of gas per gram at standard temperature (25°C) and pressure (1 bar); E, estimated; ft, feet; gal/min, gallons per minute; hh:mm, hour:minute; LSD, land surface datum; M, presence verified but not quantified; mg/L, milligrams per liter; mm/dd/yyyy, month/day/year; mm Hg, millimeters of mercury; mV, millivolts; permil, parts per thousand; pCi/L, picocuries per liter; SHE, standard hydrogen electrode; SiO₂, silicon dioxide; U, analyzed for but not detected; μg/L, micrograms per liter; μS/cm, microsiemens per centimeter; °C, degrees Celcius; <, less than; —, no data]

Station name	Station number	Date (mm/dd/ yyyy)	Time (hh:mm)	Orthophosphate, dissolved, (mg/L as P)	Aluminum, dissolved, (µg/L)	Aluminum, total, (µg/L)	Barium, dissolved, (µg/L)	Beryllium, dissolved, (µg/L)	Cadmium, dissolved, (µg/L)	Chromium, dissolved, (µg/L)	Chromium, total, (µg/L)	Cobalt, dissolved, (µg/L)
Reference Spring North	373550109341701	06/21/2007		—	—	—	—	—	—	—	—	—
(D–38–22)23cda–1 South well	372756109280901	09/11/2007		—								
		12/11/2007	15:00	—	<1 6	8	82	E0 01	<0 04	0 44	<0 40	E0 01
		03/11/2008	15:05	—	<1 6	<4	78	<0 01	<0 04	<0 12	<0 40	<0 02
		11/12/2008	10:05	0 008	<4 0	10	82	<0 02	<0 02	<0 12	<0 40	E0 01
(D–38–22)23acb–1 North well	372817109275701	09/11/2007	11:00	—								
		12/11/2007	12:00	—	<1 6	<4	83	E0 01	<0 04	<0 12	<0 40	E0 02
		03/11/2008	11:20	—	<1 6	<4	79	<0 01	E0 04	E0 08	<0 40	E0 01
		11/11/2008	09:41	E0 008	<4 0	<6	77	<0 02	<0 02	<0 12	<0 40	E0 02
(D–38–22)23bba–S1 Right Hand Fork Seep	372832109282001	03/12/2008	17:00	—	—	—	—	—	—	—	—	—
(D–38–22) 8dcd–1 West well	372930109310701	06/21/2007		—	—	—	—	—	—	—	—	—
		09/11/2007	15:30	—	—	—	—	—	—	—	—	—
		12/13/2007	13:00	—	<6 4	72	25	<0 03	0 52	6 5	13 3	2 6
		03/13/2008	11:40	—	<4 8	15	21	<0 02	0 27	0 66	2 5	1 2
		09/16/2008	14:15	E0 005	<4 8	18	25	<0 02	0 41	0 49	3 1	0 76
		11/13/2008	10:14	0 013	<12 0	37	31	<0 06	0 62	3 8	8 5	0 64
		12/08/2008	14:50	—	—	—	—	—	—	—	—	—
		12/08/2008	14:55	—	—	—	—	—	—	—	—	—
		12/08/2008	15:00	—	—	—	—	—	—	—	—	—
		04/21/2009	11:00	0 019	<12 0	<18	47	<0 06	0 32	3 6	4 1	4 5
		09/22/2009	10:30	0 011	<12 0	<18	15	<0 06	0 26	1 2	2 1	0 61
		10/19/2009	15:00	—	—	—	—	—	—	—	—	—
		10/19/2009	15:05	—	—	—	—	—	—	—	—	—
		10/19/2009	15:10	—	—	—	—	—	—	—	—	—
(D–38–22)10cbc Anasazi Pond near spillway	372943109293201	09/18/2008	10:50	0 091	2 4	1,680	121	<0 01	<0 04	<0 12	1 2	1 1
(D–38–22)10bcc–1 East well	372954109293601	06/21/2007										
		09/11/2007	13:30	—	—	—	—	—	—	—	—	—
		12/14/2007	10:45	—	2 7	783	14	<0 01	0 06	0 15	14 1	0 16
		03/13/2008	17:00	—	3 6	69	16	<0 01	0 04	0 31	14	0 08
		09/16/2008	10:45	E0 004	E1 4	45	7	<0 01	0 04	E0 09	1 7	0 1
		11/13/2008	14:00	0 031	E2 6	65	16	<0 02	0 06	0 46	3	0 03
(D–38–22)10bcc–1 East well		04/21/2009	15:50	0 056	E3 5	31	20	<0 02	0 07	0 68	1 7	0 09
		09/22/2009	15:40	0 085	<12 0	228	11	<0 06	E0 03	<0 36	1 6	E0 04
(D–38–22) 8bad–S1 Ruin Spring	373006109312301	06/01/2007		—	—	—	—	—	—	—	—	—
		09/11/2007	16:00	—	—	—	—	—	—	—	—	—
		12/13/2007	09:30	—	<1 6	<4	30	<0 01	0 07	E0 08	<0 40	0 05
		03/13/2008	12:20	—	<1 6	<4	22	<0 01	<0 04	<0 12	<0 40	<0 02
		06/18/2008	15:20	0 013	<1 6	<4	26	<0 01	0 05	<0 12	<0 40	0 08
		09/17/2008	12:20	E0 004	<1 6	<4	28	<0 01	0 05	<0 12	<0 40	0 06
		11/11/2008	13:45	0 011	<4 0	E6	28	<0 02	0 06	<0 12	<0 40	0 07

Appendix 1. Field and laboratory data for water samples collected near the White Mesa uranium mill, San Juan County, Utah, June 2007–October 2009.—Continued

[**Abbreviations**: ANC, acid neutralizing capacity; $CaCO_3$, calcium carbonate; cm^3/g at STP, cubic centimeters of gas per gram at standard temperature (25°C) and pressure (1 bar); E, estimated; ft, feet; gal/min, gallons per minute; hh:mm, hour:minute; LSD, land surface datum; M, presence verified but not quantified; mg/L, milligrams per liter; mm/dd/yyyy, month/day/year; mm Hg, millimeters of mercury; mV, millivolts; permil, parts per thousand; pCi/L, picocuries per liter; SHE, standard hydrogen electrode; SiO_2, silicon dioxide; U, analyzed for but not detected; µg/L, micrograms per liter; µS/cm, microsiemens per centimeter; °C, degrees Celcius; <, less than; —, no data]

Station name	Station number	Date (mm/dd/yyyy)	Time (hh:mm)	Orthophosphate, dissolved, (mg/L as P)	Aluminum, dissolved, (µg/L)	Aluminum, total, (µg/L)	Barium, dissolved, (µg/L)	Beryllium, dissolved, (µg/L)	Cadmium, dissolved, (µg/L)	Chromium, dissolved, (µg/L)	Chromium, total, (µg/L)	Cobalt, dissolved, (µg/L)
		04/22/2009	10:15	0 013	<4 0	11	30	<0 02	0 08	0 25	<0 40	0 29
		09/23/2009	13:30	0 012	<12 0	<6	28	<0 06	E0 05	<0 36	<0 40	0 17
(D–38–22) 4adb South Mill Pond	373052109294901	03/12/2008	14:40	—	8 2	3,470	54	<0 01	<0 04	<0 12	2 3	0 44
(D–37–22)32ddc–1 MW3A	373116109305601	12/09/2008	09:15	—	—	—	—	—	—	—	—	—
		12/09/2008	09:20	—	—	—	—	—	—	—	—	—
		10/20/2009	08:20	—	—	—	—	—	—	—	—	—
		10/20/2009	08:25	—	—	—	—	—	—	—	—	—
(D–37–22)31dcb–S1 Cow Camp Spring	373122109321501	09/18/2007	17:00	—	—	—	—	—	—	—	—	—
		09/19/2007	19:00	—	—	—	—	—	—	—	—	—
		03/12/2008	13:25	—	<1 6	38	34	<0 01	<0 04	E0 07	E0 20	0 05
		06/18/2008	14:20	0 011	3 8	368	40	<0 01	<0 04	E0 08	E0 35	0 11
		09/17/2008	13:35	E0 005	E1 0	499	35	<0 01	<0 04	E0 06	0 48	0 06
		11/13/2008	15:55	0 009	<4 0	808	46	<0 02	<0 02	E0 07	0 61	0 08
		04/22/2009	11:15	0 01	<4 0	522	43	<0 02	<0 02	0 23	E0 38	0 21
		09/23/2009	10:15	E0 008	161	307	34	<0 06	E0 03	<0 36	E0 22	0 16
(D–37–22)32bab–S1 Mill Spring	373158109312601	03/12/2008	11:55	—	E0 9	118	49	<0 01	<0 04	E0 06	E0 26	0 63
		09/18/2008	09:20	E0 004	<1 6	472	64	<0 01	E0 02	<0 12	0 52	1 2
		11/12/2008	15:00	E0 007	<4 0	426	53	<0 02	E0 02	E0 07	E0 37	0 26
		04/23/2009	08:55	0 009	<4 0	13	31	<0 02	<0 02	0 19	<0 40	0 41
		09/24/2009	11:10	0 015	<12 0	1,980	46	<0 06	E0 03	E0 19	1 5	1 5
		10/19/2009	17:00	—	—	—	—	—	—	—	—	—
(D–37–22)27ccc–S1 Entrance Spring	373202109293401	06/21/2007		—	—	—	—	—	—	—	—	—
		09/20/2007	14:00	—	—	—	—	—	—	—	—	—
		12/13/2007	15:30	—	1 7	80	156	M	<0 04	E0 10	<0 40	0 42
		03/13/2008	16:10	—	<1 6	330	143	<0 01	<0 04	E0 08	0 43	0 23
		06/19/2008	09:15	0 012	E1 0	707	141	<0 01	<0 04	E0 08	0 63	0 21
		07/22/2008	12:00	—	—	116	—	—	—	—	0 49	—
		09/17/2008	10:00	E0 005	<1 6	142	148	<0 01	<0 04	<0 12	<0 40	0 17
		11/11/2008	12:45	E0 004	<4 0	66	132	<0 02	E0 01	E0 06	<0 40	0 18
		04/22/2009	09:30	E0 007	<4 0	2,930	152	<0 02	E0 02	0 23	2	0 57
		09/23/2009	09:00	E0 007	<12 0	52	118	<0 06	<0 06	<0 36	<0 40	0 31
(D–37–22)28acc–1 MW18	373233109301001	12/09/2008	08:45	—	—	—	—	—	—	—	—	—
		12/09/2008	08:50	—	—	—	—	—	—	—	—	—
		12/09/2008	08:55	—	—	—	—	—	—	—	—	—
		10/20/2009	08:55	—	—	—	—	—	—	—	—	—
		10/20/2009	09:00	—	—	—	—	—	—	—	—	—
		10/20/2009	09:05	—	—	—	—	—	—	—	—	—
(D–37–22)10cdc–1 Lyman well	373442109291501	12/12/2007	11:00	—	6 3	9	87	0 04	0 31	E0 07	<0 40	0 28

Appendix 1.   Field and laboratory data for water samples collected near the White Mesa uranium mill, San Juan County, Utah, June 2007—October 2009.—Continued

[**Abbreviations**: ANC, acid neutralizing capacity; CaCO$_3$, calcium carbonate; cm^3/g at STP, cubic centimeters of gas per gram at standard temperature (25°C) and pressure (1 bar); E, estimated; ft, feet; gal/min, gallons per minute; hh:mm, hour:minute; LSD, land surface datum; M, presence verified but not quantified; mg/L, milligrams per liter; mm/dd/yyyy, month/day/year; mm Hg, millimeters of mercury; mV, millivolts; permil, parts per thousand; pCi/L, picocuries per liter; SHE, standard hydrogen electrode; SiO$_2$, silicon dioxide; U, analyzed for but not detected; μg/L, micrograms per liter; μS/cm, microsiemens per centimeter; °C, degrees Celcius; <, less than; —, no data]

Station name	Station number	Date (mm/dd/yyyy)	Time (hh:mm)	Orthophosphate, dissolved, (mg/L as P)	Aluminum, dissolved, (µg/L)	Aluminum, total, (µg/L)	Barium, dissolved, (µg/L)	Beryllium, dissolved, (µg/L)	Cadmium, dissolved, (µg/L)	Chromium, dissolved, (µg/L)	Chromium, total, (µg/L)	Cobalt, dissolved, (µg/L)
(D–37–22) 8dba–1 Millview well	373501109310801	09/18/2007	12:00	—	—	—	—	—	—	—	—	—
(D–37–22) 2aad–1 Bayless well	373612109273201	12/12/2007	13:30	—	<1 6	E4	137	M	0 24	0 16	0 5	2 5
(D–36–22)19aad–S1 Oasis Spring	373850109315301	09/19/2007	15:55	—	—	—	—	—	—	—	—	—
		09/18/2008	14:00	0 064	E1 1	70	128	<0 01	<0 04	<0 12	<0 40	0 39
		11/12/2008	12:15	E0 005	<4 0	178	100	<0 02	<0 02	<0 12	<0 40	0 13
		04/23/2009	10:45	E0 005	<4 0	70	100	<0 02	<0 02	0 13	<0 40	0 2
		09/24/2009	09:10	0 091	<12 0	28	176	<0 06	<0 06	<0 36	0 46	1
(D–36–22)12dbc Recapture Reservoir	374002109263501	04/23/2009	12:40	<0 008	<4 0	24	85	<0 02	<0 02	0 12	<0 40	0 12

Appendix 1. Field and laboratory data for water samples collected near the White Mesa uranium mill, San Juan County, Utah, June 2007–October 2009.—Continued

[**Abbreviations**: ANC, acid neutralizing capacity; CaCO$_3$, calcium carbonate; cm^3/g at STP, cubic centimeters of gas per gram at standard temperature (25°C) and pressure (1 bar); E, estimated; ft, feet; gal/min, gallons per minute; hh:mm, hour:minute; LSD, land surface datum; M, presence verified but not quantified; mg/L, milligrams per liter; mm/dd/yyyy, month/day/year; mm Hg, millimeters of mercury; mV, millivolts; permil, parts per thousand; pCi/L, picocuries per liter; SHE, standard hydrogen electrode; SiO$_2$, silicon dioxide; U, analyzed for but not detected; µg/L, micrograms per liter; µS/cm, microsiemens per centimeter; °C, degrees Celcius; <, less than; —, no data]

Station name	Station number	Date (mm/dd/yyyy)	Time (hh:mm)	Copper, dissolved, (µg/L)	Copper, total, (µg/L)	Ferrous iron, dissolved, field, (mg/L)	Iron, dissolved, (µg/L)	Iron, total, (µg/L)	Lead, dissolved, (µg/L)	Lead, total, (µg/L)	Lithium, dissolved, (µg/L)	Manganese, dissolved, (µg/L)
Reference Spring North	373550109341701	06/21/2007		—	—	—	—	—	—	—	—	—
(D–38–22)23cda–1 South well	372756109280901	09/11/2007		—	—	—	—	—	—	—	—	—
		12/11/2007	15:00	<1 0	3 8	<0 200	254	251	E0 05	0 42	54 4	10 9
		03/11/2008	15:05	<1 0	E0 65	<0 200	228	238	<0 08	0 08	33 1	8 3
		11/12/2008	10:05	<1 0	<4 0	0 32	198	232	<0 06	0 19	42 5	9 1
(D–38–22)23acb–1 North well	372817109275701	09/11/2007	11:00	—	—	—	—	—	—	—	—	—
		12/11/2007	12:00	<1 0	<1 2	<0 200	161	160	0 11	0 14	51 3	6 6
		03/11/2008	11:20	<1 0	<1 2	<0 200	219	216	E0 05	0 19	30 6	5 4
		11/11/2008	09:41	E0 86	E3 0	<0 200	245	233	<0 06	0 43	49 7	6 4
(D–38–22)23bba–S1 Right Hand Fork Seep	372832109282001	03/12/2008	17:00	—	—	—	—	—	—	—	—	—
(D–38–22) 8dcd–1 West well	372930109310701	06/21/2007		—	—	—	—	—	—	—	—	—
		09/11/2007	15:30	—	—	—	—	—	—	—	—	—
		12/13/2007	13:00	37 6	45 6	—	93	313	1 2	5 23	414	124
		03/13/2008	11:40	9	26 4	1 89	3,050	4,090	0 34	3 07	429	374
		09/16/2008	14:15	E2 0	4 3	<0 200	32	390	<0 24	1 7	349	414
		11/13/2008	10:14	8 5	E11 9	<0 200	136	171	0 7	3 43	299	305
		12/08/2008	14:50	—	—	—	—	210	—	—	—	—
		12/08/2008	14:55	—	—	—	—	98	—	—	—	—
		12/08/2008	15:00	—	—	—	—	73,800	—	—	—	—
		04/21/2009	11:00	E2 4	13	0 32	215	219	1 16	4 74	290	344
		09/22/2009	10:30	E2 1	<12 0	—	131	140	0 46	1	161	176
		10/19/2009	15:00	—	—	—	—	<28	—	—	—	—
		10/19/2009	15:05	—	—	—	—	509	—	—	—	—
		10/19/2009	15:10	—	—	—	—	28,600	—	—	—	—
(D–38–22)10cbc Anasazi Pond near spillway	372943109293201	09/18/2008	10:50	1 5	2 7		29	1,490	<0 08	2 94	2 3	550
(D–38–22)10bcc–1 East well	372954109293601	06/21/2007		—	—	—	—	—	—	—	—	—
		09/11/2007	13:30	—	—	—	—	—	—	—	—	—
		12/14/2007	10:45	1 6	109	<0 200	<8	406	1 05	24 9	63 9	11 9
		03/13/2008	17:00	4 7	19 2	<0 20	<8	309	0 15	7 63	46 9	8 2
		09/16/2008	10:45	E0 86	8 7	<0 200	17	2,200	<0 08	1 78	50 6	29 1
		11/13/2008	14:00	2 3	8 2	<0 200	E3	510	E0 03	1 51	49 7	1 3
		04/21/2009	15:50	1 7	4 1	<0 200	E3	138	<0 06	0 67	45 8	8 8
		09/22/2009	15:40	<3 0	6 4	—	<4	381	<0 18	1 35	48 8	4 9
(D–38–22) 8bad–S1 Ruin Spring	373006109312301	06/01/2007		—	—	—	—	—	—	—	—	—
		09/11/2007	16:00	—	—	—	—	—	—	—	—	—
		12/13/2007	09:30	E0 61	<1 2	—	<8	<6	<0 08	<0 06	63 6	E0 2
		03/13/2008	12:20	<1 0	<1 2	—	<8	<6	<0 08	<0 06	—	<0 2
		06/18/2008	15:20	<1 0	<1 2	—	<8	E4	<0 08	<0 06	53 3	<0 2
		09/17/2008	12:20	<1 0	<1 2	—	<8	E5	<0 08	<0 06	62 5	<0 2
		11/11/2008	13:45	<1 0	<4 0	—	5	<14	<0 06	0 11	61 8	E0 1
		04/22/2009	10:15	<1 0	<4 0	—	<4	E14	<0 06	<0 10	66 8	E0 2
		09/23/2009	13:30	<3 0	<4 0	—	<4	<14	<0 18	<0 10	58 7	<0 6

Appendix 1.   Field and laboratory data for water samples collected near the White Mesa uranium mill, San Juan County, Utah, June 2007–October 2009.—Continued

[**Abbreviations**: ANC, acid neutralizing capacity; CaCO$_3$, calcium carbonate; cm^3/g at STP, cubic centimeters of gas per gram at standard temperature (25°C) and pressure (1 bar); E, estimated; ft, feet; gal/min, gallons per minute; hh:mm, hour:minute; LSD, land surface datum; M, presence verified but not quantified; mg/L, milligrams per liter; mm/dd/yyyy, month/day/year; mm Hg, millimeters of mercury; mV, millivolts; permil, parts per thousand; pCi/L, picocuries per liter; SHE, standard hydrogen electrode; SiO$_2$, silicon dioxide; U, analyzed for but not detected; µg/L, micrograms per liter; µS/cm, microsiemens per centimeter; °C, degrees Celcius; <, less than; —, no data]

Station name	Station number	Date (mm/dd/yyyy)	Time (hh:mm)	Copper, dissolved, (µg/L)	Copper, total, (µg/L)	Ferrous iron, dissolved, field, (mg/L)	Iron, dissolved, (µg/L)	Iron, total, (µg/L)	Lead, dissolved, (µg/L)	Lead, total, (µg/L)	Lithium, dissolved, (µg/L)	Manganese, dissolved, (µg/L)
(D–38–22) 4adb South Mill Pond	373052109294901	03/12/2008	14:40	2 7	4 5	—	10	2,380	<0 08	3 04	E0 9	0 5
(D–37–22)32ddc–1 MW3A	373116109305601	12/09/2008	09:15	—	—	—	—	<70	—	—	—	—
		12/09/2008	09:20	—	—	—	—	E36	—	—	—	—
		10/20/2009	08:20	—	—	—	—	<46	—	—	—	—
		10/20/2009	08:25	—	—	—	—	<46	—	—	—	—
(D–37–22)31dcb–S1 Cow Camp Spring	373122109321501	09/18/2007	17:00	—	—	—	—	—	—	—	—	—
		09/19/2007	19:00	—	—	—	—	—	—	—	—	—
		03/12/2008	13:25	<1 0	<1 2	—	<8	29	<0 08	E0 05	60 9	0 9
		06/18/2008	14:20	<1 0	E0 74	—	E6	259	<0 08	0 51	74 2	2 4
		09/17/2008	13:35	<1 0	<1 2	—	<8	322	<0 08	0 5	71 1	4 6
		11/13/2008	15:55	<1 0	<4 0	—	E3	474	0 07	0 94	73 8	3 6
		04/22/2009	11:15	<1 0	<4 0	—	5	310	E0 04	0 68	76 7	8 6
		09/23/2009	10:15	E2 7	<4 0	—	6	205	0 32	0 36	63 1	0 9
(D–37–22)32bab–S1 Mill Spring	373158109312601	03/12/2008	11:55	<1 0	<1 2	—	95	686	<0 08	0 11	60	95 1
		09/18/2008	09:20	<1 0	<1 2	—	16	1,090	<0 08	0 58	125	136
		11/12/2008	15:00	1 3	<4 0	—	22	294	<0 06	0 43	89 2	39 9
		04/23/2009	08:55	<1 0	<4 0	—	18	116	<0 06	<0 10	58 8	57 8
		09/24/2009	11:10	<3 0	<8 0	—	26	1,860	<0 18	1 76	195	83 1
		10/19/2009	17:00	—	—	—	—	—	—	—	—	—
(D–37–22)27ccc–S1 Entrance Spring	373202109293401	06/21/2007		—	—	—	—	—	—	—	—	—
		09/20/2007	14:00	—	—	—	—	—	—	—	—	—
		12/13/2007	15:30	E0 79	<1 2	—	E8	54	E0 05	0 24	44 6	94 7
		03/13/2008	16:10	1 2	1 5	—	E5	205	<0 08	0 46	30 1	30 8
		06/19/2008	09:15	<1 0	1 3	—	14	686	<0 08	1 57	32 8	24 6
		07/22/2008	12:00	—	2 3	—	—	96	—	2 18	—	—
		09/17/2008	10:00	<1 0	<1 2	—	E6	96	<0 08	0 33	31 9	56 2
		11/11/2008	12:45	<1 0	<4 0	—	6	43	E0 05	0 11	34	52 7
		04/22/2009	09:30	<1 0	4 2	—	11	2,050	E0 03	5 59	30 8	144
		09/23/2009	09:00	<3 0	<4 0	—	7	46	<0 18	0 28	35 8	4 1
(D–37–22)28acc–1 MW18	373233109301001	12/09/2008	08:45	—	—	—	—	54	—	—	—	—
		12/09/2008	08:50	—	—	—	—	1,840	—	—	—	—
		12/09/2008	08:55	—	—	—	—	2,700	—	—	—	—
		10/20/2009	08:55	—	—	—	—	1,800	—	—	—	—
		10/20/2009	09:00	—	—	—	—	2,470	—	—	—	—
		10/20/2009	09:05	—	—	—	—	1,190	—	—	—	—
(D–37–22)10cdc–1 Lyman well	373442109291501	12/12/2007	11:00	1 1	<1 2	<0 200	E5	9	4 41	3 69	25 6	17 5
(D–37–22) 8dba–1 Millview well	373501109310801	09/18/2007	12:00	—	—	—	—	—	—	—	—	—
(D–37–22) 2aad–1 Bayless well	373612109273201	12/12/2007	13:30	2 4	2 2	<0 200	E5	372	0 5	0 77	21 1	3,150

Appendix 1. Field and laboratory data for water samples collected near the White Mesa uranium mill, San Juan County, Utah, June 2007–October 2009.—Continued

[**Abbreviations**: ANC, acid neutralizing capacity; CaCO₃, calcium carbonate; cm³/g at STP, cubic centimeters of gas per gram at standard temperature (25°C) and pressure (1 bar); E, estimated; ft, feet; gal/min, gallons per minute; hh:mm, hour:minute; LSD, land surface datum; M, presence verified but not quantified; mg/L, milligrams per liter; mm/dd/yyyy, month/day/year; mm Hg, millimeters of mercury; mV, millivolts; permil, parts per thousand; pCi/L, picocuries per liter; SHE, standard hydrogen electrode; SiO₂, silicon dioxide; U, analyzed for but not detected; μg/L, micrograms per liter; μS/cm, microsiemens per centimeter; °C, degrees Celcius; <, less than; —, no data]

Station name	Station number	Date (mm/dd/yyyy)	Time (hh:mm)	Copper, dissolved, (μg/L)	Copper, total, (μg/L)	Ferrous iron, dissolved, field, (mg/L)	Iron, dissolved, (μg/L)	Iron, total, (μg/L)	Lead, dissolved, (μg/L)	Lead, total, (μg/L)	Lithium, dissolved, (μg/L)	Manganese, dissolved, (μg/L)
(D–36–22)19aad–S1 Oasis Spring	373850109315301	09/19/2007	15:55	—	—	—	—	—	—	—	—	—
		09/18/2008	14:00	<1 0	E0 75	—	29	199	<0 08	0 07	10	353
		11/12/2008	12:15	<1 0	<4 0	—	10	120	<0 06	0 2	8 9	49 8
		04/23/2009	10:45	<1 0	<4 0	—	E4	47	E0 04	E0 08	9 2	18 6
		09/24/2009	09:10	<3 0	<4 0	—	94	175	<0 18	E0 10	6 2	1,140
(D–36–22)12dbc Recapture Reservoir	374002109263501	04/23/2009	12:40	<1 0	<4 0	—	<4	19	<0 06	<0 10	3 2	3 4

Appendix 1.   Field and laboratory data for water samples collected near the White Mesa uranium mill, San Juan County, Utah, June 2007–October 2009.—Continued

[**Abbreviations**: ANC, acid neutralizing capacity; $CaCO_3$, calcium carbonate; cm^3/g at STP, cubic centimeters of gas per gram at standard temperature (25°C) and pressure (1 bar); E, estimated; ft, feet; gal/min, gallons per minute; hh:mm, hour:minute; LSD, land surface datum; M, presence verified but not quantified; mg/L, milligrams per liter; mm/dd/yyyy, month/day/year; mm Hg, millimeters of mercury; mV, millivolts; permil, parts per thousand; pCi/L, picocuries per liter; SHE, standard hydrogen electrode; SiO_2, silicon dioxide; U, analyzed for but not detected; µg/L, micrograms per liter; µS/cm, microsiemens per centimeter; °C, degrees Celcius; <, less than; —, no data]

Station name	Station number	Date (mm/dd/yyyy)	Time (hh:mm)	Manganese, total, (µg/L)	Molybdenum, dissolved, (µg/L)	Molybdenum, total, (µg/L)	Nickel, dissolved, (µg/L)	Nickel, total, (µg/L)	Silver, dissolved, (µg/L)	Strontium, dissolved, (µg/L)	Thallium, dissolved, (µg/L)	Vanadium, dissolved, (µg/L)
Reference Spring North	373550109341701	06/21/2007		—	—	—	—	—	—	—	—	—
(D–38–22)23cda–1 South well	372756109280901	09/11/2007		—	—	—	—	—	—	—	—	—
		12/11/2007	15:00	9 6	0 9	0 8	E0 19	E0 07	<0 1	2,300	<0 04	0 06
		03/11/2008	15:05	8	0 8	0 9	E0 11	E0 10	<0 1	2,010	<0 04	E0 02
		11/12/2008	10:05	8 6	0 9	0 9	0 14	<0 20	M	2,050	<0 04	<0 16
(D–38–22)23acb–1 North well	372817109275701	09/11/2007	11:00	—	—	—	—	—	—	—	—	—
		12/11/2007	12:00	5 5	0 9	1	E0 15	E0 06	<0 1	2,180	<0 04	E0 03
		03/11/2008	11:20	5 2	0 9	0 9	E0 11	E0 08	<0 1	1,850	<0 04	<0 04
		11/11/2008	09:41	6 5	1 2	1 3	E0 10	<0 20	<0 008	1,760	<0 04	<0 16
(D–38–22)23bba–S1 Right Hand Fork Seep	372832109282001	03/12/2008	17:00	—	—	—	—	—	—	—	—	—
(D–38–22) 8dcd–1 West well	372930109310701	06/21/2007		—	—	—	—	—	—	—	—	—
		09/11/2007	15:30	—	—	—	—	—	—	—	—	—
		12/13/2007	13:00	139	10 7	40 5	12 5	14 4	<0 4	8,860	0 35	0 4
		03/13/2008	11:40	348	33 4	36 9	7 7	7 4	<0 3	8,980	0 17	E0 12
		09/16/2008	14:15	399	32	33 6	9 3	9 5	<0 3	9,490	0 35	0 26
		11/13/2008	10:14	281	32 9	38 6	11	11 1	M	8,560	0 52	0 6
		12/08/2008	14:50	—	—	—	—	—	—	—	—	—
		12/08/2008	14:55	—	—	—	—	—	—	—	—	—
		12/08/2008	15:00	—	—	—	—	—	—	—	—	—
		04/21/2009	11:00	322	48 8	43 5	13 7	10	M	8,680	0 54	1 6
		09/22/2009	10:30	326	24 1	43 2	5 3	6 1	<0 024	4,980	0 32	E0 40
		10/19/2009	15:00	—	—	—	—	—	—	—	—	—
		10/19/2009	15:05	—	—	—	—	—	—	—	—	—
		10/19/2009	15:10	—	—	—	—	—	—	—	—	—
(D–38–22)10cbc Anasazi Pond near spillway	372943109293201	09/18/2008	10:50	586	1	0 9	1	1 9	<0 1	147	<0 04	5 5
(D–38–22)10bcc–1 East well	372954109293601	06/21/2007		—	—	—	—	—	—	—	—	—
		09/11/2007	13:30	—	—	—	—	—	—	—	—	—
		12/14/2007	10:45	12	18 4	16 9	7 4	6 7	<0 1	203	<0 04	0 89
		03/13/2008	17:00	8 1	17 6	18 4	1 2	1 3	<0 1	209	<0 04	0 52
		09/16/2008	10:45	27 2	15 3	15 6	1 3	1 9	<0 1	100	<0 04	0 19
		11/13/2008	14:00	9 4	16 4	17	1 9	2 8	<0 008	149	<0 04	0 5
		04/21/2009	15:50	9 2	18 6	17 2	2	2 1	<0 008	205	<0 04	0 63
		09/22/2009	15:40	7 4	16 6	18 2	0 67	1	<0 024	150	<0 12	0 92
(D–38–22) 8bad–S1 Ruin Spring	373006109312301	06/01/2007		—	—	—	—	—	—	—	—	—
		09/11/2007	16:00	—	—	—	—	—	—	—	—	—
		12/13/2007	09:30	E0 2	19 7	19 4	0 45	E0 12	<0 1	1,530	<0 04	0 33
		03/13/2008	12:20	<0 4	18 7	19 4	—	0 18	<0 1	—	<0 04	0 24
		06/18/2008	15:20	<0 4	18 5	19	0 53	0 22	<0 1	1,340	<0 04	0 26
		09/17/2008	12:20	<0 4	19 5	19 9	0 43	0 13	<0 1	1,480	E0 03	0 33
		11/11/2008	13:45	<0 4	19 4	19 9	0 69	<0 20	M	1,410	<0 04	0 34
		04/22/2009	10:15	6 8	16 5	17 3	0 26	0 25	M	1,750	<0 04	0 56
		09/23/2009	13:30	<0 4	17 6	20 1	0 78	0 52	<0 024	1,380	<0 12	0 53

Appendix 1. Field and laboratory data for water samples collected near the White Mesa uranium mill, San Juan County, Utah, June 2007–October 2009.—Continued

[**Abbreviations**: ANC, acid neutralizing capacity; CaCO₃, calcium carbonate; cm³/g at STP, cubic centimeters of gas per gram at standard temperature (25°C) and pressure (1 bar); E, estimated; ft, feet; gal/min, gallons per minute; hh:mm, hour:minute; LSD, land surface datum; M, presence verified but not quantified; mg/L, milligrams per liter; mm/dd/yyyy, month/day/year; mm Hg, millimeters of mercury; mV, millivolts; permil, parts per thousand; pCi/L, picocuries per liter; SHE, standard hydrogen electrode; SiO₂, silicon dioxide; U, analyzed for but not detected; µg/L, micrograms per liter; µS/cm, microsiemens per centimeter; °C, degrees Celcius; <, less than; —, no data]

Station name	Station number	Date (mm/dd/ yyyy)	Time (hh:mm)	Manganese, total, (µg/L)	Molybdenum, dissolved, (µg/L)	Molybdenum, total, (µg/L)	Nickel, dissolved, (µg/L)	Nickel, total, (µg/L)	Silver, dissolved, (µg/L)	Strontium, dissolved, (µg/L)	Thallium, dissolved, (µg/L)	Vanadium, dissolved, (µg/L)
(D–38–22) 4adb South Mill Pond	373052109294901	03/12/2008	14:40	84 5	0 5	0 4	0 72	2 8	<0 1	123	<0 04	6 4
(D–37–22)32ddc–1 MW3A	373116109305601	12/09/2008	09:15	—	—	—	—	—	—	—	—	—
		12/09/2008	09:20	—	—	—	—	—	—	—	—	—
		10/20/2009	08:20	—	—	—	—	—	—	—	—	—
		10/20/2009	08:25	—	—	—	—	—	—	—	—	—
(D–37–22)31dcb–S1 Cow Camp Spring	373122109321501	09/18/2007	17:00	—	—	—	—	—	—	—	—	—
		09/19/2007	19:00	—	—	—	—	—	—	—	—	—
		03/12/2008	13:25	1 6	1 7	1 8	0 32	E0 12	<0 1	2,970	<0 04	0 9
		06/18/2008	14:20	13 6	1 7	1 7	0 66	0 43	<0 1	3,500	<0 04	1 5
		09/17/2008	13:35	14 4	1 7	1 6	0 29	0 43	<0 1	3,120	<0 04	0 94
		11/13/2008	15:55	32 7	1 8	1 8	0 45	0 62	M	3,070	<0 04	1 3
		04/22/2009	11:15	21 7	1 8	1 7	0 2	0 5	M	3,420	<0 04	2 4
		09/23/2009	10:15	5 2	1 6	1 9	0 51	0 65	0 1	2,710	<0 12	1
(D–37–22)32bab–S1 Mill Spring	373158109312601	03/12/2008	11:55	89 4	1 4	1 5	1 2	1 1	<0 1	1,640	<0 04	0 3
		09/18/2008	09:20	158	6 3	6 5	1 8	1 7	<0 1	3,000	<0 04	0 59
		11/12/2008	15:00	56 3	3 5	3 9	1 3	0 96	M	2,340	<0 04	0 29
		04/23/2009	08:55	59 4	0 8	0 8	0 57	0 47	<0 008	1,530	<0 04	0 34
		09/24/2009	11:10	96 5	19 2	22 7	3 3	3 5	<0 024	4,300	<0 12	3 2
		10/19/2009	17:00	—	—	—	—	—	—	—	—	—
(D–37–22)27ccc–S1 Entrance Spring	373202109293401	06/21/2007		—	—	—	—	—	—	—	—	—
		09/20/2007	14:00	—	—	—	—	—	—	—	—	—
		12/13/2007	15:30	67 5	4 2	4	1 2	0 94	<0 1	1,320	<0 04	4 6
		03/13/2008	16:10	37	5 5	5 8	0 94	1	<0 1	1,090	<0 04	4 6
		06/19/2008	09:15	41 8	4 9	4 6	0 91	1 2	<0 1	1,330	<0 04	3 2
		07/22/2008	12:00	69 3	—	1 4	—	2	—	—	—	—
		09/17/2008	10:00	59 1	4 4	4 4	0 71	0 64	<0 1	1,200	<0 04	6 1
		11/11/2008	12:45	51	3 6	3 9	0 72	0 41	<0 008	1,080	<0 04	4 9
		04/22/2009	09:30	204	4 7	4 1	1 2	3	<0 008	1,090	<0 04	6 5
		09/23/2009	09:00	5 4	3 8	3 9	0 76	0 64	<0 024	1,180	<0 12	4
(D–37–22)28acc–1 MW18	373233109301001	12/09/2008	08:45	—	—	—	—	—	—	—	—	—
		12/09/2008	08:50	—	—	—	—	—	—	—	—	—
		12/09/2008	08:55	—	—	—	—	—	—	—	—	—
		10/20/2009	08:55	—	—	—	—	—	—	—	—	—
		10/20/2009	09:00	—	—	—	—	—	—	—	—	—
		10/20/2009	09:05	—	—	—	—	—	—	—	—	—
(D–37–22)10cdc–1 Lyman well	373442109291501	12/12/2007	11:00	16 3	3 6	3 5	11	9 2	<0 1	975	0 66	0 76
(D–37–22) 8dba–1 Millview well	373501109310801	09/18/2007	12:00	—	—	—	—	—	—	—	—	—
(D–37–22) 2aad–1 Bayless well	373612109273201	12/12/2007	13:30	3320	7 2	7 4	7 4	9 4	<0 1	862	0 29	0 43

Appendix 1.   Field and laboratory data for water samples collected near the White Mesa uranium mill, San Juan County, Utah, June 2007–October 2009.—Continued

[**Abbreviations**: ANC, acid neutralizing capacity; $CaCO_3$, calcium carbonate; cm^3/g at STP, cubic centimeters of gas per gram at standard temperature (25°C) and pressure (1 bar); E, estimated; ft, feet; gal/min, gallons per minute; hh:mm, hour:minute; LSD, land surface datum; M, presence verified but not quantified; mg/L, milligrams per liter; mm/dd/yyyy, month/day/year; mm Hg, millimeters of mercury; mV, millivolts; permil, parts per thousand; pCi/L, picocuries per liter; SHE, standard hydrogen electrode; SiO_2, silicon dioxide; U, analyzed for but not detected; µg/L, micrograms per liter; µS/cm, microsiemens per centimeter; °C, degrees Celcius; <, less than; —, no data]

Station name	Station number	Date (mm/dd/yyyy)	Time (hh:mm)	Manganese, total, (µg/L)	Molybdenum, dissolved, (µg/L)	Molybdenum, total, (µg/L)	Nickel, dissolved, (µg/L)	Nickel, total, (µg/L)	Silver, dissolved, (µg/L)	Strontium, dissolved, (µg/L)	Thallium, dissolved, (µg/L)	Vanadium, dissolved, (µg/L)
(D–36–22)19aad–S1 Oasis Spring	373850109315301	09/19/2007	15:55	—	—	—	—	—	—	—	—	—
		09/18/2008	14:00	333	1 1	1 1	0 75	0 62	<0 1	493	<0 04	0 97
		11/12/2008	12:15	48 7	1 4	1 4	0 51	0 34	<0 008	436	<0 04	1 1
		04/23/2009	10:45	20	1 8	1 6	0 31	0 31	<0 008	472	<0 04	1 3
		09/24/2009	09:10	1,130	2 5	2 7	0 99	1	<0 024	407	<0 12	1
(D–36–22)12dbc Recapture Reservoir	374002109263501	04/23/2009	12:40	6 6	1 8	1 7	0 41	0 39	<0 008	434	<0 04	0 66

Appendix 1. Field and laboratory data for water samples collected near the White Mesa uranium mill, San Juan County, Utah, June 2007–October 2009.—Continued

[**Abbreviations**: ANC, acid neutralizing capacity; CaCO$_3$, calcium carbonate; cm^3/g at STP, cubic centimeters of gas per gram at standard temperature (25°C) and pressure (1 bar); E, estimated; ft, feet; gal/min, gallons per minute; hh:mm, hour:minute; LSD, land surface datum; M, presence verified but not quantified; mg/L, milligrams per liter; mm/dd/yyyy, month/day/year; mm Hg, millimeters of mercury; mV, millivolts; permil, parts per thousand; pCi/L, picocuries per liter; SHE, standard hydrogen electrode; SiO$_2$, silicon dioxide; U, analyzed for but not detected; µg/L, micrograms per liter; µS/cm, microsiemens per centimeter; °C, degrees Celcius; <, less than; —, no data]

Station name	Station number	Date (mm/dd/yyyy)	Time (hh:mm)	Vanadium, total, (µg/L)	Zinc, dissolved, (µg/L)	Zinc, total, (µg/L)	Antimony, dissolved, (µg/L)	Arsenic, dissolved, (µg/L)	Arsenic, total, (µg/L)	Boron, dissolved, (µg/L)	Selenium, dissolved, (µg/L)	Selenium, total, (µg/L)
Reference Spring North	373550109341701	06/21/2007		—	—	—	—	—	—	—	—	—
(D–38–22)23cda–1 South well	372756109280901	09/11/2007		—	—	—	—	—	—	—	—	—
		12/11/2007	15:00	E0 05	6 9	7 3	<0 14	8 3	8 9	25	<0 04	<0 08
		03/11/2008	15:05	<1 0	4 5	6 7	<0 14	8 2	8 7	20	<0 04	<0 08
		11/12/2008	10:05	<1 6	E1 4	10 8	E0 04	8 6	8 2	24	<0 06	<0 12
(D–38–22)23acb–1 North well	372817109275701	09/11/2007	11:00	—	—	—	—	—	—	—	—	—
		12/11/2007	12:00	E0 07	3 8	10 2	<0 14	8 6	9 6	22	<0 04	<0 08
		03/11/2008	11:20	0 28	3 2	4 5	<0 14	8	8 5	17	<0 04	<0 08
		11/11/2008	09:41	<1 6	E1 3	51 2	E0 02	9 9	9 8	24	<0 06	<0 12
(D–38–22)23bba–S1 Right Hand Fork Seep	372832109282001	03/12/2008	17:00	—	—	—	—	—	—	—	—	—
(D–38–22) 8dcd–1 West well	372930109310701	06/21/2007		—	—	—	—	—	—	—	—	—
		09/11/2007	15:30	—	—	—	—	—	—	—	—	—
		12/13/2007	13:00	0 68	556	557	0 66	0 26	<1 8	62	1 2	0 88
		03/13/2008	11:40	<0 30	42 5	46 1	<0 42	0 22	<1 8	62	0 53	0 36
		09/16/2008	14:15	0 6	25 6	23 4	<0 42	E0 14	<1 8	58	0 63	0 34
		11/13/2008	10:14	<4 8	48 9	41	0 28	0 27	E0 57	47	0 77	0 42
		12/08/2008	14:50	<4 8	—	—	—	—	—	—	—	—
		12/08/2008	14:55	<4 8	—	—	—	—	—	—	—	—
		12/08/2008	15:00	<4 8	—	—	—	—	—	—	—	—
		04/21/2009	11:00	<4 8	28 8	24 8	0 27	0 39	2 5	70	1	0 52
		09/22/2009	10:30	<4 8	13 7	19 5	E0 11	E0 15	4 8	26	0 57	0 6
		10/19/2009	15:00	<4 8	—	—	—	—	—	—	—	—
		10/19/2009	15:05	<4 8	—	—	—	—	—	—	—	—
		10/19/2009	15:10	<4 8	—	—	—	—	—	—	—	—
(D–38–22)10cbc Anasazi Pond near spillway	372943109293201	09/18/2008	10:50	8 2	<1 8	4 9	E0 11	2 8	3 3	24	0 2	0 22
(D–38–22)10bcc–1 East well	372954109293601	06/21/2007		—	—	—	—	—	—	—	—	—
		09/11/2007	13:30	—	—	—	—	—	—	—	—	—
		12/14/2007	10:45	1 9	6 2	16 9	0 24	0 79	1 1	99	7 7	6 3
		03/13/2008	17:00	0 91	21 7	40 2	0 32	0 53	0 61	85	5 8	5 5
		09/16/2008	10:45	0 69	4 7	25	0 34	0 31	0 8	123	4 7	4 3
		11/13/2008	14:00	<1 6	3 8	10 4	0 42	0 49	0 62	106	4 4	4 2
		04/21/2009	15:50	<1 6	4 3	5 2	0 43	0 58	0 64	122	3 8	3 8
		09/22/2009	15:40	<1 6	<6 0	7 1	0 32	0 79	0 98	114	4 8	4 5
(D–38–22) 8bad–S1 Ruin Spring	373006109312301	06/01/2007		—	—	—	—	—	—	—	—	—
		09/11/2007	16:00	—	—	—	—	—	—	—	—	—
		12/13/2007	09:30	0 33	<1 8	<2 0	E0 08	0 45	E0 49	56	11 3	10 3
		03/13/2008	12:20	<1 0	<1 8	<2 0	<0 14	0 38	E0 55	—	7 8	7 1
		06/18/2008	15:20	0 34	<1 8	<2 0	<0 14	0 41	<1 0	60	9 5	8 9
		09/17/2008	12:20	0 39	<1 8	<2 0	E0 07	0 46	E0 54	67	9 8	8 7
		11/11/2008	13:45	<1 6	<2 0	2 6	0 09	0 45	0 49	68	10 1	9 4
		04/22/2009	10:15	<1 6	<2 0	<2 0	0 08	0 55	0 45	74	11 7	10 3
		09/23/2009	13:30	<1 6	<6 0	E1 5	E0 07	0 4	2	67	10 6	10 3

Appendix 1.   Field and laboratory data for water samples collected near the White Mesa uranium mill, San Juan County, Utah, June 200–October 2009.—Continued

[**Abbreviations**: ANC, acid neutralizing capacity; $CaCO_3$, calcium carbonate; cm³/g at STP, cubic centimeters of gas per gram at standard temperature (25°C) and pressure (1 bar); E, estimated; ft, feet; gal/min, gallons per minute; hh:mm, hour:minute; LSD, land surface datum; M, presence verified but not quantified; mg/L, milligrams per liter; mm/dd/yyyy, month/day/year; mm Hg, millimeters of mercury; mV, millivolts; permil, parts per thousand; pCi/L, picocuries per liter; SHE, standard hydrogen electrode; SiO_2, silicon dioxide; U, analyzed for but not detected; µg/L, micrograms per liter; µS/cm, microsiemens per centimeter; °C, degrees Celcius; <, less than; —, no data]

Station name	Station number	Date (mm/dd/yyyy)	Time (hh:mm)	Vanadium, total, (µg/L)	Zinc, dissolved, (µg/L)	Zinc, total, (µg/L)	Antimony, dissolved, (µg/L)	Arsenic, dissolved, (µg/L)	Arsenic, total, (µg/L)	Boron, dissolved, (µg/L)	Selenium, dissolved, (µg/L)	Selenium, total, (µg/L)
(D–38–22) 4adb South Mill Pond	373052109294901	03/12/2008	14:40	9 9	<1 8	11 9	E0 10	1 1	1 7	12	0 14	0 17
(D–37–22)32ddc–1 MW3A	373116109305601	12/09/2008	09:15	<4 8	—	—	—	—	—	—	—	—
		12/09/2008	09:20	<4 8	—	—	—	—	—	—	—	—
		10/20/2009	08:20	<4 8	—	—	—	—	—	—	—	—
		10/20/2009	08:25	<4 8	—	—	—	—	—	—	—	—
(D–37–22)31dcb–S1 Cow Camp Spring	373122109321501	09/18/2007	17:00	—	—	—	—	—	—	—	—	—
		09/19/2007	19:00	—	—	—	—	—	—	—	—	—
		03/12/2008	13:25	1	<1 8	<2 0	E0 12	1 8	2	58	1 9	1 7
		06/18/2008	14:20	2	<1 8	E1 2	E0 10	2 2	2 4	66	1 8	1 7
		09/17/2008	13:35	1 6	<1 8	E1 0	E0 11	1 9	2	60	2	1 7
		11/13/2008	15:55	2 1	<2 0	2 3	0 14	2	2	63	1 7	1 5
		04/22/2009	11:15	2 3	<2 0	<2 0	0 13	2 3	2 1	71	1 8	1 4
		09/23/2009	10:15	1 9	<6 0	E1 0	E0 11	1 7	2 7	56	<0 18	1 6
(D–37–22)32bab–S1 Mill Spring	373158109312601	03/12/2008	11:55	0 76	<1 8	<2 0	<0 14	0 88	1 5	61	0 45	0 47
		09/18/2008	09:20	2 1	<1 8	E1 6	<0 14	1 6	3	98	0 53	0 43
		11/12/2008	15:00	E0 84	<2 0	E1 5	0 05	0 75	1	82	0 42	0 34
		04/23/2009	08:55	<1 6	<2 0	<2 0	E0 02	0 74	0 86	76	0 32	0 27
		09/24/2009	11:10	6 9	<6 0	E3 6	0 29	1 7	6 7	130	8 7	8
		10/19/2009	17:00	—	—	—	—	—	—	—	—	—
(D–37–22)27ccc–S1 Entrance Spring	373202109293401	06/21/2007		—	—	—	—	—	—	—	—	—
		09/20/2007	14:00	—	—	—	—	—	—	—	—	—
		12/13/2007	15:30	4 4	E1 3	E2 0	E0 12	1 9	1 8	70	9 3	9 1
		03/13/2008	16:10	6 6	<1 8	E1 6	E0 14	1 7	2	79	9 8	9
		06/19/2008	09:15	6 9	<1 8	3 9	E0 12	1 6	2 1	105	11 7	10 3
		07/22/2008	12:00	11 9	—	536	—	—	1 7	—	—	1 3
		09/17/2008	10:00	7 8	<1 8	<2 0	E0 13	2 8	2 9	90	11	10 1
		11/11/2008	12:45	5 3	<2 0	<2 0	0 12	2 2	2 3	76	9 8	8 9
		04/22/2009	09:30	17 2	<2 0	11 3	0 13	3 8	4 9	95	13 4	11 9
		09/23/2009	09:00	5 2	<6 0	2 1	E0 11	1 2	2 3	93	8 7	8 1
(D–37–22)28acc–1 MW18	373233109301001	12/09/2008	08:45	E2 0	—	—	—	—	—	—	—	—
		12/09/2008	08:50	<3 2	—	—	—	—	—	—	—	—
		12/09/2008	08:55	<3 2	—	—	—	—	—	—	—	—
		10/20/2009	08:55	E1 6	—	—	—	—	—	—	—	—
		10/20/2009	09:00	<3 2	—	—	—	—	—	—	—	—
		10/20/2009	09:05	<3 2	—	—	—	—	—	—	—	—
(D–37–22)10cdc–1 Lyman well	373442109291501	12/12/2007	11:00	0 71	6 2	4 6	E0 13	0 78	0 81	55	0 37	0 24
(D–37–22) 8dba–1 Millview well	373501109310801	09/18/2007	12:00	—	—	—	—	—	—	—	—	—
(D–37–22) 2aad–1 Bayless well	373612109273201	12/12/2007	13:30	0 5	74 2	102	<0 14	2 4	2 9	49	0 46	0 43

Appendix 1. Field and laboratory data for water samples collected near the White Mesa uranium mill, San Juan County, Utah, June 2007–October 2009.—Continued

[**Abbreviations**: ANC, acid neutralizing capacity; $CaCO_3$, calcium carbonate; cm^3/g at STP, cubic centimeters of gas per gram at standard temperature (25°C) and pressure (1 bar); E, estimated; ft, feet; gal/min, gallons per minute; hh:mm, hour:minute; LSD, land surface datum; M, presence verified but not quantified; mg/L, milligrams per liter; mm/dd/yyyy, month/day/year; mm Hg, millimeters of mercury; mV, millivolts; permil, parts per thousand; pCi/L, picocuries per liter; SHE, standard hydrogen electrode; SiO_2, silicon dioxide; U, analyzed for but not detected; μg/L, micrograms per liter; μS/cm, microsiemens per centimeter; °C, degrees Celcius; <, less than; —, no data]

Station name	Station number	Date (mm/dd/yyyy)	Time (hh:mm)	Vanadium, total, (μg/L)	Zinc, dissolved, (μg/L)	Zinc, total, (μg/L)	Antimony, dissolved, (μg/L)	Arsenic, dissolved, (μg/L)	Arsenic, total, (μg/L)	Boron, dissolved, (μg/L)	Selenium, dissolved, (μg/L)	Selenium, total, (μg/L)
(D–36–22)19aad–S1 Oasis Spring	373850109315301	09/19/2007	15:55	—	—	—	—	—	—	—	—	—
		09/18/2008	14:00	1 2	<1 8	<2 0	<0 14	3 1	3 3	44	0 46	0 42
		11/12/2008	12:15	E1 2	<2 0	<2 0	0 06	1	1	26	2 2	2 1
		04/23/2009	10:45	E1 1	<2 0	<2 0	0 05	1	0 93	41	1 8	1 9
		09/24/2009	09:10	<1 6	<6 0	<2 0	E0 12	4	5	37	0 63	0 65
(D–36–22)12dbc Recapture Reservoir	374002109263501	04/23/2009	12:40	<1 6	<2 0	<2 0	0 09	1 5	1 4	18	0 3	0 28

Appendix 1.   Field and laboratory data for water samples collected near the White Mesa uranium mill, San Juan County, Utah, June 2007–October 2009.—Continued

[**Abbreviations**: ANC, acid neutralizing capacity; CaCO₃, calcium carbonate; cm³/g at STP, cubic centimeters of gas per gram at standard temperature (25°C) and pressure (1 bar); E, estimated; ft, feet; gal/min, gallons per minute; hh:mm, hour:minute; LSD, land surface datum; M, presence verified but not quantified; mg/L, milligrams per liter; mm/dd/yyyy, month/day/year; mm Hg, millimeters of mercury; mV, millivolts; permil, parts per thousand; pCi/L, picocuries per liter; SHE, standard hydrogen electrode; SiO₂, silicon dioxide; U, analyzed for but not detected; µg/L, micrograms per liter; µS/cm, microsiemens per centimeter; °C, degrees Celcius; <, less than; —, no data]

Station name	Station number	Date (mm/dd/yyyy)	Time (hh:mm)	Uranium, dissolved, (µg/L)	Uranium, total, (µg/L)	Uranium-234/Uranium-238, ratio, water	Uranium-234/Uranium-235, ratio, water	Uranium-235/Uranium-238, ratio, water	δ^2H, water, (permil)	δ^{18}O, water, (permil)	δ^{34}S, sulfate in water, (permil)
Reference Spring North	373550109341701	06/21/2007		8 1	—	0 000096	—	0 00715	—	—	—
(D–38–22)23cda–1 South well	372756109280901	09/11/2007		—			—		—	–121	–16 60	—
		12/11/2007	15:00	<0 02	<0 020	—	—	—	–117	–15 92	–6 03
		03/11/2008	15:05	E0 01	E0 011				–16	–15 83	—
		11/12/2008	10:05	0 01	E0 013				–117	–15 92	–4 71
(D–38–22)23acb–1 North well	372817109275701	09/11/2007	11:00	—				—	–122	–16 70	—
		12/11/2007	12:00	E0 01	E0 016				–119	–16 11	–14 87
		03/11/2008	11:20	<0 02	<0 020				–117	–16 05	—
		11/11/2008	09:41	0 02	E0 019				–118	–16 11	–13 41
(D–38–22)23bba–S1 Right Hand Fork Seep	372832109282001	03/12/2008	17:00	—	—	—	—	—	–93 7	–12 16	—
(D–38–22) 8dcd–1 West well	372930109310701	06/21/2007		15 1		0 00016	—	0 00717	—	—	—
		09/11/2007	15:30	—					–101	–13 10	—
		12/13/2007	13:00	11 1	16 5	0 000163	—		–109	–13 96	9 40
		03/13/2008	11:40	11 8	15 1	0 000176			–108	–13 90	—
		09/16/2008	14:15	12 7	14	—	—		–110	–14 01	—
		11/13/2008	10:14	13 4	16 3				–108	–14 00	9 38
		12/08/2008	14:50	—	13 6						
		12/08/2008	14:55	—	13 7						
		12/08/2008	15:00	—	11 4						
		04/21/2009	11:00	16 1	18						
		09/22/2009	10:30	7 46	16 4	—	0 0237	—	—	—	—
		10/19/2009	15:00	—	14 5						
		10/19/2009	15:05	—	13 4						
		10/19/2009	15:10	—	11 7						
(D–38–22)10cbc Anasazi Pond near spillway	372943109293201	09/18/2008	10:50	0 37	0 544	—	—	—	–88 5	–11 14	—
(D–38–22)10bcc–1 East well	372954109293601	06/21/2007		3 00	—	0 000144	—	0 00709	—	—	—
		09/11/2007	13:30	—					–103	–13 70	—
		12/14/2007	10:45	3 01	3 03	0 000154			–103	–13 46	7 92
		03/13/2008	17:00	3 03	3 29	0 000158			–101	–13 14	—
		09/16/2008	10:45	1 84	2 12	—	—		–101	–13 12	—
		11/13/2008	14:00	2 4	2 64				–100	–13 00	8 06
		04/21/2009	15:50	4 03	4 08						
		09/22/2009	15:40	2 81	2 78						
(D–38–22) 8bad–S1 Ruin Spring	373006109312301	06/01/2007		7 4	—	0 000102	—	0 00723	—	—	—
		09/11/2007	16:00	—					–114	–14 40	—
		12/13/2007	09:30	8 61	9 39	0 00009			–98 9	–12 71	12 24
		03/13/2008	12:20	—	8 49	0 000101			–98 7	–12 93	—
		06/18/2008	15:20	9 02	9 7	—	—		–98 2	–12 74	—
		09/17/2008	12:20	8 24	9 24				–98 7	–12 77	—
		11/11/2008	13:45	8 62	10				–98 4	–12 79	12 00
		04/22/2009	10:15	11	10 8						
		09/23/2009	13:30	7 81	10 2	—	0 0141	—	—	—	—

Appendix 1. Field and laboratory data for water samples collected near the White Mesa uranium mill, San Juan County, Utah, June 2007–October 2009.—Continued

[**Abbreviations**: ANC, acid neutralizing capacity; CaCO$_3$, calcium carbonate; cm^3/g at STP, cubic centimeters of gas per gram at standard temperature (25°C) and pressure (1 bar); E, estimated; ft, feet; gal/min, gallons per minute; hh:mm, hour:minute; LSD, land surface datum; M, presence verified but not quantified; mg/L, milligrams per liter; mm/dd/yyyy, month/day/year; mm Hg, millimeters of mercury; mV, millivolts; permil, parts per thousand; pCi/L, picocuries per liter; SHE, standard hydrogen electrode; SiO$_2$, silicon dioxide; U, analyzed for but not detected; µg/L, micrograms per liter; µS/cm, microsiemens per centimeter; °C, degrees Celcius; <, less than; —, no data]

Station name	Station number	Date (mm/dd/yyyy)	Time (hh:mm)	Uranium, dissolved, (µg/L)	Uranium, total, (µg/L)	Uranium-234/Uranium-238, ratio, water	Uranium-234/Uranium-235, ratio, water	Uranium-235/Uranium-238, ratio, water	δ²H, water, (permil)	δ¹⁸O, water, (permil)	δ³⁴S, sulfate in water, (permil)
(D–38–22) 4adb South Mill Pond	373052109294901	03/12/2008	14:40	0 4	0 499	0 000059	—	—	-111	-14 15	—
(D–37–22)32ddc–1 MW3A	373116109305601	12/09/2008	09:15	—	17 6	—	—	—	—	—	—
		12/09/2008	09:20	—	18 8	—	—	—	—	—	—
		10/20/2009	08:20	—	19	—	—	—	—	—	—
		10/20/2009	08:25	—	22 9	—	—	—	—	—	—
(D–37–22)31dcb–S1 Cow Camp Spring	373122109321501	09/18/2007	17:00	—	—	—	—	—	-89 7	-11 99	—
		09/19/2007	19:00	—	—	—	—	—	-93 3	-12 30	—
		03/12/2008	13:25	7 71	9 03	0 00018	—	—	-91 6	-12 02	—
		06/18/2008	14:20	8 12	9 41	—	—	—	-90 4	-11 96	—
		09/17/2008	13:35	8 21	9 18	—	—	—	-90 6	-12 00	—
		11/13/2008	15:55	8 38	10 2	—	—	—	-90 1	-12 01	7 35
		04/22/2009	11:15	8 54	9 91	—	—	—	—	—	—
		09/23/2009	10:15	7 64	10 9	—	0 0242	—	—	—	—
(D–37–22)32bab–S1 Mill Spring	373158109312601	03/12/2008	11:55	3 98	4 48	0 000117	—	—	-103	-13 37	—
		09/18/2008	09:20	25 8	29 3	—	—	—	-97 7	-12 64	—
		11/12/2008	15:00	8 35	10 5	—	—	—	-102	-12 92	9 45
		04/23/2009	08:55	1 57	1 89	—	—	—	—	—	—
		09/24/2009	11:10	75 6	114	—	0 0173	—	—	—	—
		10/19/2009	17:00								
(D–37–22)27ccc–S1 Entrance Spring	373202109293401	06/21/2007		20 9	—	0 000083	—	0 00711	—	—	—
		09/20/2007	14:00	—	—	—	—	—	-80 4	-9 54	—
		12/13/2007	15:30	33 2	41 8	0 000085	—	—	-84 3	-10 01	6 77
		03/13/2008	16:10	48 4	55 8	0 000069	—	—	-84 4	-10 08	—
		06/19/2008	09:15	26 6	28 1	—	—	—	-80 6	-9 68	—
		07/22/2008	12:00	—	2 95	—	—	—	—	—	—
		09/17/2008	10:00	21 3	24 6	—	—	—	-80 1	-9 37	—
		11/11/2008	12:45	21 9	25 7	—	—	—	-79 0	-9 17	6 66
		04/22/2009	09:30	23 5	27 5	0 000094	0 0129	—	—	—	—
		09/23/2009	09:00	16 9	20 2	—	0 0134	—	—	—	—
(D–37–22)28acc–1 MW18	373233109301001	12/09/2008	08:45	—	27 2	—	—	—	—	—	—
		12/09/2008	08:50	—	27 7	—	—	—	—	—	—
		12/09/2008	08:55	—	38 4	—	—	—	—	—	—
		10/20/2009	08:55	—	20 2	—	—	—	—	—	—
		10/20/2009	09:00	—	36 5	—	—	—	—	—	—
		10/20/2009	09:05	—	44 5	—	—	—	—	—	—
(D–37–22)10cdc–1 Lyman well	373442109291501	12/12/2007	11:00	5 36	5 42	0 000115	—	—	-75 4	-8 45	4 88
(D–37–22) 8dba–1 Millview well	373501109310801	09/18/2007	12:00	—	—	—	—	—	-97 7	-12 53	—
(D–37–22) 2aad–1 Bayless well	373612109273201	12/12/2007	13:30	3 1	3 34	0 000108	—	—	-82 2	-9 6	2 98

Appendix 1.   Field and laboratory data for water samples collected near the White Mesa uranium mill, San Juan County, Utah, June 2007–October 2009.—Continued

[**Abbreviations**: ANC, acid neutralizing capacity; CaCO$_3$, calcium carbonate; cm^3/g at STP, cubic centimeters of gas per gram at standard temperature (25°C) and pressure (1 bar); E, estimated; ft, feet; gal/min, gallons per minute; hh:mm, hour:minute; LSD, land surface datum; M, presence verified but not quantified; mg/L, milligrams per liter; mm/dd/yyyy, month/day/year; mm Hg, millimeters of mercury; mV, millivolts; permil, parts per thousand; pCi/L, picocuries per liter; SHE, standard hydrogen electrode; SiO$_2$, silicon dioxide; U, analyzed for but not detected; µg/L, micrograms per liter; µS/cm, microsiemens per centimeter; °C, degrees Celcius; <, less than; —, no data]

Station name	Station number	Date (mm/dd/yyyy)	Time (hh:mm)	Uranium, dissolved, (µg/L)	Uranium, total, (µg/L)	Uranium-234/ Uranium-238, ratio, water	Uranium-234/ Uranium-235, ratio, water	Uranium-235/ Uranium-238, ratio, water	δ^{2}H, water, (permil)	δ^{18}O, water, (permil)	δ^{34}S, sulfate in water, (permil)
(D–36–22)19aad–S1 Oasis Spring	373850109315301	09/19/2007	15:55	—	—	—	—	—	–96 9	–12 49	—
		09/18/2008	14:00	2 6	2 85	—	—	—	–94 6	–12 29	—
		11/12/2008	12:15	6 06	6 86	0 000112	—	0 00720	–98 6	–12 75	2 64
		04/23/2009	10:45	7 05	7 16	0 000113	0 0155	—	—	—	—
		09/24/2009	09:10	4 14	4 7	—	0 0149	—	—	—	—
(D–36–22)12dbc Recapture Reservoir	374002109263501	04/23/2009	12:40	0 61	0 627	0 000132	0 0182	—	–81 1	–9 75	—

Appendix 1. Field and laboratory data for water samples collected near the White Mesa uranium mill, San Juan County, Utah, June 2007–October 2009.—Continued

[**Abbreviations**: ANC, acid neutralizing capacity; CaCO₃, calcium carbonate; cm³/g at STP, cubic centimeters of gas per gram at standard temperature (25°C) and pressure (1 bar); E, estimated; ft, feet; gal/min, gallons per minute; hh:mm, hour:minute; LSD, land surface datum; M, presence verified but not quantified; mg/L, milligrams per liter; mm/dd/yyyy, month/day/year; mm Hg, millimeters of mercury; mV, millivolts; permil, parts per thousand; pCi/L, picocuries per liter; SHE, standard hydrogen electrode; SiO₂, silicon dioxide; U, analyzed for but not detected; μg/L, micrograms per liter; μS/cm, microsiemens per centimeter; °C, degrees Celcius; <, less than; —, no data]

Station name	Station number	Date (mm/dd/yyyy)	Time (hh:mm)	$\delta^{18}O$, sulfate in water, (permil)	Helium-4, dissolved, (cm³/g at STP)	Krypton, dissolved, (cm³/g at STP)	Neon, dissolved, (cm³/g at STP)	Xenon, dissolved, (cm³/g at STP)	Argon, dissolved, (cm³/g at STP)	Tritium, total, (pCi/L)	Tritium, total, (tritium units)
Reference Spring North	373550109341701	06/21/2007		—	—	—	—	—	—	—	—
(D–38–22)23cda–1 South well	372756109280901	09/11/2007	—	—	1 38E–07	8 36E–08	2 69E–07	1 21E–08	3 92E–04	—	<0 1
		12/11/2007	15:00	—	—	—	—	—	—	—	—
		03/11/2008	15:05	—	—	—	—	—	—	—	—
		11/12/2008	10:05	–2 92	—	—	—	—	—	—	—
(D–38–22)23acb–1 North well	372817109275701	09/11/2007	11:00	—	1 24E–07	8 45E–08	2 54E–07	1 23E–08	3 88E–04	—	<0 1
		12/11/2007	12:00	—	—	—	—	—	—	—	—
		03/11/2008	11:20	—	—	—	—	—	—	—	—
		11/11/2008	09:41	–3 44	—	—	—	—	—	—	—
(D–38–22)23bba–S1 Right Hand Fork Seep	372832109282001	03/12/2008	17:00	—	—	—	—	—	—	—	—
(D–38–22) 8dcd–1 West well	372930109310701	06/21/2007		—	—	—	—	—	—	—	—
		09/11/2007	15:30	—	3 37E–08	5 96E–08	1 49E–07	8 20E–09	2 56E–04	—	0 5
		12/13/2007	13:00	—	—	—	—	—	—	—	—
		03/13/2008	11:40	—	—	—	—	—	—	—	—
		09/16/2008	14:15	—	—	—	—	—	—	—	—
		11/13/2008	10:14	–5 23	—	—	—	—	—	—	—
		12/08/2008	14:50	—	—	—	—	—	—	—	—
		12/08/2008	14:55	—	—	—	—	—	—	—	—
		12/08/2008	15:00	—	—	—	—	—	—	—	—
		04/21/2009	11:00	—	—	—	—	—	—	—	—
		09/22/2009	10:30	—	—	—	—	—	—	—	—
		10/19/2009	15:00	—	—	—	—	—	—	—	—
		10/19/2009	15:05	—	—	—	—	—	—	—	—
		10/19/2009	15:10	—	—	—	—	—	—	—	—
(D–38–22)10cbc Anasazi Pond near spillway	372943109293201	09/18/2008	10:50	—	—	—	—	—	—	—	—
(D–38–22)10bcc–1 East well	372954109293601	06/21/2007		—	—	—	—	—	—	—	—
		09/11/2007	13:30	—	3 90E–08	6 07E–08	1 77E–07	8 61E–09	2 76E–04	—	0 1
		12/14/2007	10:45	—	—	—	—	—	—	—	—
		03/13/2008	17:00	—	—	—	—	—	—	—	—
		09/16/2008	10:45	—	—	—	—	—	—	—	—
		11/13/2008	14:00	1 10	—	—	—	—	—	—	—
		04/21/2009	15:50	—	—	—	—	—	—	—	—
		09/22/2009	15:40	—	—	—	—	—	—	—	—
(D–38–22) 8bad–S1 Ruin Spring	373006109312301	06/01/2007		—	—	—	—	—	—	—	—
		09/11/2007	16:00	—	4 18E–08	6 33E–08	1 87E–07	8 21E–09	2 82E–04	—	<0 1
		12/13/2007	09:30	—	—	—	—	—	—	—	—
		03/13/2008	12:20	—	—	—	—	—	—	—	—
		06/18/2008	15:20	—	—	—	—	—	—	—	—
		09/17/2008	12:20	—	—	—	—	—	—	—	—
		11/11/2008	13:45	–4 20	—	—	—	—	—	—	—

Appendix 1. Field and laboratory data for water samples collected near the White Mesa uranium mill, San Juan County, Utah, June 2007–October 2009.—Continued

[**Abbreviations**: ANC, acid neutralizing capacity; CaCO₃, calcium carbonate; cm³/g at STP, cubic centimeters of gas per gram at standard temperature (25°C) and pressure (1 bar); E, estimated; ft, feet; gal/min, gallons per minute; hh:mm, hour:minute; LSD, land surface datum; M, presence verified but not quantified; mg/L, milligrams per liter; mm/dd/yyyy, month/day/year; mm Hg, millimeters of mercury; mV, millivolts; permil, parts per thousand; pCi/L, picocuries per liter; SHE, standard hydrogen electrode; SiO₂, silicon dioxide; U, analyzed for but not detected; μg/L, micrograms per liter; μS/cm, microsiemens per centimeter; °C, degrees Celcius; <, less than; —, no data]

Station name	Station number	Date (mm/dd/yyyy)	Time (hh:mm)	$\delta^{18}O$, sulfate in water, (permil)	Helium-4, dissolved, (cm³/g at STP)	Krypton, dissolved, (cm³/g at STP)	Neon, dissolved, (cm³/g at STP)	Xenon, dissolved, (cm³/g at STP)	Argon, dissolved, (cm³/g at STP)	Tritium, total, (pCi/L)	Tritium, total, (tritium units)
		04/22/2009	10:15	—	—	—	—	—	—	—	—
		09/23/2009	13:30	—	—	—	—	—	—	—	—
(D–38–22) 4adb South Mill Pond	373052109294901	03/12/2008	14:40	—	—	—	—	—	—	—	—
(D–37–22)32ddc–1 MW3A	373116109305601	12/09/2008	09:15	—	—	—	—	—	—	—	—
		12/09/2008	09:20	—	—	—	—	—	—	—	—
		10/20/2009	08:20	—	—	—	—	—	—	—	—
		10/20/2009	08:25	—	—	—	—	—	—	—	—
(D–37–22)31dcb–S1 Cow Camp Spring	373122109321501	09/18/2007	17:00	—	4 27E–08	4 30E–08	1 61E–07	3 22E–09	0 00034	—	5 3
		09/19/2007	19:00	—	3 57E–08	6 47E–08	1 57E–07	9 46E–09	2 88E–04	—	5 6
		03/12/2008	13:25	—	—	—	—	—	—	—	—
		06/18/2008	14:20	—	—	—	—	—	—	—	—
		09/17/2008	13:35	—	—	—	—	—	—	—	—
		11/13/2008	15:55	5 40	—	—	—	—	—	—	—
		04/22/2009	11:15	—	—	—	—	—	—	—	—
		09/23/2009	10:15	—	—	—	—	—	—	—	—
(D–37–22)32bab–S1 Mill Spring	373158109312601	03/12/2008	11:55	—	—	—	—	—	—	—	—
		09/18/2008	09:20	—	—	—	—	—	—	—	—
		11/12/2008	15:00	1 15	—	—	—	—	—	—	—
		04/23/2009	08:55	—	—	—	—	—	—	—	—
		09/24/2009	11:10	—	—	—	—	—	—	—	—
		10/19/2009	17:00	—	—	—	—	—	—	1 7	0 5
(D–37–22)27ccc–S1 Entrance Spring	373202109293401	06/21/2007		—	—	—	—	—	—	—	—
		09/20/2007	14:00	—	3 68E–08	2 98E–08	1 35E–07	2 26E–09	0 000253	—	4 2
		12/13/2007	15:30	—	—	—	—	—	—	—	—
		03/13/2008	16:10	—	—	—	—	—	—	—	—
		06/19/2008	09:15	—	—	—	—	—	—	—	—
		07/22/2008	12:00	—	—	—	—	—	—	—	—
		09/17/2008	10:00	—	—	—	—	—	—	—	—
		11/11/2008	12:45	0 44	—	—	—	—	—	—	—
		04/22/2009	09:30	—	—	—	—	—	—	—	—
		09/23/2009	09:00	—	—	—	—	—	—	—	—
(D–37–22)28acc–1 MW18	373233109301001	12/09/2008	08:45	—	—	—	—	—	—	—	—
		12/09/2008	08:50	—	—	—	—	—	—	—	—
		12/09/2008	08:55	—	—	—	—	—	—	—	—
		10/20/2009	08:55	—	—	—	—	—	—	—	—
		10/20/2009	09:00	—	—	—	—	—	—	—	—
		10/20/2009	09:05	—	—	—	—	—	—	—	—
(D–37–22)10cdc–1 Lyman well	373442109291501	12/12/2007	11:00	—	—	—	—	—	—	—	—

Appendix 1. Field and laboratory data for water samples collected near the White Mesa uranium mill, San Juan County, Utah, June 2007–October 2009.—Continued

[**Abbreviations**: ANC, acid neutralizing capacity; CaCO$_3$, calcium carbonate; cm^3/g at STP, cubic centimeters of gas per gram at standard temperature (25°C) and pressure (1 bar); E, estimated; ft, feet; gal/min, gallons per minute; hh:mm, hour:minute; LSD, land surface datum; M, presence verified but not quantified; mg/L, milligrams per liter; mm/dd/yyyy, month/day/year; mm Hg, millimeters of mercury; mV, millivolts; permil, parts per thousand; pCi/L, picocuries per liter; SHE, standard hydrogen electrode; SiO$_2$, silicon dioxide; U, analyzed for but not detected; µg/L, micrograms per liter; µS/cm, microsiemens per centimeter; °C, degrees Celcius; <, less than; —, no data]

Station name	Station number	Date (mm/dd/yyyy)	Time (hh:mm)	δ^{18}O, sulfate in water, (permil)	Helium-4, dissolved, (cm^3/g at STP)	Krypton, dissolved, (cm^3/g at STP)	Neon, dissolved, (cm^3/g at STP)	Xenon, dissolved, (cm^3/g at STP)	Argon, dissolved, (cm^3/g at STP)	Tritium, total, (pCi/L)	Tritium, total, (tritium units)
(D–37–22) 8dba–1 Millview well	373501109310801	09/18/2007	12:00	—	3 90E–08	3 73E–08	1 46E–07	2 65E–09	0 000284	—	0 3
(D–37–22) 2aad–1 Bayless well	373612109273201	12/12/2007	13:30	—	—	—	—	—	—	—	—
(D–36–22)19aad–S1 Oasis Spring	373850109315301	09/19/2007	15:55	—	3 28E–08	3 20E–08	1 29E–07	2 18E–09	0 000245	—	3 6
		09/18/2008	14:00	—	—	—	—	—	—	—	—
		11/12/2008	12:15	–2 97	—	—	—	—	—	—	—
		04/23/2009	10:45	—	—	—	—	—	—	—	—
		09/24/2009	09:10	—	—	—	—	—	—	—	—
(D–36–22)12dbc Recapture Reservoir	374002109263501	04/23/2009	12:40	—	—	—	—	—	—	—	—

Appendix 2

Appendix 2. Chemical composition of fine sediment from dry ephemeral streams near the White Mesa uranium mill, San Juan County, Utah, June 2008.

[Analyses are by total digestion and reported as dry weight of bed sediment, except as noted. **Abbreviations**: mm/dd/yyyy, month/day/year; µg/g, micrograms per gram; <, less than]

Field ID	Station number	Sample date (mm/dd/yyyy)	Aluminum, (percent)	Calcium, (percent)	Iron, (percent)	Potassium, (percent)	Magnesium, (percent)	Sodium, (percent)	Sulfur, (percent)
WM2-S1	373159109311601	06/17/2008	2.73	1.51	1.14	1.02	0.35	0.28	0.02
WM2-S2	373201109311901	06/17/2008	2.02	0.95	0.77	0.6	0.2	0.15	0.02
WM2-S3	373159109312201	06/17/2008	4.25	0.99	1.87	1.93	0.54	0.67	0.02
WM2-S5	373159109312801	06/17/2008	4.14	1.41	1.81	1.85	0.54	0.76	0.3
WM2-S6	373214109310001	06/17/2008	4.91	0.68	1.34	1.59	0.48	0.46	0.02
WM2-S7	373201109311201	06/17/2008	4.24	0.62	1.6	1.96	0.52	0.66	0.02
WM2-S9	373204109314201	06/17/2008	4.76	1.64	1.59	2.02	0.64	0.61	0.05
WM2-S10	373109109300901	06/17/2008	5.37	0.62	2.09	2.22	0.78	0.65	0.02
WM2-S11	373106109302001	06/17/2008	4.59	0.4	1.97	2.02	0.59	0.69	0.02
WM2-S12	373110109301701	06/17/2008	4.38	0.37	1.66	1.99	0.56	0.68	0.02
WM2-S13	373102109304001	06/17/2008	4.99	1.62	1.98	2.08	0.71	0.67	0.02
WM2-S14	373051109304601	06/17/2008	4.13	2.78	2.01	1.8	0.63	0.62	0.02
WM2-S15	373044109304601	06/17/2008	4.05	1.54	1.53	1.9	0.54	0.64	0.01
WM2-S16	373056109294601	06/18/2008	4.59	0.65	1.72	2.06	0.57	0.64	0.02
WM2-S17	373113109312501	06/18/2008	4.62	0.97	1.74	2.03	0.62	0.68	0.02
WM2-S18	373125109311801	06/18/2008	4.54	0.9	1.82	1.96	0.58	0.65	0.02
WM2-S19	373048109310401	06/18/2008	4.98	0.43	2.11	2.1	0.71	0.67	0.01
WM2-S20	373109109294701	06/18/2008	4.67	0.41	1.91	2.1	0.63	0.69	0.02
WM2-S21	373045109303401	06/18/2008	4.84	2.01	1.77	1.96	0.76	1.04	0.55
WMS-1A	373205109293701	06/17/2008	4.95	0.61	1.92	2.09	0.64	0.67	0.02
WMS-2A	373202109293402	06/17/2008	4.43	1.07	1.5	1.96	0.58	0.72	0.04
WMS-3A	373202109292301	06/17/2008	5.87	1.13	2.17	2.02	0.69	0.49	0.07
WMS-4A	373154109293601	06/17/2008	4.77	1.55	1.86	2	0.62	0.69	0.02
WMS-5A	373151109292401	06/17/2008	5.24	0.96	2.07	2.04	0.62	0.6	0.03
WMS-6A	373152109292001	06/17/2008	4.58	0.56	1.73	1.91	0.61	0.66	0.02
WMS-7A	373146109294001	06/18/2008	4.87	0.53	1.81	2.16	0.64	0.72	0.02
WMS-8A	373145109293601	06/18/2008	4.61	0.71	1.92	2.02	0.59	0.72	0.02
WMS-9A	373146109292401	06/18/2008	4.54	0.73	1.66	2.02	0.58	0.7	0.02
WMS-10A	373147109291201	06/18/2008	5.37	1.09	1.91	2.03	0.63	0.47	0.21
WMS-30	373503109310401	06/18/2008	3.97	1.28	1.34	1.84	0.52	0.52	0.01
WMS-31	373458109311201	06/18/2008	4.06	2.06	1.5	1.75	0.63	0.52	0.02
WMS-32	373457109313101	06/18/2008	4.57	1.26	1.74	1.89	0.59	0.64	0.01

Appendix 2. Chemical composition of fine sediment from dry ephemeral streams near the White Mesa uranium mill, San Juan County, Utah, June 2008.—Continued

[Analyses are by total digestion and reported as dry weight of bed sediment, except as noted. **Abbreviations**: mm/dd/yyyy, month/day/year; μg/g, micrograms per gram; <, less than]

Field ID	Station number	Sample date (mm/dd/yyyy)	Titanium, (percent)	Silver, (μg/g)	Arsenic, (μg/g)	Barium, (μg/g)	Beryllium, (μg/g)	Bismuth, (μg/g)	Cadmium, (μg/g)
WM2-S1	373159109311601	06/17/2008	0.16	<1	6	331	0.9	0.16	0.1
WM2-S2	373201109311901	06/17/2008	0.08	<1	8	220	0.7	0.12	0.1
WM2-S3	373159109312201	06/17/2008	0.31	<1	5	489	1.5	0.19	<0.1
WM2-S5	373159109312801	06/17/2008	0.31	<1	5	466	1.2	0.2	0.1
WM2-S6	373214109310001	06/17/2008	0.23	<1	8	565	1.8	0.4	<0.1
WM2-S7	373201109311201	06/17/2008	0.22	<1	5	506	1.6	0.15	0.1
WM2-S9	373204109314201	06/17/2008	0.23	<1	5	949	2	0.2	0.2
WM2-S10	373109109300901	06/17/2008	0.26	<1	6	506	2	0.29	0.2
WM2-S11	373106109302001	06/17/2008	0.29	<1	5	483	1.8	0.26	0.1
WM2-S12	373110109301701	06/17/2008	0.22	<1	5	488	1.7	0.19	0.1
WM2-S13	373102109304001	06/17/2008	0.26	<1	6	523	1.8	0.44	0.2
WM2-S14	373051109304601	06/17/2008	0.31	<1	5	499	1.7	0.19	0.2
WM2-S15	373044109304601	06/17/2008	0.24	<1	4	482	1.2	0.19	0.1
WM2-S16	373056109294601	06/18/2008	0.22	<1	5	497	1.6	0.18	0.1
WM2-S17	373113109312501	06/18/2008	0.23	<1	6	512	1.7	0.2	0.1
WM2-S18	373125109311801	06/18/2008	0.25	<1	6	513	2	0.18	0.1
WM2-S19	373048109310401	06/18/2008	0.28	<1	5	498	2.2	0.26	0.2
WM2-S20	373109109294701	06/18/2008	0.26	<1	5	477	1.5	0.19	0.2
WM2-S21	373045109303401	06/18/2008	0.23	<1	7	534	1.7	0.22	0.1
WMS-1A	373205109293701	06/17/2008	0.26	<1	5	522	2.1	0.27	0.2
WMS-2A	373202109293402	06/17/2008	0.23	<1	6	475	1.6	0.26	0.2
WMS-3A	373202109292301	06/17/2008	0.26	<1	6	493	1.6	0.31	0.2
WMS-4A	373154109293601	06/17/2008	0.26	<1	5	515	1.6	0.42	0.1
WMS-5A	373151109292401	06/17/2008	0.26	<1	6	634	2.1	0.22	0.2
WMS-6A	373152109292001	06/17/2008	0.25	<1	6	485	1.5	0.3	0.2
WMS-7A	373146109294001	06/18/2008	0.24	<1	6	497	1.7	0.34	0.1
WMS-8A	373145109293601	06/18/2008	0.29	<1	5	492	1.8	0.2	0.4
WMS-9A	373146109292401	06/18/2008	0.22	<1	5	505	1.7	0.23	0.2
WMS-10A	373147109291201	06/18/2008	0.23	<1	5	592	2.4	0.21	0.1
WMS-30	373503109310401	06/18/2008	0.17	<1	5	465	1.5	0.27	0.1
WMS-31	373458109311201	06/18/2008	0.23	<1	5	454	1.7	0.34	0.1
WMS-32	373457109313101	06/18/2008	0.24	<1	6	502	1.5	0.3	0.2

Appendix 2. Chemical composition of fine sediment from dry ephemeral streams near the White Mesa uranium mill, San Juan County, Utah, June 2008.—Continued

[Analyses are by total digestion and reported as dry weight of bed sediment, except as noted. **Abbreviations**: mm/dd/yyyy, month/day/year; μg/g, micrograms per gram; <, less than]

Field ID	Station number	Sample date (mm/dd/yyyy)	Cerium, (μg/kg)	Cobalt, (μg/g)	Chromium, (μg/g)	Cesium, (μg/g)	Copper, (μg/g)	Gallium, (μg/g)	Indium, (μg/g)
WM2-S1	373159109311601	06/17/2008	45.4	4.2	16	<5	4.4	6.94	0.02
WM2-S2	373201109311901	06/17/2008	35.9	3.2	11	<5	2.9	4.72	<0.02
WM2-S3	373159109312201	06/17/2008	70.9	6	30	<5	12.1	10.7	0.03
WM2-S5	373159109312801	06/17/2008	56.7	5.4	28	<5	10	9.99	0.03
WM2-S6	373214109310001	06/17/2008	62.7	4.3	24	<5	9.3	13.2	0.05
WM2-S7	373201109311201	06/17/2008	50.7	5.3	23	<5	10.9	10.2	0.02
WM2-S9	373204109314201	06/17/2008	61.1	5.5	26	<5	9.8	11.4	0.03
WM2-S10	373109109300901	06/17/2008	65.6	7.4	41	<5	16.5	13.3	0.04
WM2-S11	373106109302001	06/17/2008	65.2	6.2	29	<5	13.4	11.4	0.03
WM2-S12	373110109301701	06/17/2008	53.7	6	26	<5	12.4	10.8	0.03
WM2-S13	373102109304001	06/17/2008	66	6.9	30	<5	13.9	12.5	0.03
WM2-S14	373051109304601	06/17/2008	74	6.4	33	<5	11.4	10.4	0.03
WM2-S15	373044109304601	06/17/2008	49.8	5.1	27	<5	9.7	9.81	0.02
WM2-S16	373056109294601	06/18/2008	58.8	5.7	25	<5	11.4	11	0.03
WM2-S17	373113109312501	06/18/2008	52.9	6	30	<5	12	11.2	0.03
WM2-S18	373125109311801	06/18/2008	53.6	6.2	33	<5	10.8	11.7	0.03
WM2-S19	373048109310401	06/18/2008	63.6	7.2	32	<5	14.4	12.5	0.03
WM2-S20	373109109294701	06/18/2008	58.3	6.4	28	<5	13.1	11.3	0.03
WM2-S21	373045109303401	06/18/2008	51.1	6.9	28	<5	16.8	11.5	0.04
WMS-1A	373205109293701	06/17/2008	62.3	7.3	31	<5	14.1	12.5	0.04
WMS-2A	373202109293402	06/17/2008	49.6	5.8	25	<5	15.3	10.4	0.03
WMS-3A	373202109292301	06/17/2008	45.4	5.7	27	<5	14.3	14.9	0.04
WMS-4A	373154109293601	06/17/2008	59.9	6.6	28	<5	14.3	11.6	0.03
WMS-5A	373151109292401	06/17/2008	41.6	6.7	33	<5	16.2	13.5	0.04
WMS-6A	373152109292001	06/17/2008	56.3	6.2	26	<5	14.3	11.3	0.03
WMS-7A	373146109294001	06/18/2008	53.1	6.2	28	<5	14	11.7	0.03
WMS-8A	373145109293601	06/18/2008	59.1	6.1	28	<5	13.5	11.2	0.03
WMS-9A	373146109292401	06/18/2008	49.5	5.8	28	<5	13	10.9	0.03
WMS-10A	373147109291201	06/18/2008	59.5	6.2	34	<5	10.7	13.2	0.04
WMS-30	373503109310401	06/18/2008	61.1	5.2	22	<5	8.4	9.74	0.03
WMS-31	373458109311201	06/18/2008	64	5.4	25	<5	10	9.23	0.03
WMS-32	373457109313101	06/18/2008	64.4	5.4	28	<5	10.6	11.1	0.03

Appendix 2. Chemical composition of fine sediment from dry ephemeral streams near the White Mesa uranium mill, San Juan County, Utah, June 2008.—Continued

[Analyses are by total digestion and reported as dry weight of bed sediment, except as noted. **Abbreviations**: mm/dd/yyyy, month/day/year; µg/g, micrograms per gram; <, less than]

Field ID	Station number	Sample date (mm/dd/yyyy)	Lanthanum, (µg/g)	Lithium, (µg/g)	Manganese, (µg/g)	Molybdenum, (µg/g)	Niobium, (µg/g)	Nickel, (µg/g)	Phosphorus, (µg/g)
WM2-S1	373159109311601	06/17/2008	21.8	16	244	0.97	3.9	7.2	200
WM2-S2	373201109311901	06/17/2008	18	13	171	2.7	2.8	6.1	170
WM2-S3	373159109312201	06/17/2008	35.1	21	376	0.71	7	11.6	360
WM2-S5	373159109312801	06/17/2008	27.2	21	378	0.91	6.1	10.6	370
WM2-S6	373214109310001	06/17/2008	32.3	35	216	1.31	6.9	9.6	400
WM2-S7	373201109311201	06/17/2008	25.3	20	337	0.79	5.5	10.8	350
WM2-S9	373204109314201	06/17/2008	30.4	22	371	1.04	6	12.5	300
WM2-S10	373109109300901	06/17/2008	32	27	541	0.87	6.5	20.5	610
WM2-S11	373106109302001	06/17/2008	35.2	23	460	0.71	5.9	12.6	470
WM2-S12	373110109301701	06/17/2008	26	21	401	0.69	5.2	12.5	480
WM2-S13	373102109304001	06/17/2008	32.1	26	467	0.78	5.3	13.5	540
WM2-S14	373051109304601	06/17/2008	36	21	496	0.67	4.4	13.1	530
WM2-S15	373044109304601	06/17/2008	23.3	20	391	0.56	4.9	11.9	470
WM2-S16	373056109294601	06/18/2008	28	22	376	0.59	5.1	11.3	460
WM2-S17	373113109312501	06/18/2008	26.7	23	417	0.64	4.8	13	500
WM2-S18	373125109311801	06/18/2008	27.6	23	377	0.65	5.7	14.2	440
WM2-S19	373048109310401	06/18/2008	30.9	24	579	0.7	6.1	14	560
WM2-S20	373109109294701	06/18/2008	28.1	22	491	0.72	5.9	12.2	570
WM2-S21	373045109303401	06/18/2008	24.9	25	408	1.43	6.6	13.8	470
WMS-1A	373205109293701	06/17/2008	30.5	25	497	0.71	6.4	13.7	540
WMS-2A	373202109293402	06/17/2008	24.6	21	320	1.02	5.4	12.7	390
WMS-3A	373202109292301	06/17/2008	22.9	28	239	1.27	8.5	10.8	300
WMS-4A	373154109293601	06/17/2008	29.8	23	389	0.9	6.2	12.6	480
WMS-5A	373151109292401	06/17/2008	21	23	322	1.08	7.5	14.1	450
WMS-6A	373152109292001	06/17/2008	28.7	21	326	0.82	5.3	11.3	490
WMS-7A	373146109294001	06/18/2008	26.1	24	425	0.7	5.3	12.1	560
WMS-8A	373145109293601	06/18/2008	28.9	23	429	0.69	6	11.8	510
WMS-9A	373146109292401	06/18/2008	24.3	22	369	0.82	4.9	12.1	460
WMS-10A	373147109291201	06/18/2008	28.4	22	248	0.85	7.4	15.5	320
WMS-30	373503109310401	06/18/2008	30.3	20	237	0.56	4.1	10.1	240
WMS-31	373458109311201	06/18/2008	32.4	21	266	0.58	4.4	10.3	320
WMS-32	373457109313101	06/18/2008	32	24	325	0.7	5.3	11.7	380

Appendix 2. Chemical composition of fine sediment from dry ephemeral streams near the White Mesa uranium mill, San Juan County, Utah, June 2008.—Continued

[Analyses are by total digestion and reported as dry weight of bed sediment, except as noted. **Abbreviations**: mm/dd/yyyy, month/day/year; µg/g, micrograms per gram; <, less than]

Field ID	Station number	Sample date (mm/dd/yyyy)	Lead, (µg/g)	Rubidium, (µg/g)	Antimony, (µg/g)	Scandium, (µg/g)	Tin, (µg/g)	Strontium, (µg/g)	Tellurium, (µg/g)
WM2-S1	373159109311601	06/17/2008	12.4	42.7	0.37	3.6	1.1	75.2	<0.1
WM2-S2	373201109311901	06/17/2008	11.1	25.3	0.36	2.5	1.1	54.4	<0.1
WM2-S3	373159109312201	06/17/2008	16.4	74.9	0.55	6	1.9	97.8	<0.1
WM2-S5	373159109312801	06/17/2008	17	72.1	0.58	5.3	5.6	145	<0.1
WM2-S6	373214109310001	06/17/2008	22.4	63.4	0.72	5.2	7.8	93.3	<0.1
WM2-S7	373201109311201	06/17/2008	15.7	75.2	0.51	5.3	12	88.4	<0.1
WM2-S9	373204109314201	06/17/2008	15.6	93.6	0.51	6.4	4.2	219	<0.1
WM2-S10	373109109300901	06/17/2008	17.2	88.8	0.54	7.4	2.4	102	<0.1
WM2-S11	373106109302001	06/17/2008	16.1	81.6	0.52	6.1	1.2	88.8	<0.1
WM2-S12	373110109301701	06/17/2008	16.1	79.3	0.47	5.5	1.2	87	<0.1
WM2-S13	373102109304001	06/17/2008	16.4	83.2	0.47	6.5	3.2	121	<0.1
WM2-S14	373051109304601	06/17/2008	15.6	70.8	0.46	5.8	2.5	171	<0.1
WM2-S15	373044109304601	06/17/2008	13.9	71.9	0.4	4.9	4.3	127	<0.1
WM2-S16	373056109294601	06/18/2008	16.4	79.3	0.41	5.5	1.6	100	<0.1
WM2-S17	373113109312501	06/18/2008	15	77.8	0.44	5.9	1	96.9	<0.1
WM2-S18	373125109311801	06/18/2008	15.2	79.5	0.5	6.1	1.7	95.7	<0.1
WM2-S19	373048109310401	06/18/2008	16.8	85.4	0.55	6.9	1.3	92.3	<0.1
WM2-S20	373109109294701	06/18/2008	16.8	82.9	0.52	6	1.1	91.9	<0.1
WM2-S21	373045109303401	06/18/2008	18.6	75.3	0.62	6.1	2.7	154	<0.1
WMS-1A	373205109293701	06/17/2008	16.5	87.5	0.5	6.9	1.5	98.1	<0.1
WMS-2A	373202109293402	06/17/2008	19.5	72.9	0.5	5	1.8	115	<0.1
WMS-3A	373202109292301	06/17/2008	18.4	104	0.78	7.9	2.3	160	<0.1
WMS-4A	373154109293601	06/17/2008	16.8	80.8	0.51	6.3	1.3	113	<0.1
WMS-5A	373151109292401	06/17/2008	17.6	95.4	0.69	7.4	2.2	201	<0.1
WMS-6A	373152109292001	06/17/2008	19	81.9	0.57	5.5	1.2	96.5	<0.1
WMS-7A	373146109294001	06/18/2008	16.9	82.9	0.53	6	1.2	100	<0.1
WMS-8A	373145109293601	06/18/2008	16.4	77.1	0.51	5.8	1.7	98.1	<0.1
WMS-9A	373146109292401	06/18/2008	16.8	78.7	0.49	5.5	1.4	99.4	<0.1
WMS-10A	373147109291201	06/18/2008	16.9	103	0.66	7.1	6.6	150	<0.1
WMS-30	373503109310401	06/18/2008	15	74.2	0.36	4.8	1.2	92.9	<0.1
WMS-31	373458109311201	06/18/2008	15.6	66.7	0.39	4.7	1	110	<0.1
WMS-32	373457109313101	06/18/2008	16.7	74.3	0.41	5.6	1.2	102	<0.1

Appendix 2. Chemical composition of fine sediment from dry ephemeral streams near the White Mesa uranium mill, San Juan County, Utah, June 2008.—Continued

[Analyses are by total digestion and reported as dry weight of bed sediment, except as noted. **Abbreviations**: mm/dd/yyyy, month/day/year; μg/g, micrograms per gram; <, less than]

Field ID	Station number	Sample date (mm/dd/yyyy)	Thorium, (µg/g)	Thallium, (µg/g)	Uranium, (µg/g)	Vanadium, (µg/g)	Tungsten, (µg/g)	Yttrium, (µg/g)	Zinc, (µg/g)	Selenium, (µg/g)
WM2-S1	373159109311601	06/17/2008	7.3	0.3	2.2	39	0.3	9.7	26	<0.2
WM2-S2	373201109311901	06/17/2008	4.5	0.2	1.5	26	0.3	7.2	20	<0.2
WM2-S3	373159109312201	06/17/2008	11.7	0.4	3	59	0.5	17.2	37	<0.2
WM2-S5	373159109312801	06/17/2008	10.1	0.4	2.9	61	0.5	14.6	36	<0.2
WM2-S6	373214109310001	06/17/2008	10	0.4	3	51	0.7	15.2	42	<0.2
WM2-S7	373201109311201	06/17/2008	8.3	0.4	2.2	51	0.4	13.6	39	<0.2
WM2-S9	373204109314201	06/17/2008	8	0.5	2.2	51	0.5	14.7	35	<0.2
WM2-S10	373109109300901	06/17/2008	10	0.5	2.8	66	0.5	17.2	57	0.2
WM2-S11	373106109302001	06/17/2008	10.3	0.4	3.5	61	0.4	17.5	44	<0.2
WM2-S12	373110109301701	06/17/2008	8.3	0.4	2.6	53	0.4	14.8	42	<0.2
WM2-S13	373102109304001	06/17/2008	10.4	0.5	2.6	62	0.5	17	45	<0.2
WM2-S14	373051109304601	06/17/2008	11.8	0.4	3.2	66	0.4	18.6	40	<0.2
WM2-S15	373044109304601	06/17/2008	7.8	0.4	2.2	47	0.4	13.4	34	<0.2
WM2-S16	373056109294601	06/18/2008	8.9	0.4	2.2	53	0.5	14.6	41	<0.2
WM2-S17	373113109312501	06/18/2008	7.8	0.4	2.2	54	0.5	14.9	42	<0.2
WM2-S18	373125109311801	06/18/2008	9.1	0.4	2.4	59	0.4	16.4	41	<0.2
WM2-S19	373048109310401	06/18/2008	10.9	0.5	2.7	62	0.5	18.1	49	<0.2
WM2-S20	373109109294701	06/18/2008	9	0.4	2.7	58	0.5	16.5	44	<0.2
WM2-S21	373045109303401	06/18/2008	8.1	0.4	16.2	75	0.7	15.1	50	0.6
WMS-1A	373205109293701	06/17/2008	9.5	0.5	2.4	60	0.5	16.8	48	<0.2
WMS-2A	373202109293402	06/17/2008	7.6	0.5	6.6	73	0.5	12.7	46	0.4
WMS-3A	373202109292301	06/17/2008	9.9	0.5	5.9	73	0.8	11.7	39	0.2
WMS-4A	373154109293601	06/17/2008	10	0.4	5.7	71	0.5	16.2	46	<0.2
WMS-5A	373151109292401	06/17/2008	7.9	0.5	4.9	79	0.8	12.2	43	0.3
WMS-6A	373152109292001	06/17/2008	8.9	0.5	3.4	60	0.5	14	42	0.5
WMS-7A	373146109294001	06/18/2008	8.6	0.4	3.7	58	0.5	14.9	46	<0.2
WMS-8A	373145109293601	06/18/2008	9.7	0.4	3.9	66	0.5	16	46	<0.2
WMS-9A	373146109292401	06/18/2008	8.3	0.4	3.6	60	0.4	13.6	44	<0.2
WMS-10A	373147109291201	06/18/2008	8.3	0.6	2.6	56	0.8	12.7	40	<0.2
WMS-30	373503109310401	06/18/2008	6.6	0.4	1.8	42	0.3	13.6	32	<0.2
WMS-31	373458109311201	06/18/2008	8.6	0.4	2.6	50	0.3	16	36	<0.2
WMS-32	373457109313101	06/18/2008	10.5	0.4	3.6	56	0.4	15.8	40	<0.2

Appendix 3

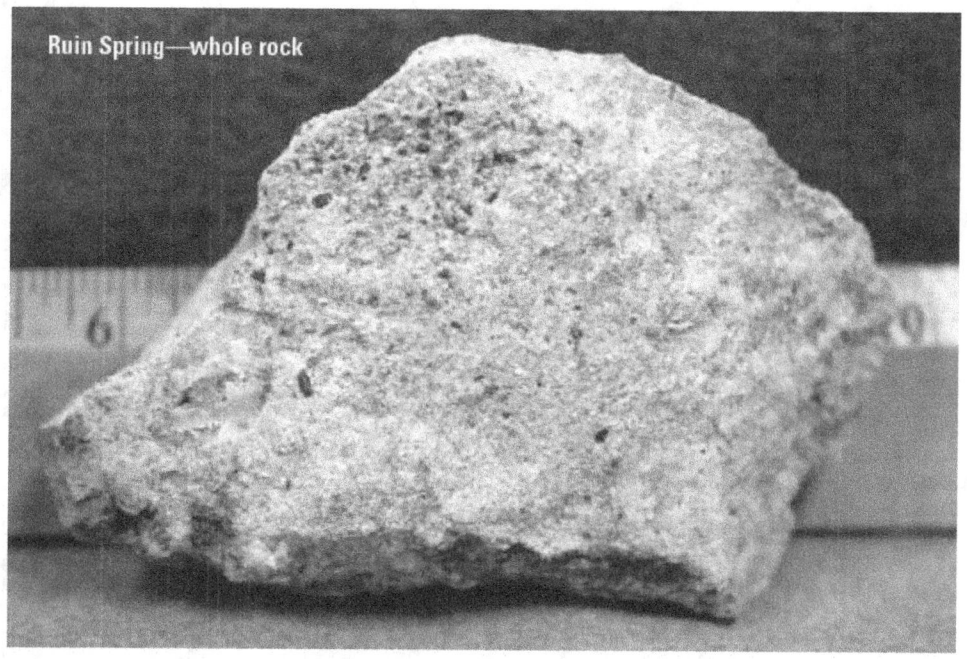

Ruin Spring—whole rock
6
9

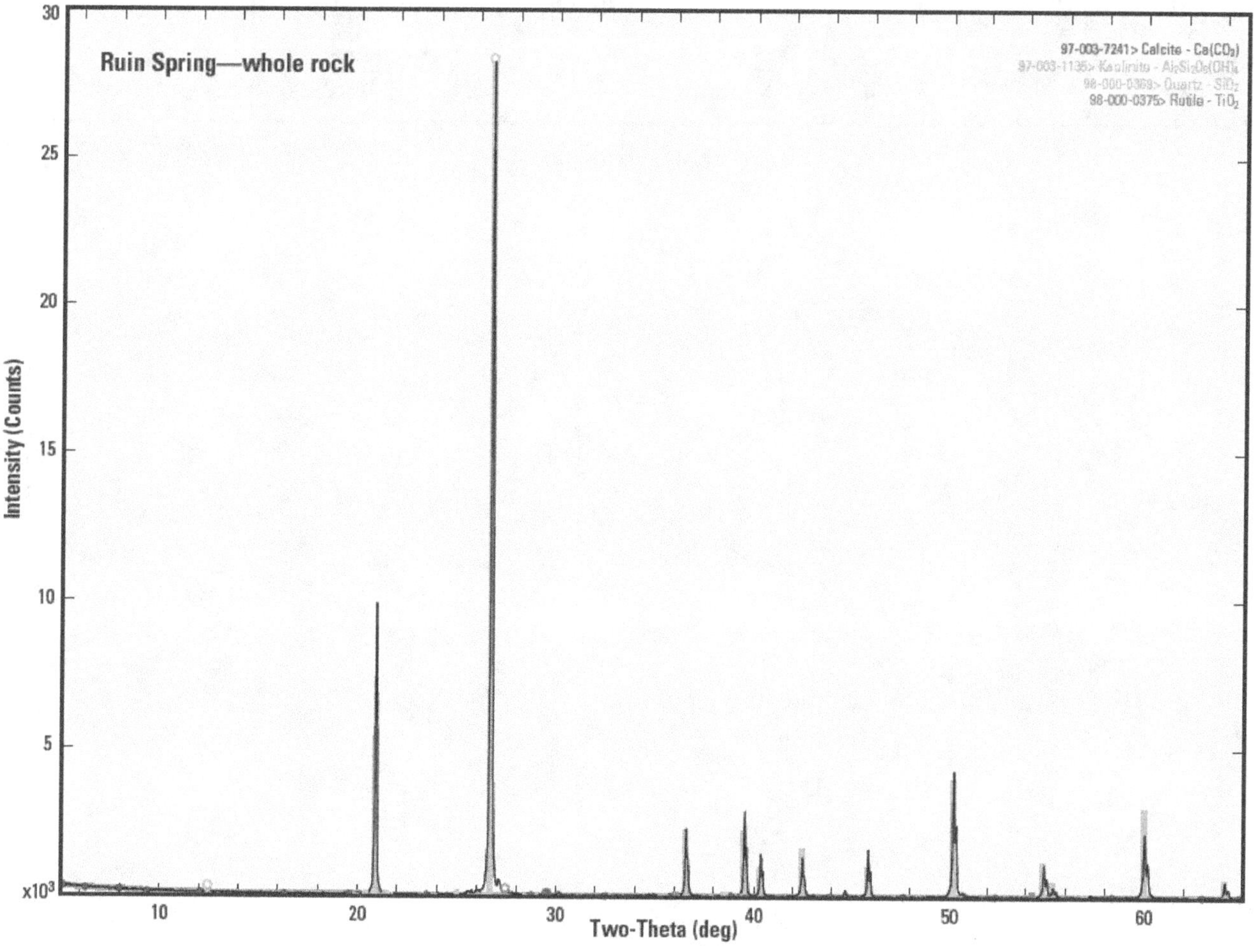
Ruin Spring—whole rock
97-003-7241> Calcite - $Ca(CO_3)$
97-003-1135> Kaolinite - $Al_2Si_2O_5(OH)_4$
98-000-0369> Quartz - SiO_2
98-000-0375> Rutile - TiO_2
Intensity (Counts)
x10³
30
25
20
15
10
5
Two-Theta (deg)
10
20
30
40
50
60

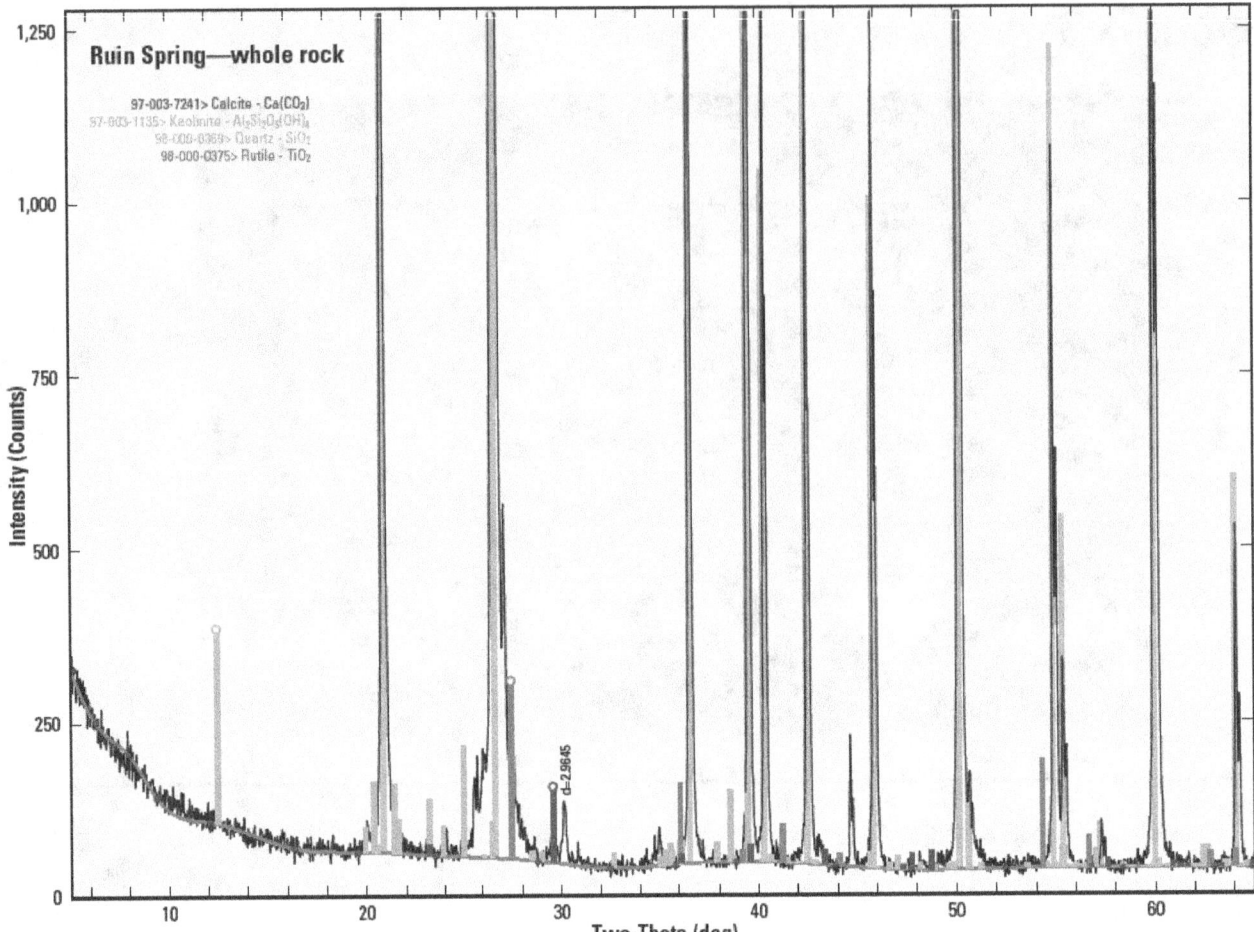

Ruin Spring—green band

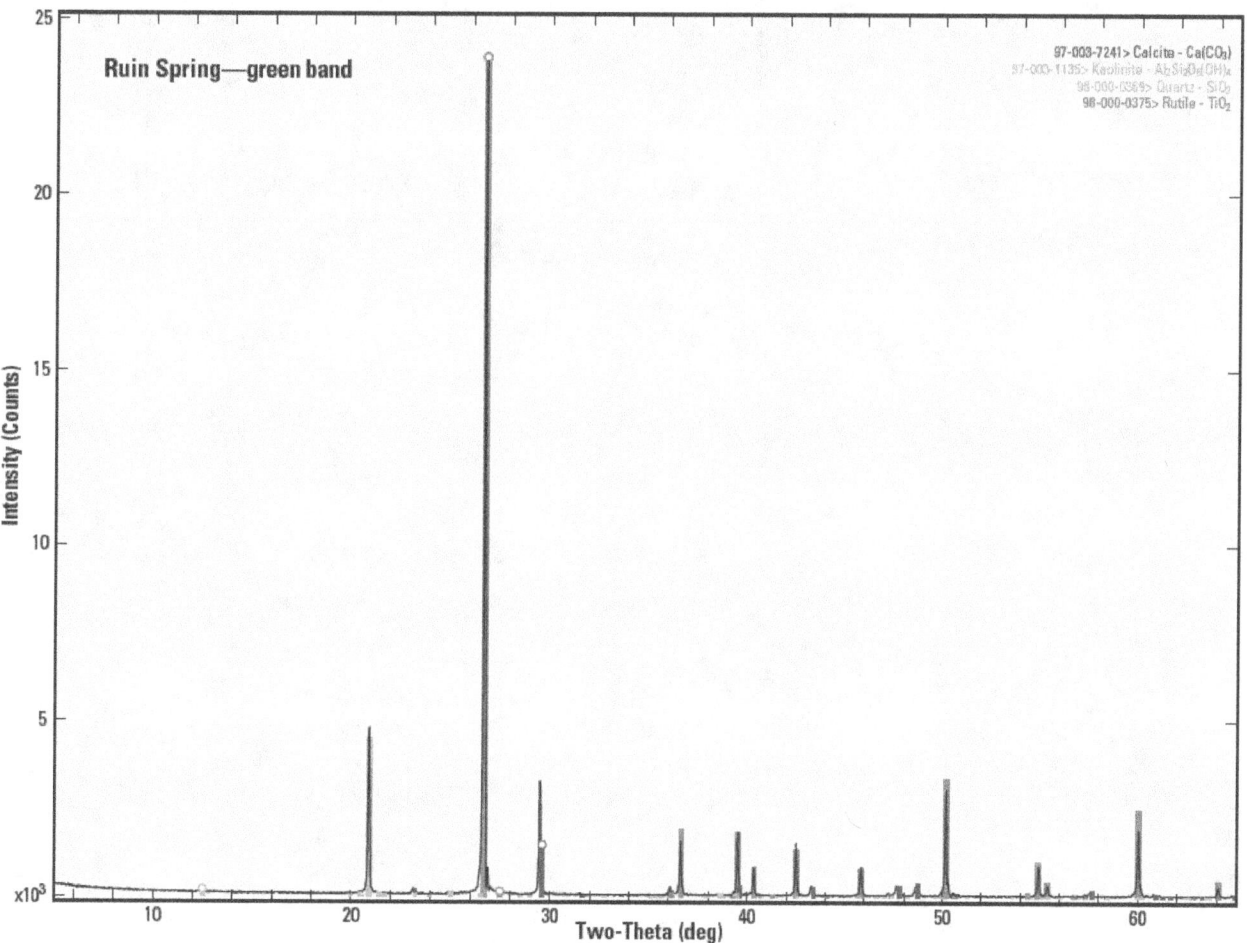

Ruin Spring—green band

97-003-7241> Calcite - Ca(CO₃)
97-000-1135> Kaolinite - Al₂Si₂O₅(OH)₄
98-000-0369> Quartz - SiO₂
98-000-0375> Rutile - TiO₂

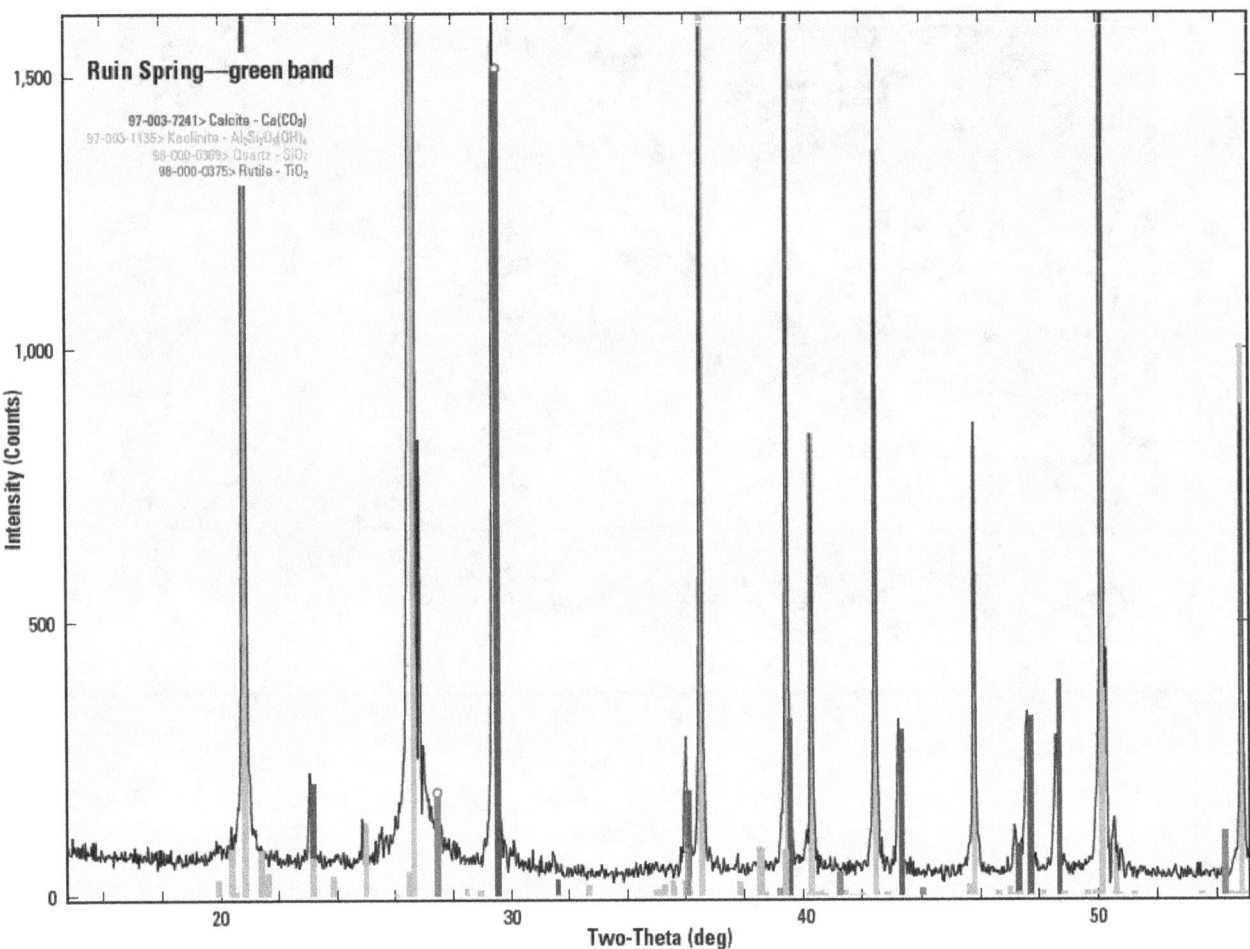

Ruin Spring—green band

97-003-7241> Calcite - Ca(CO₃)
97-003-1135> Kaolinite - Al₂Si₂O₅(OH)₄
98-000-0369> Quartz - SiO₂
98-000-0375> Rutile - TiO₂

Ruin Spring—green band

Whole Pattern Fitting and Rietveld Refinement

Phase ID (4)

- ■ Calcite - Ca(CO$_3$)
- ▨ Kaolinite 1A - Al$_2$Si$_2$O$_5$(OH)$_4$
- ▢ Quartz - SiO$_2$
- ▨ Rutile - TiO$_2$

Source	I/Ic	Wt%	#L
FIZ#37241	3.18 (0%)	11.9 (0.2)	17
FIZ#31135	1.05 (0%)	1.0 (0.2)	242
JCS#369	4.20 (0%)	85.6 (0.8)	40
JCS#375	3.40 (0%)	1.4 (0.3)	9

XRF (Wt%): TiO$_2$ = 1.4%, CaO = 6.7%, SiO$_2$ = 86.1%, Al$_2$O$_3$ = 0.4%, CO$_2$ = 5.2%

NOTE: Fitting Halted at Iteration 20(4): R = 16.65% (E = 7.85%, R/E = 2.12, P = 37, EPS = 0.5)

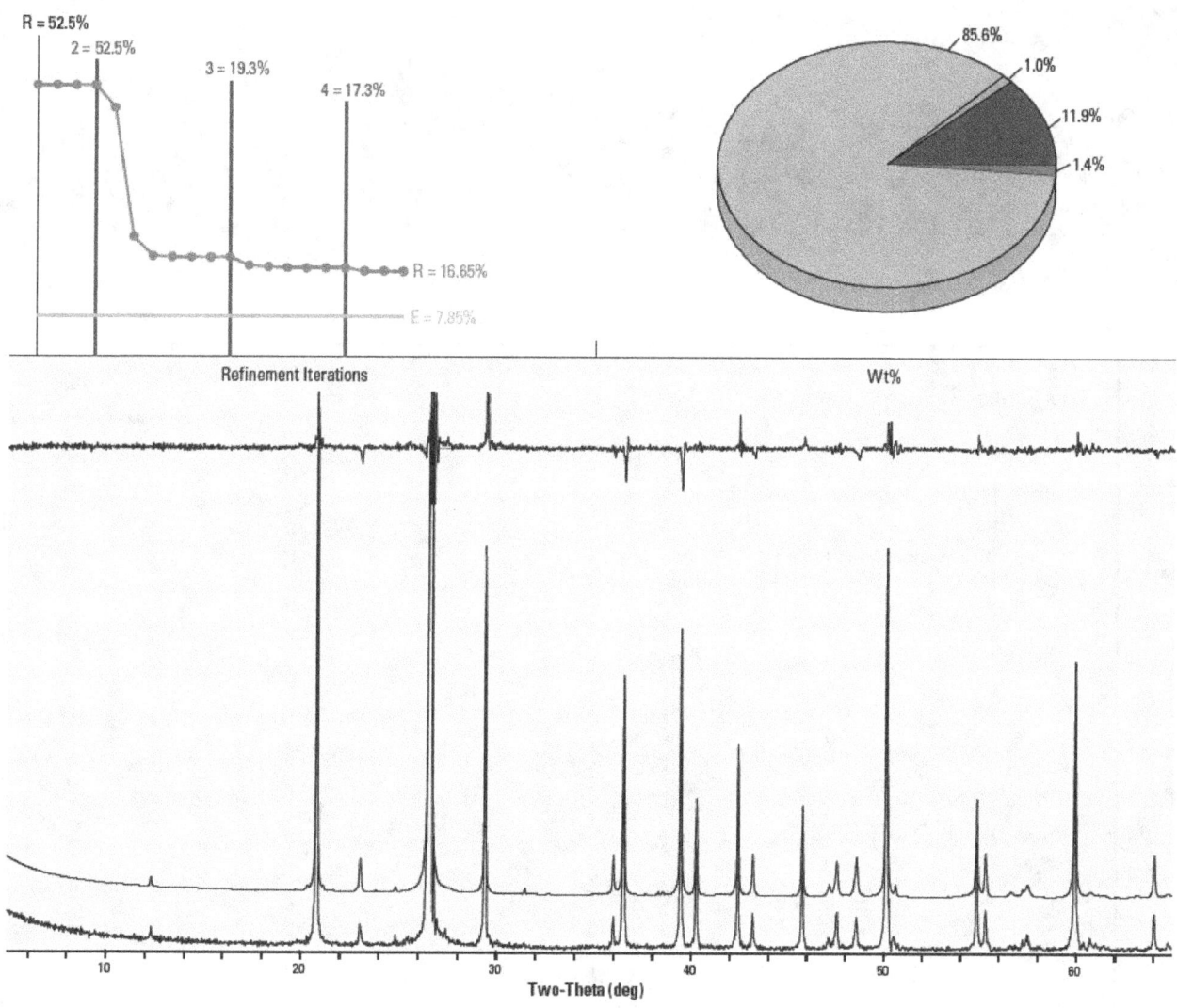

R = 52.5%

2 = 52.5%

3 = 19.3%

4 = 17.3%

R = 16.65%

E = 7.85%

Refinement Iterations

85.6%

1.0%

11.9%

1.4%

Wt%

Two-Theta (deg)

Ruin Spring—rock crust

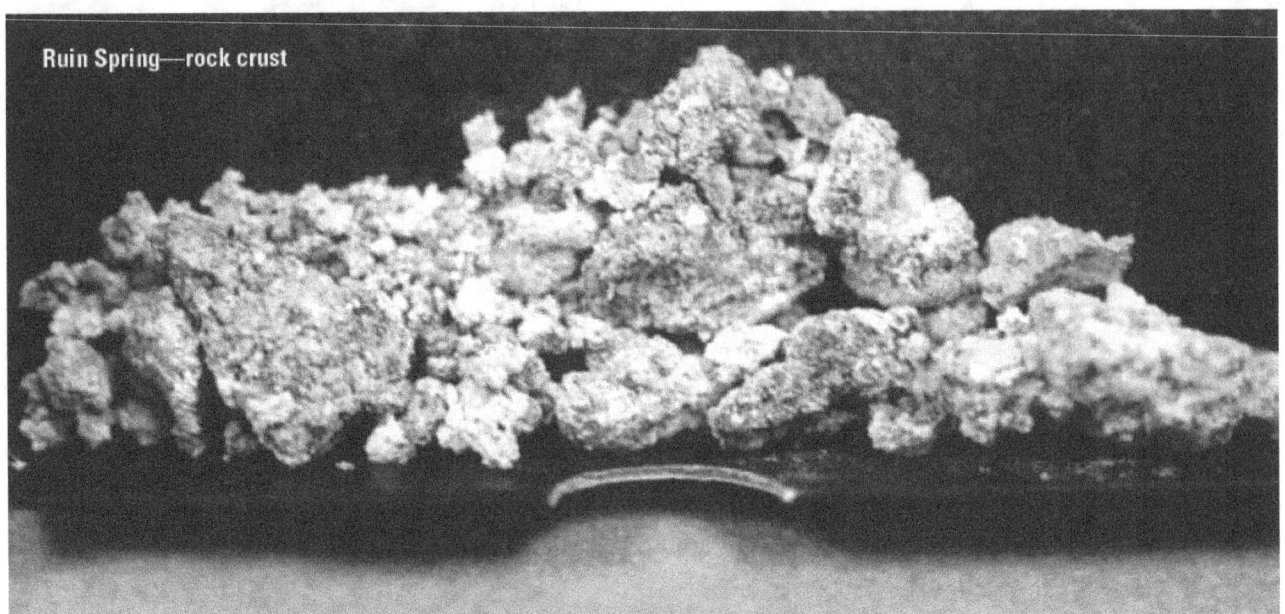

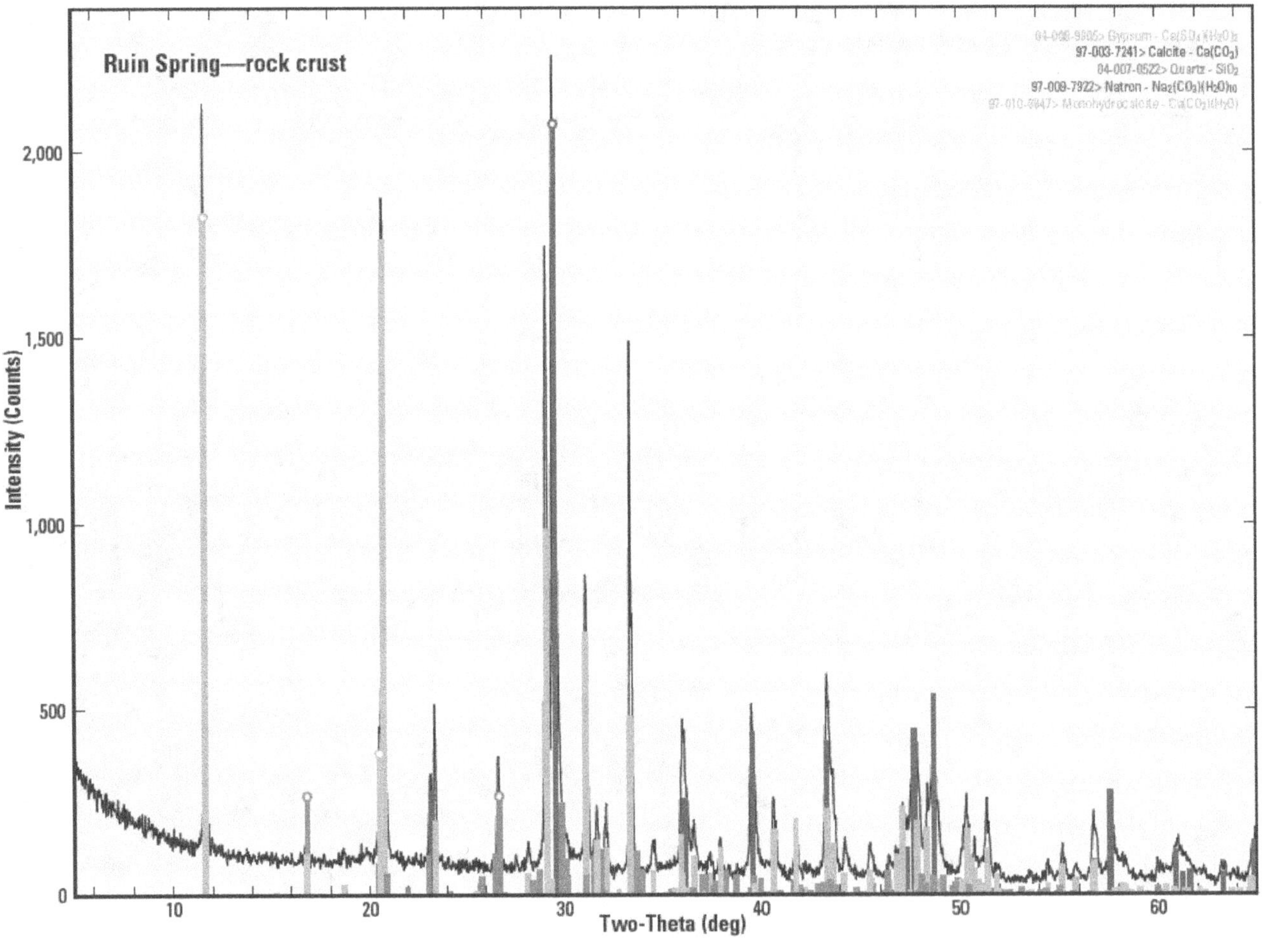
Ruin Spring—rock crust
97-003-7241> Calcite - $Ca(CO_3)$
04-007-0522> Quartz - SiO_2
97-009-7922> Natron - $Na_2(CO_3)(H_2O)_{10}$
Intensity (Counts)
0
500
1,000
1,500
2,000
10
20
30
40
50
60
Two-Theta (deg)

Ruin Spring—rock crust   # Whole Pattern Fitting and Rietveld Refinement

Phase ID (4)

	Source	I/Ic	Wt%	#L
Gypsum - $Ca(SO_4)(H_2O)_2$	04-008-9805	2.10 (0%)	39.1 (0.5)	92
Calcite - $Ca(CO_3)$	FIZ#37241	3.20 (0%)	47.4 (0.7)	17
Quartz á Fe - SiO_2	04-007-0522	1.00 (0%)	3.3 (0.2)	152
Monohydrocalcite (supercell) - $Ca(CO_3)(H_2O)$	FIZ#100847	1.32 (0%)	10.1 (0.4)	348

XRF (Wt%): CaO = 44.4%, SO_2 = 14.9%, SiO_2 = 3.3%, CO_2 = 24.6%

NOTE: Fitting Halted at Iteration 25(4): R = 15.38% (E = 8.24%, R/E = 1.87, P = 35, EPS = 0.5)

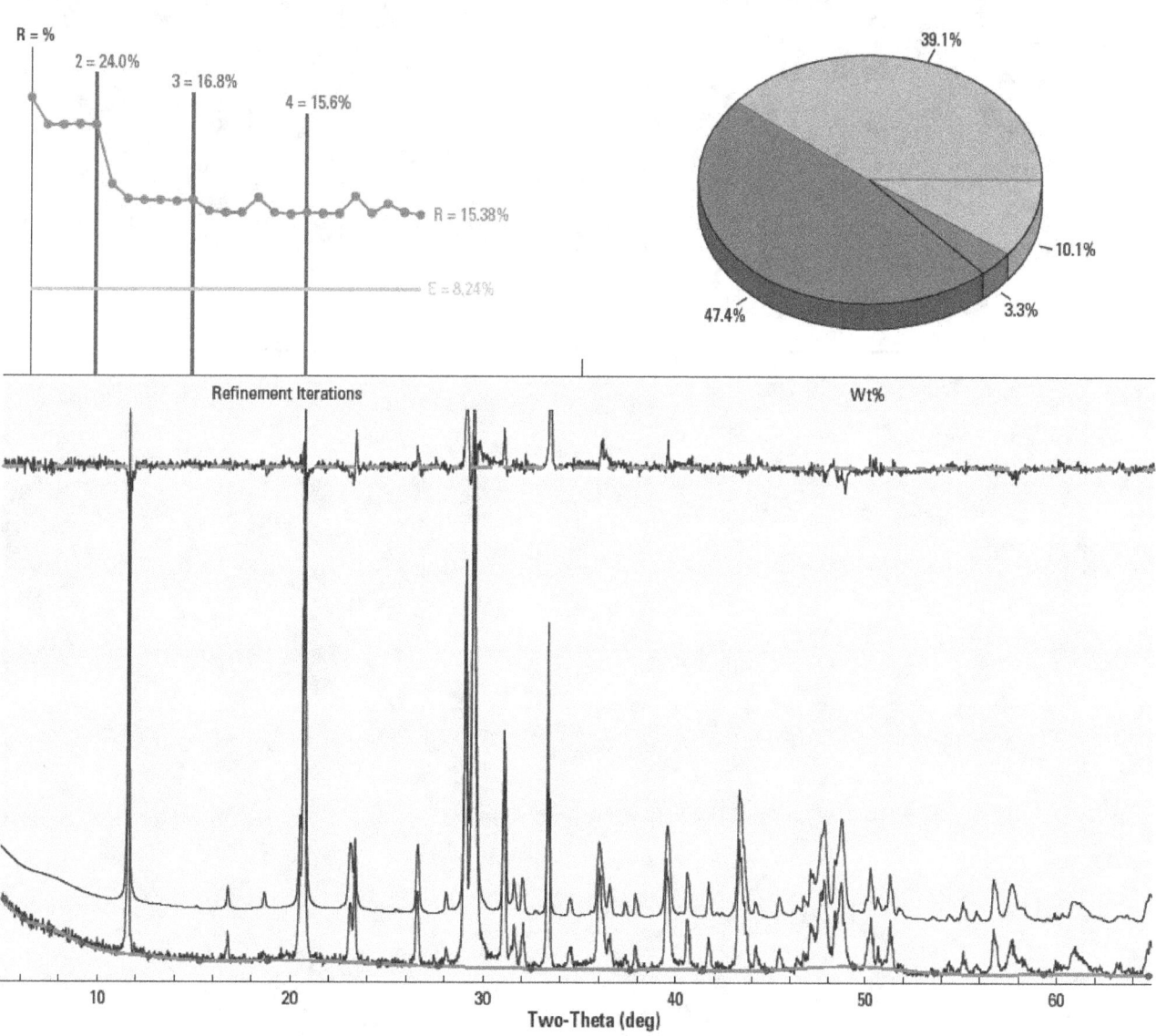

Entrance Spring

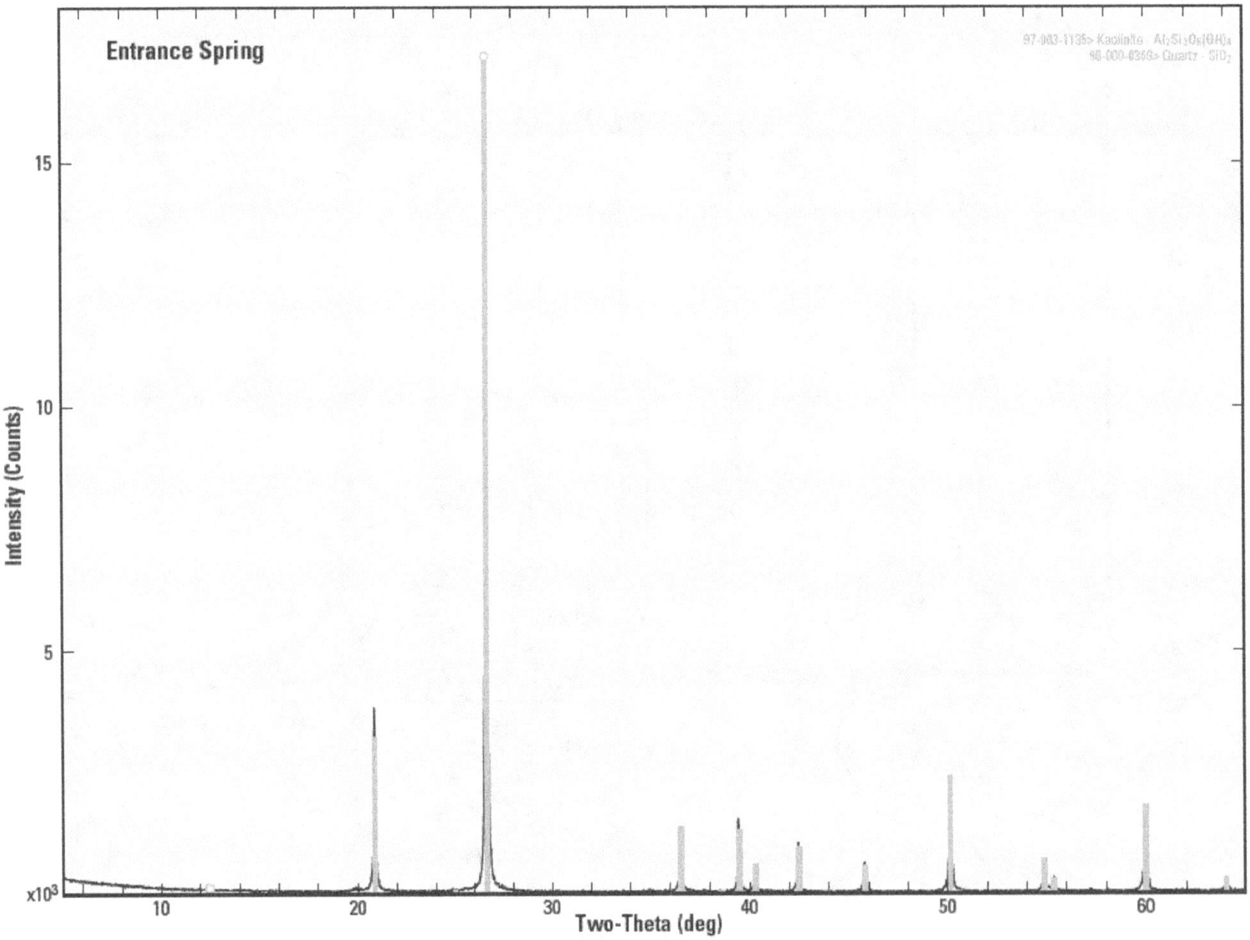

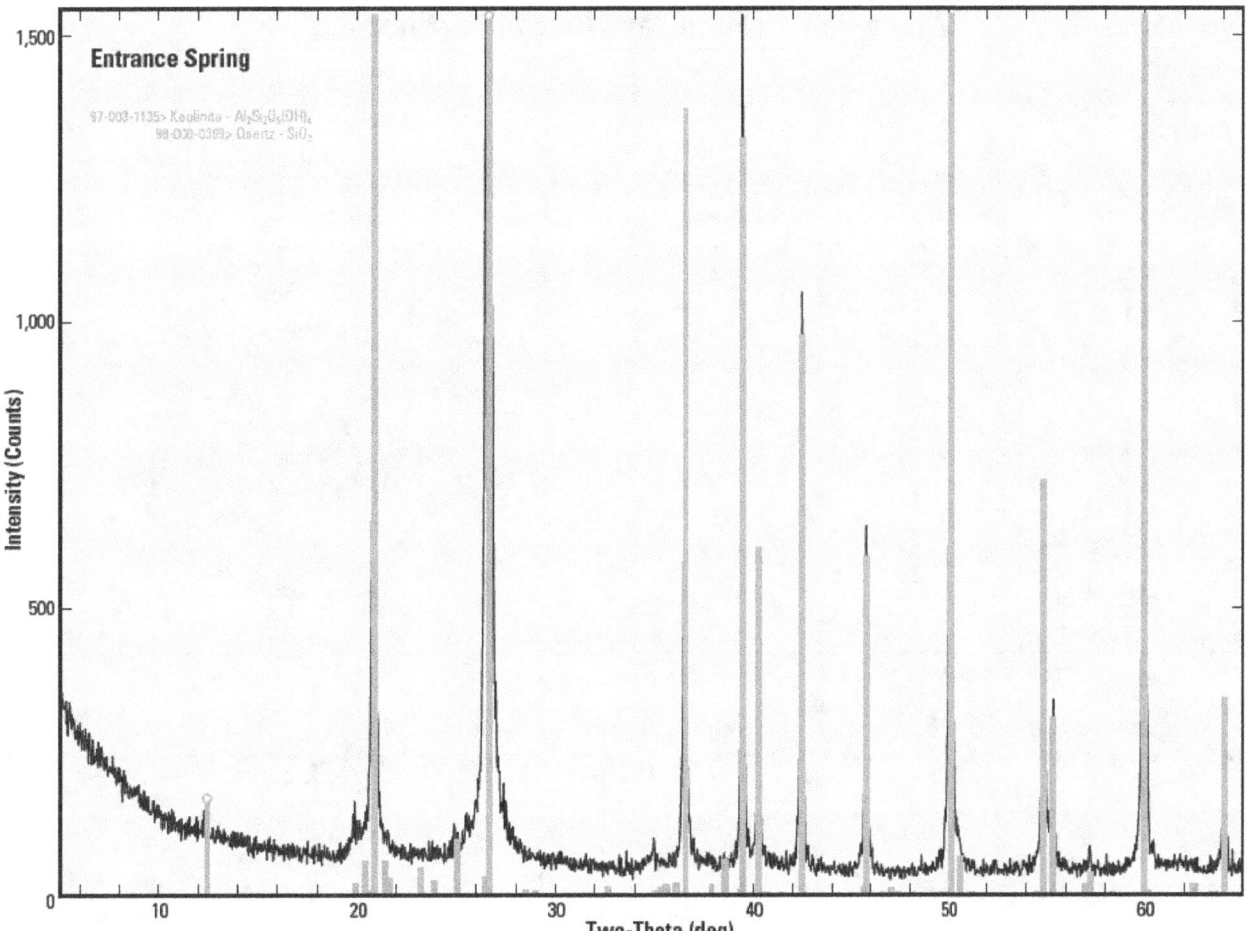

Entrance Spring

Whole Pattern Fitting and Rietveld Refinement

Phase ID (2)	Source	I/Ic	Wt%	#L
Kaolinite 1A - $Al_2Si_2O_5(OH)_4$	FIZ#31135	1.06 (0%)	1.3 (0.2)	241
Quartz - SiO_2	JCS#369	4.23 (0%)	98.7 (0.5)	40

XRF (Wt%): SiO_2 = 99.3%, Al_2O_3 = 0.5%

NOTE: Fitting Halted at Iteration 19(4): R = 11.22% (E = 6.93%, R/E = 1.62, P = 25, EPS = 0.5)

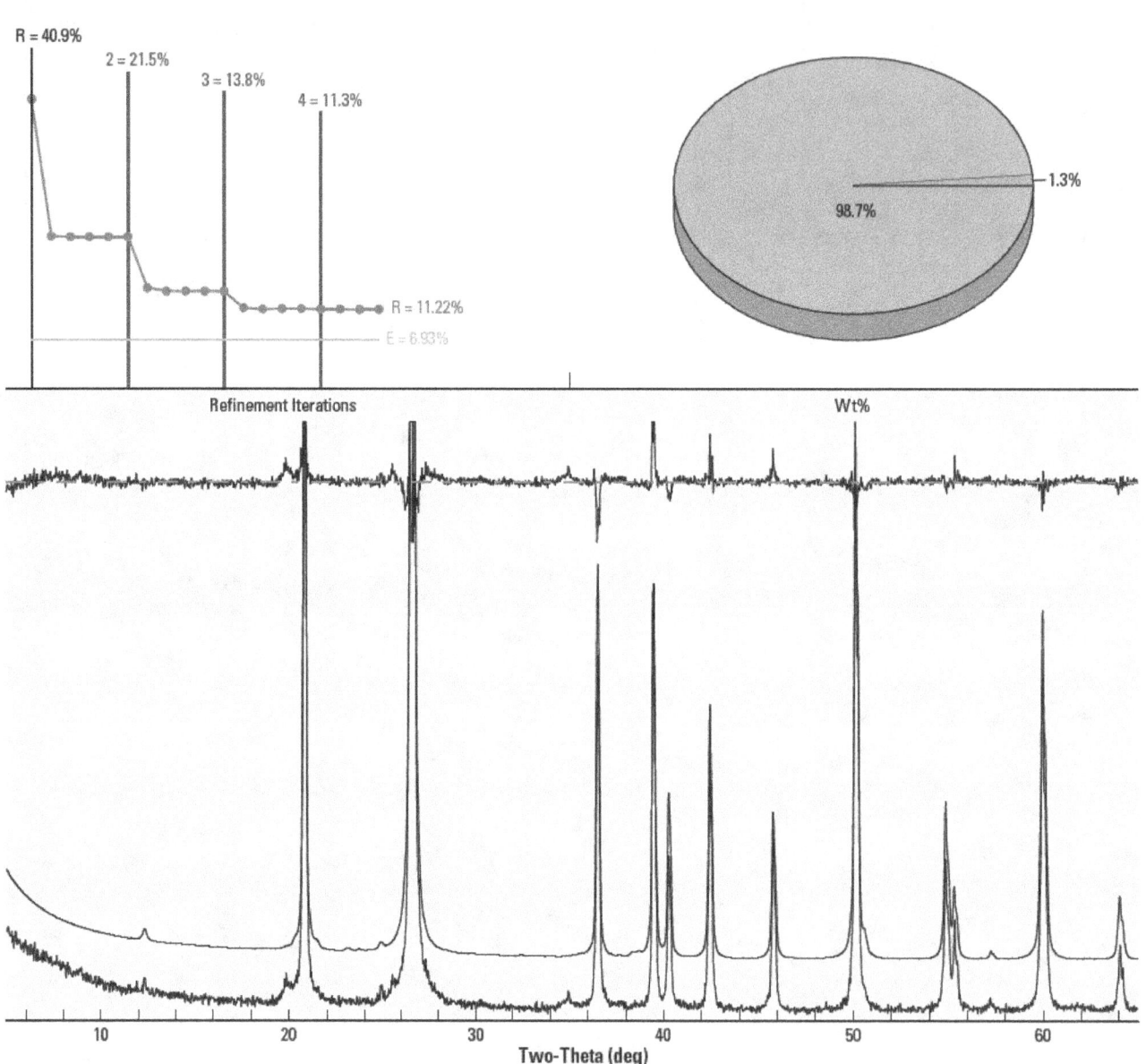

Above Oasis Spring

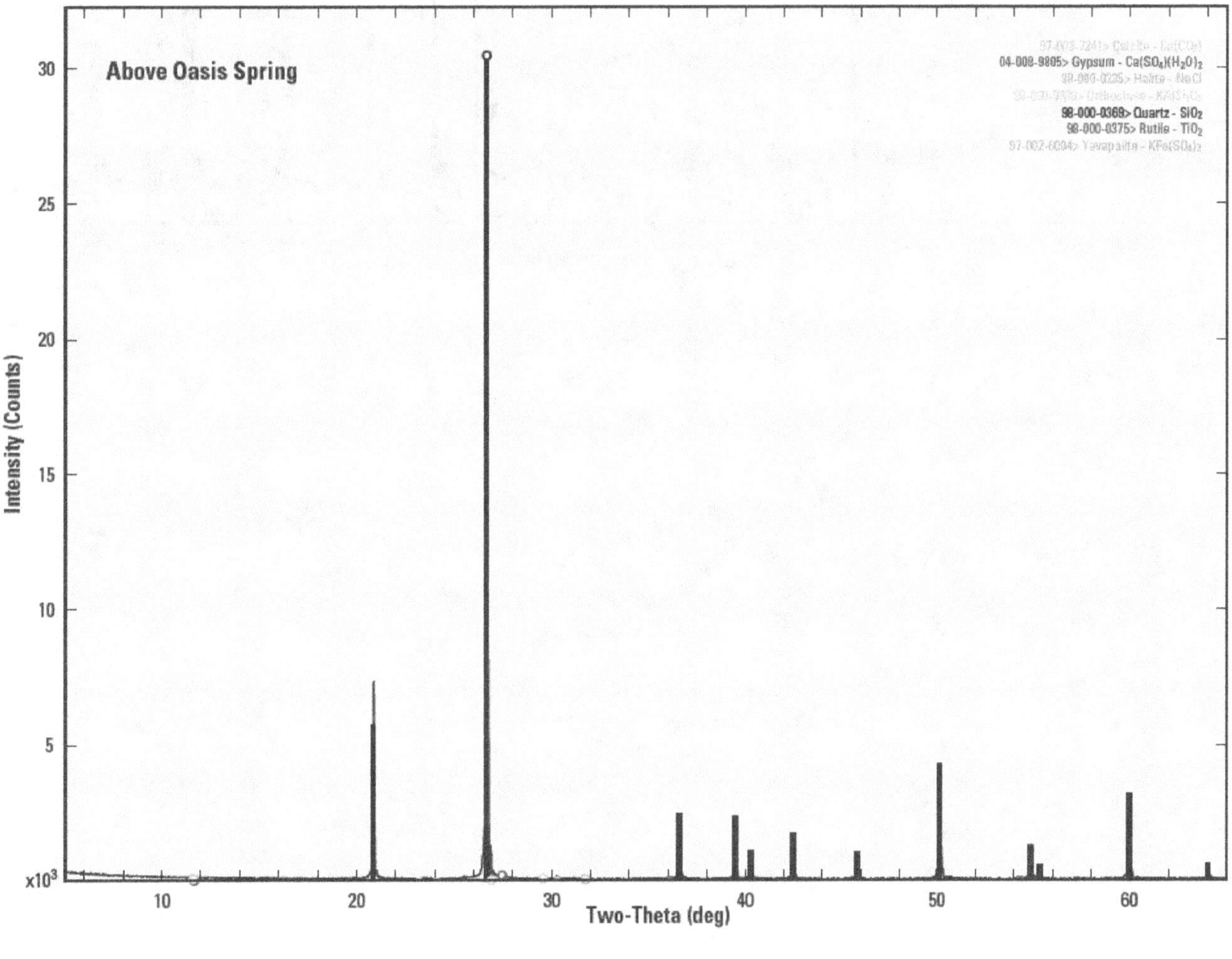

Above Oasis Spring

97-003-7241> Calcite - Ca(CO₃)
04-008-9805> Gypsum - Ca(SO₄)(H₂O)₂
98-000-0225> Halite - NaCl
98-000-3378> Orthoclase - KAlSi₃O₈
98-000-0369> Quartz - SiO₂
98-000-0375> Rutile - TiO₂
97-002-6004> Yavapaiite - KFe(SO₄)₂

Intensity (Counts)

Two-Theta (deg)

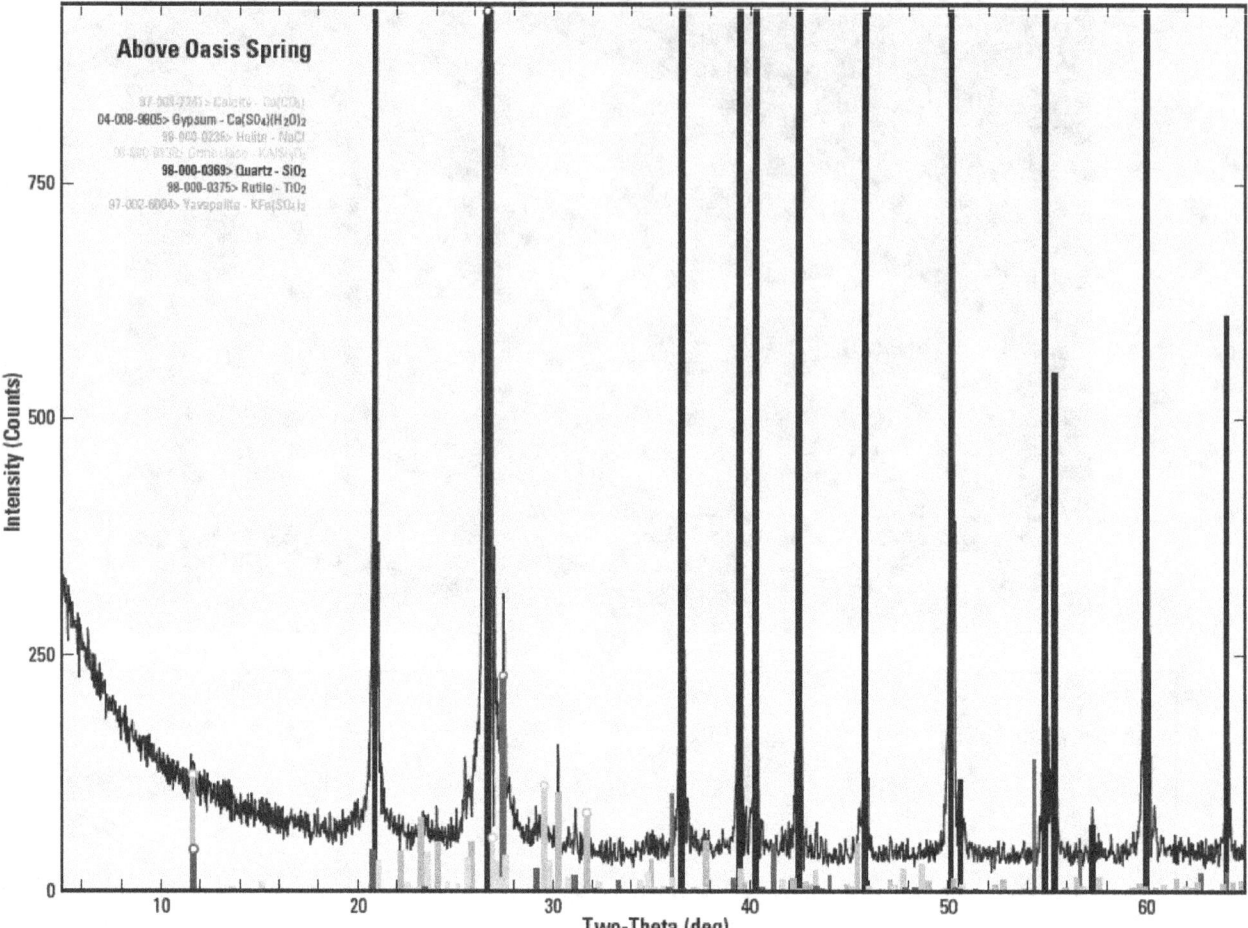

Above Oasis Spring

Whole Pattern Fitting and Rietveld Refinement

Phase ID (4)	Source	I/Ic	Wt%	#L
Calcite - Ca(CO₃)	FIZ#3724	13.19 (0%)	0.2 (?)	16
Gypsum - Ca(SO₄)(H₂O)₂	04-008-9805	2.09 (0%)	0.5 (0.4)	92
Orthoclase - KAlSi₃O₈	JCS#338	0.67 (0%)	2.9 (0.9)	137
Quartz - SiO₂	JCS#369	4.21 (0%)	96.5 (3.1)	40

XRF (Wt%): CaO = 0.2%, K₂O = 0.5%, SO₂ = 0.2%, SiO₂ = 98.3%, Al₂O₃ = 0.5%, CO₂ = 0.1%

NOTE: Fitting Halted at Iteration 0(1): R = 21.81% (E = 7.53%, R/E = 2.9, P = 28, EPS = 0.5)

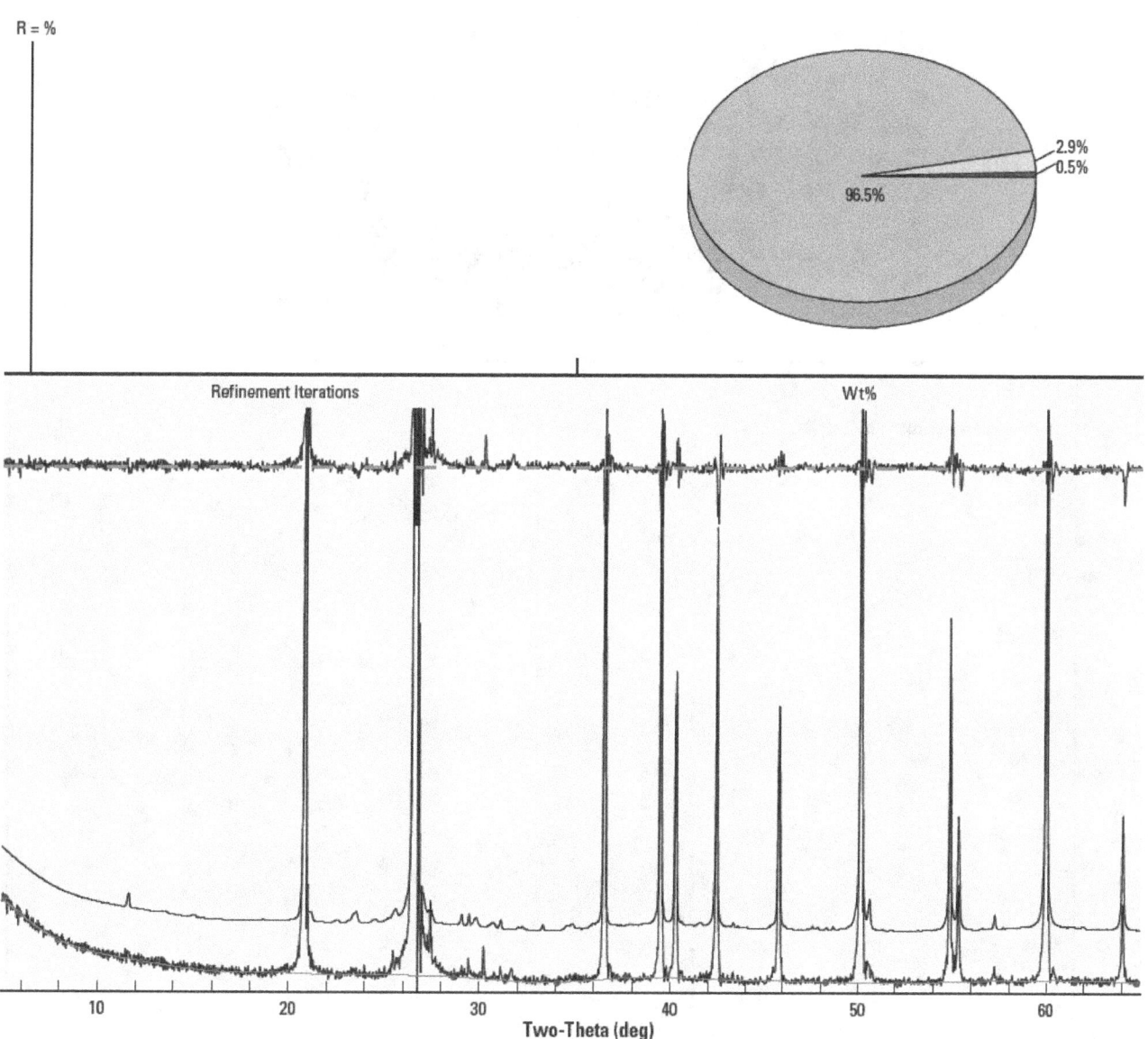

7
8
9
Below Oasis Spring

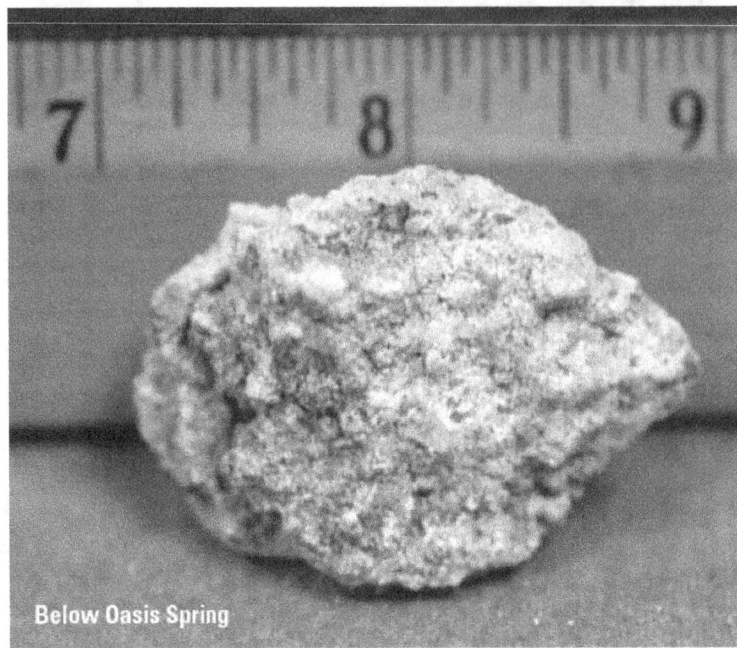

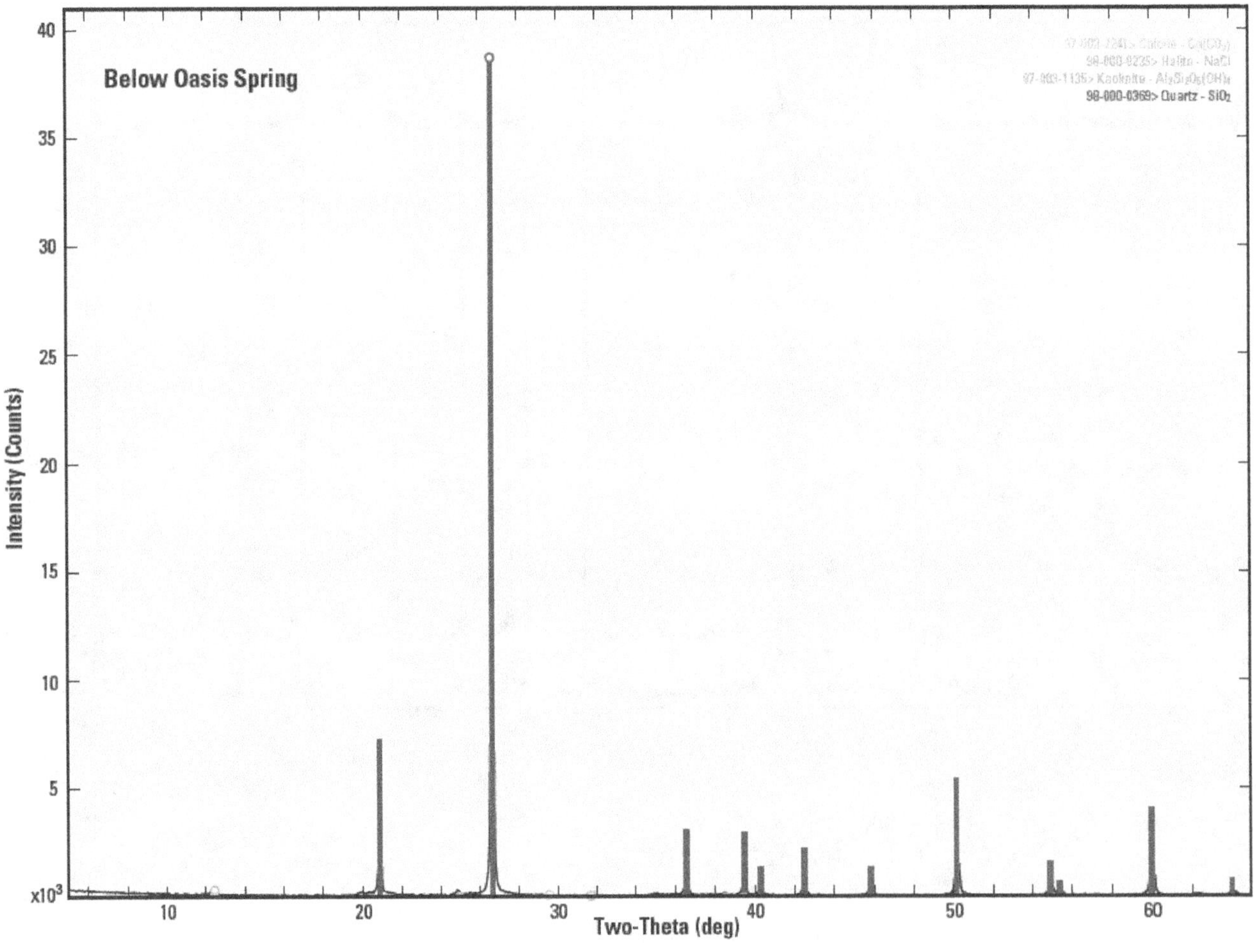
Below Oasis Spring
98-000-0369> Quartz - SiO_2
Intensity (Counts)
x10³
40
35
30
25
20
15
10
5
10
20
30
40
50
60
Two-Theta (deg)

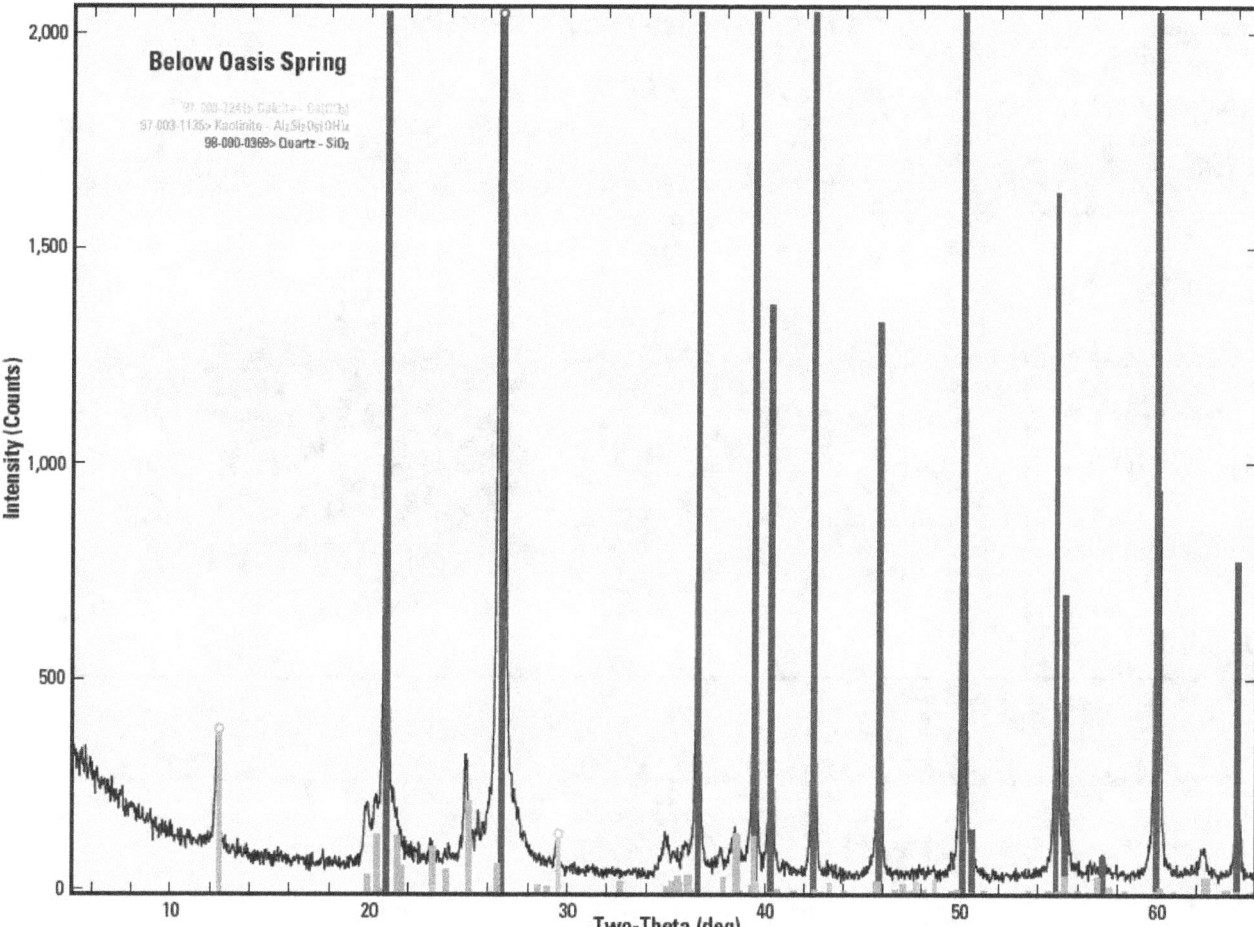

Below Oasis Spring

97-000-2241> Calcite - CaCO3
97-003-1135> Kaolinite - Al2Si2O5(OH)4
98-090-0369> Quartz - SiO2

Whole Pattern Fitting and Rietveld Refinement

Below Oasis Spring

Phase ID (3)

Calcite - Ca(CO$_3$)

Kaolinite 1A - Al$_2$Si$_2$O$_5$(OH)$_4$

Quartz - SiO$_2$

Source	I/Ic	Wt%	#L
FIZ#37241	3.20 (0%)	0.4 (0.1)	17
FIZ#31135	1.06 (0%)	6.3 (0.3)	242
JCS#369	4.23 (0%)	93.3 (0.8)	40

XRF (Wt%): CaO = 0.2%, SiO$_2$ = 96.3%, Al$_2$O$_3$ = 2.5%, CO$_2$ = 0.2%

NOTE: Fitting Halted at Iteration 20(4): R = 19.3% (E = 6.56%, R/E = 2.94, P = 29, EPS = 0.5)

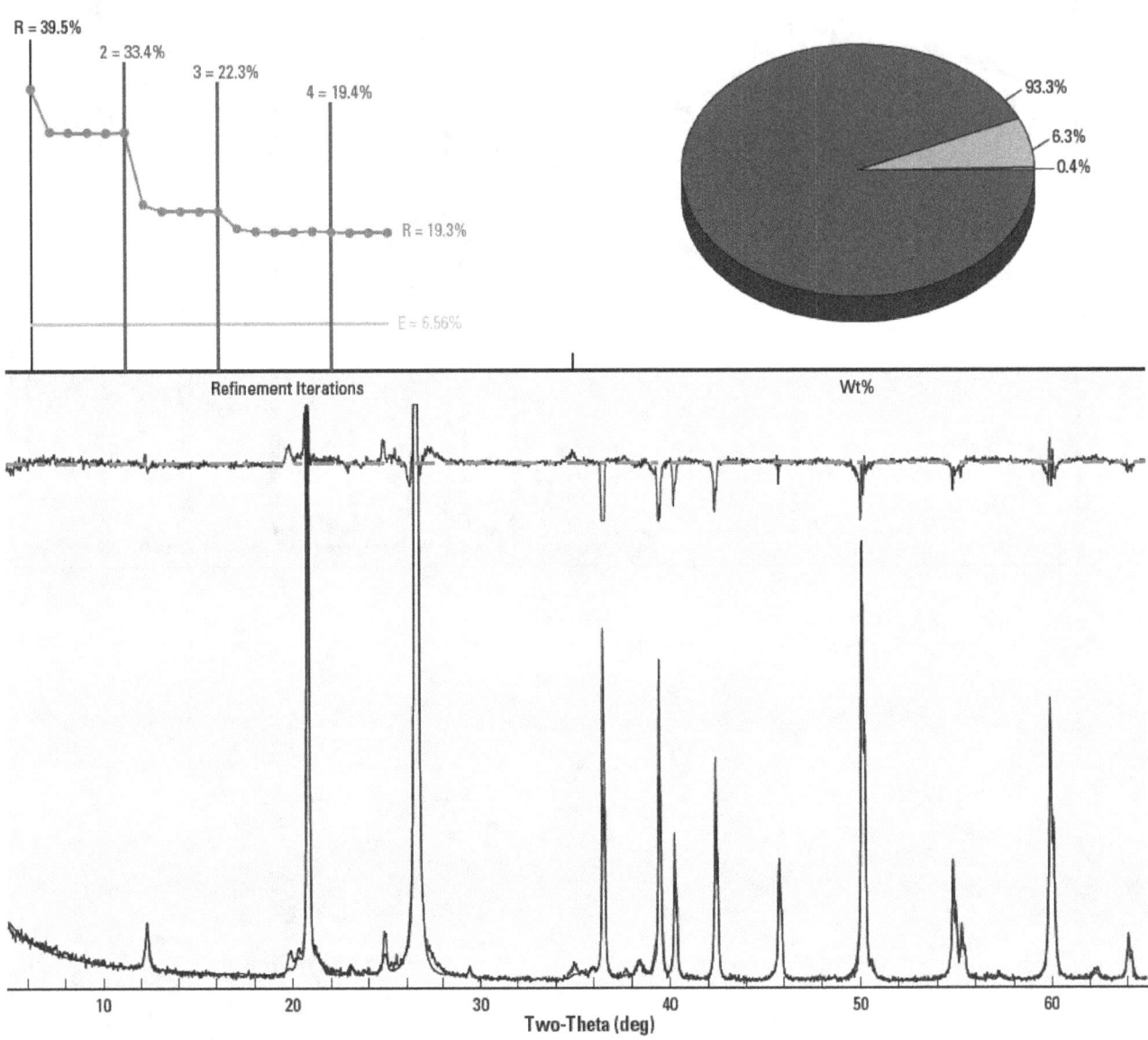

R = 39.5%

2 = 33.4%

3 = 22.3%

4 = 19.4%

R = 19.3%

E = 6.56%

Refinement Iterations

Wt%

93.3%

6.3%

0.4%

Two-Theta (deg)

Oasis Spring channel

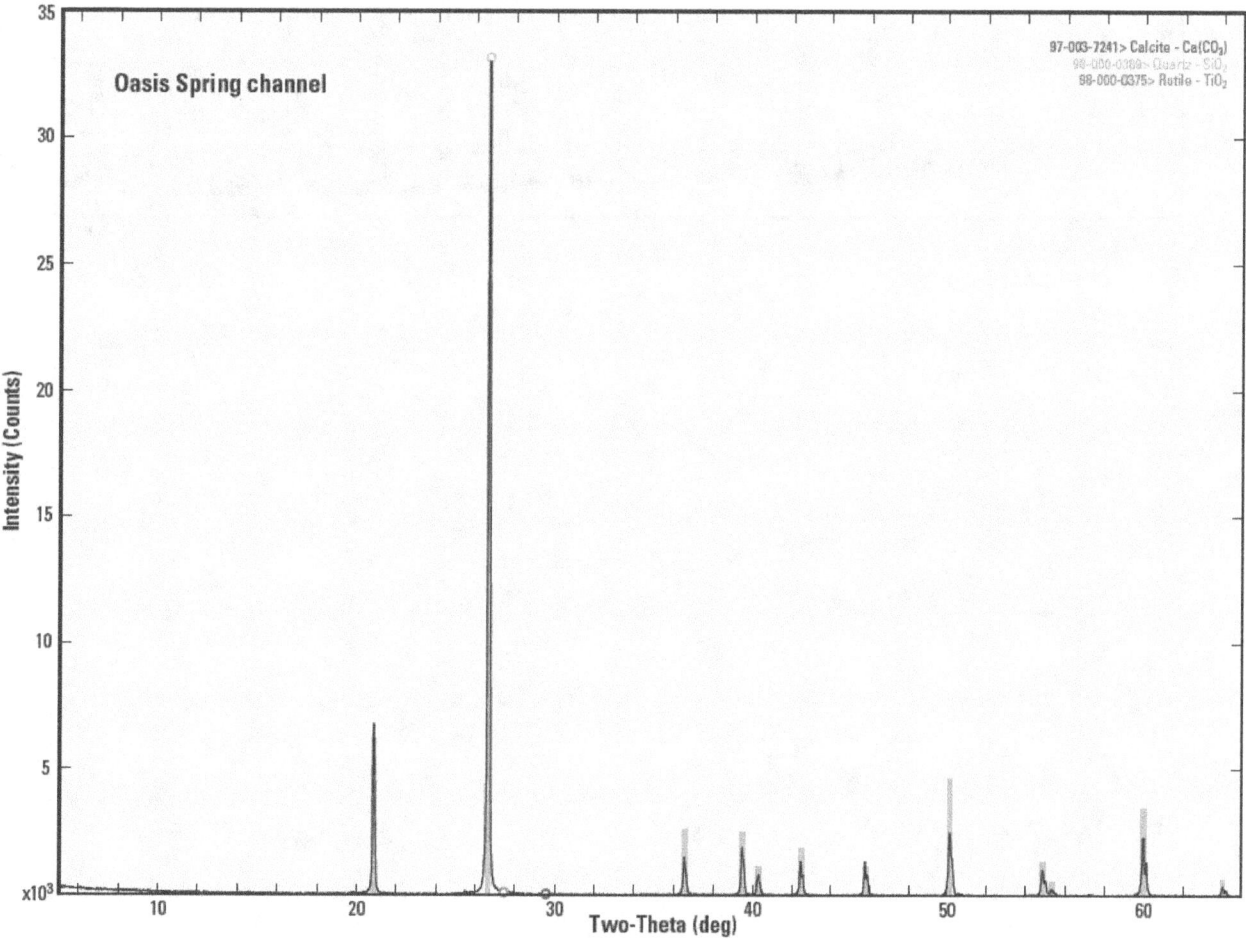

Oasis Spring channel

97-003-7241> Calcite - Ca(CO₃)
98-000-0369> Quartz - SiO₂
98-000-0375> Rutile - TiO₂

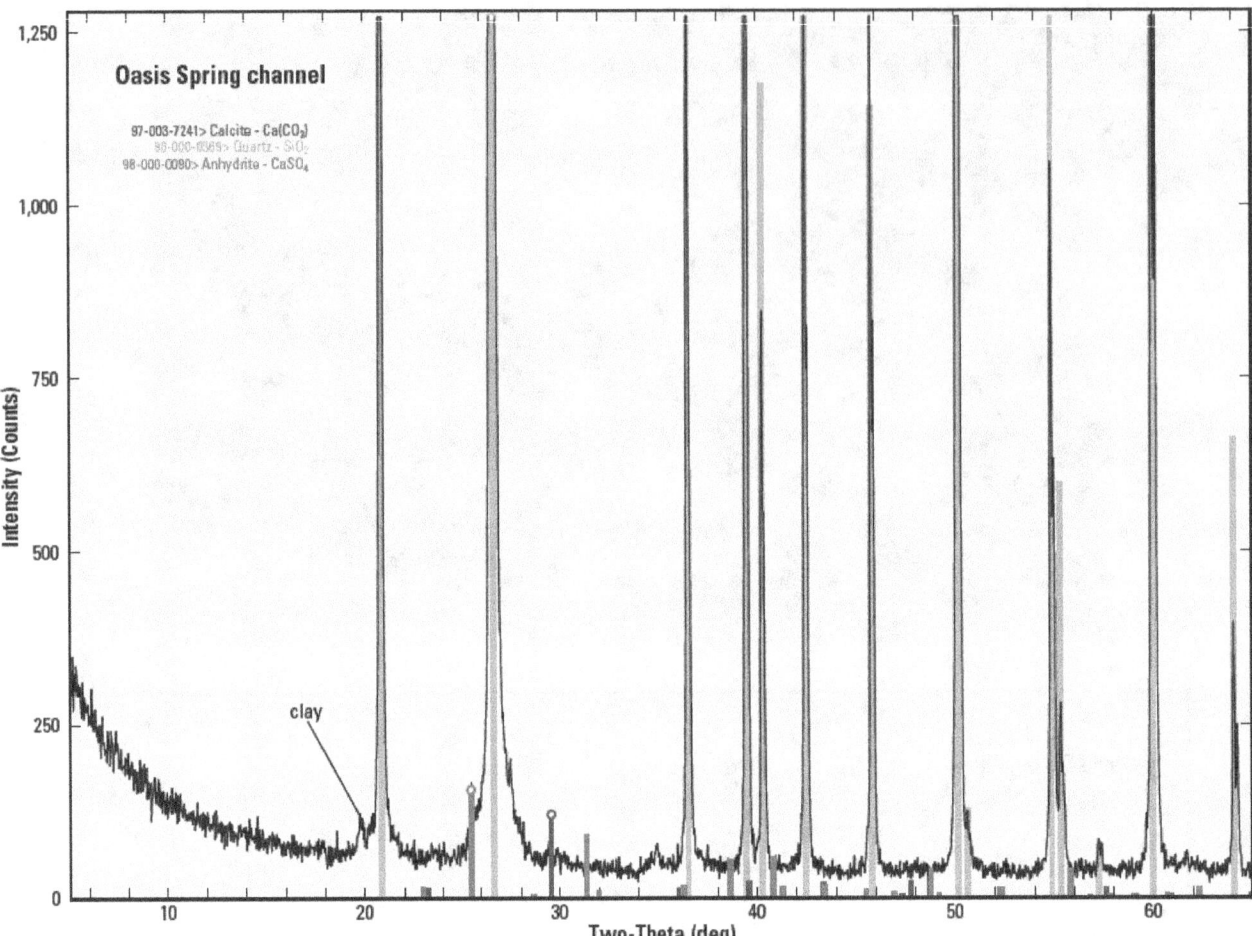

Whole Pattern Fitting and Rietveld Refinement

Oasis Spring channel

Phase ID (3)

- ■ Calcite - Ca(CO₃)
- ▢ Quartz - SiO₂
- ▢ Anhydrite - CaSO₄

Source	I/Ic	Wt%	#L
FIZ#37241	3.19 (0%)	0.1 (0.1)	17
JCS#369	4.22 (0%)	98.8 (1.1)	40
JCS#90	1.86 (0%)	1.0 (0.3)	35

XRF(Wt%): CaO = 0.5%, SO₂ = 0.5%, SiO₂ = 98.9%, CO₂ = 0.1%

NOTE: Fitting Halted at Iteration 21(4): R = 22.76% (E = 7.71%, R/E = 2.95, P = 26, EPS = 0.5)

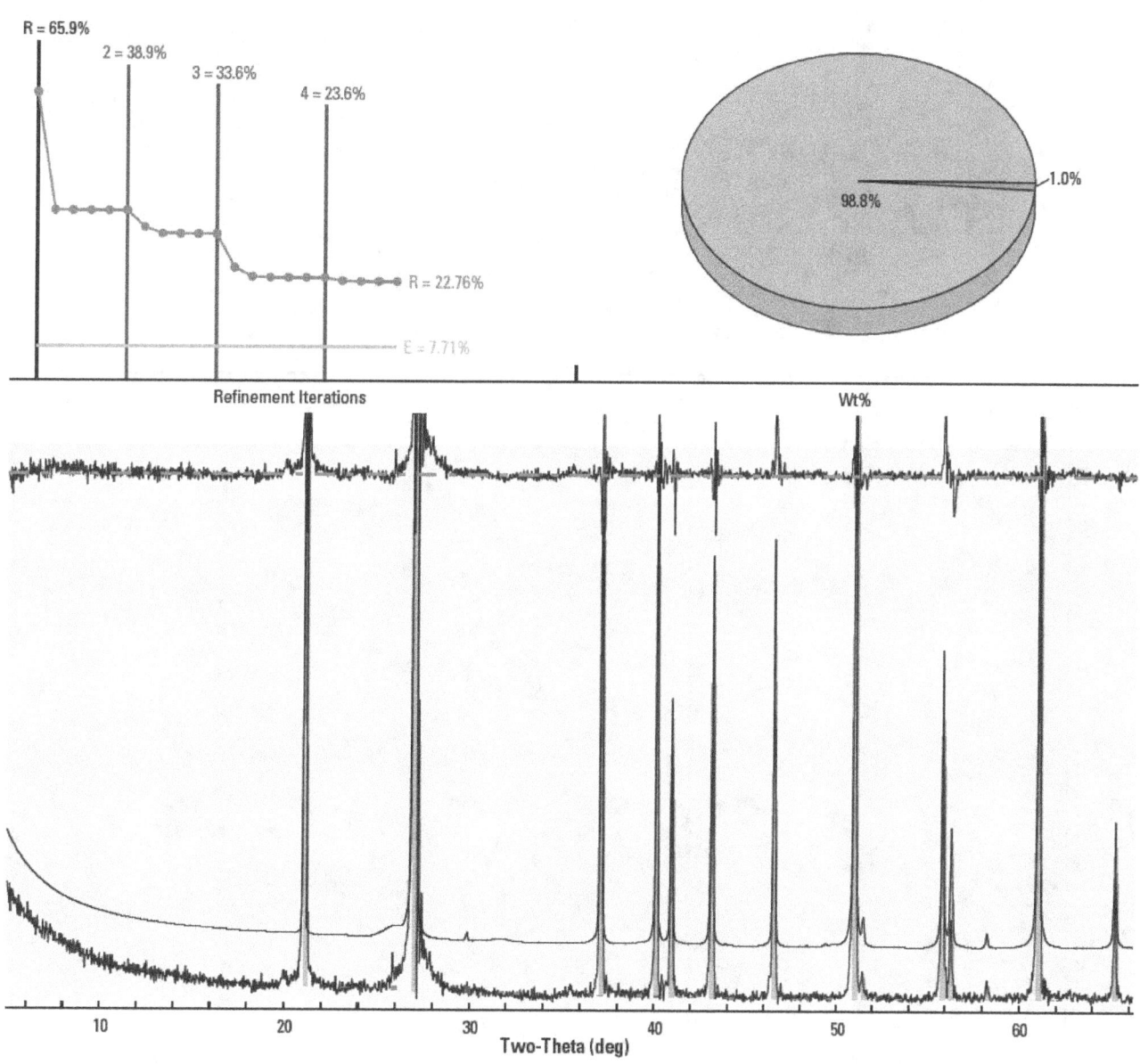

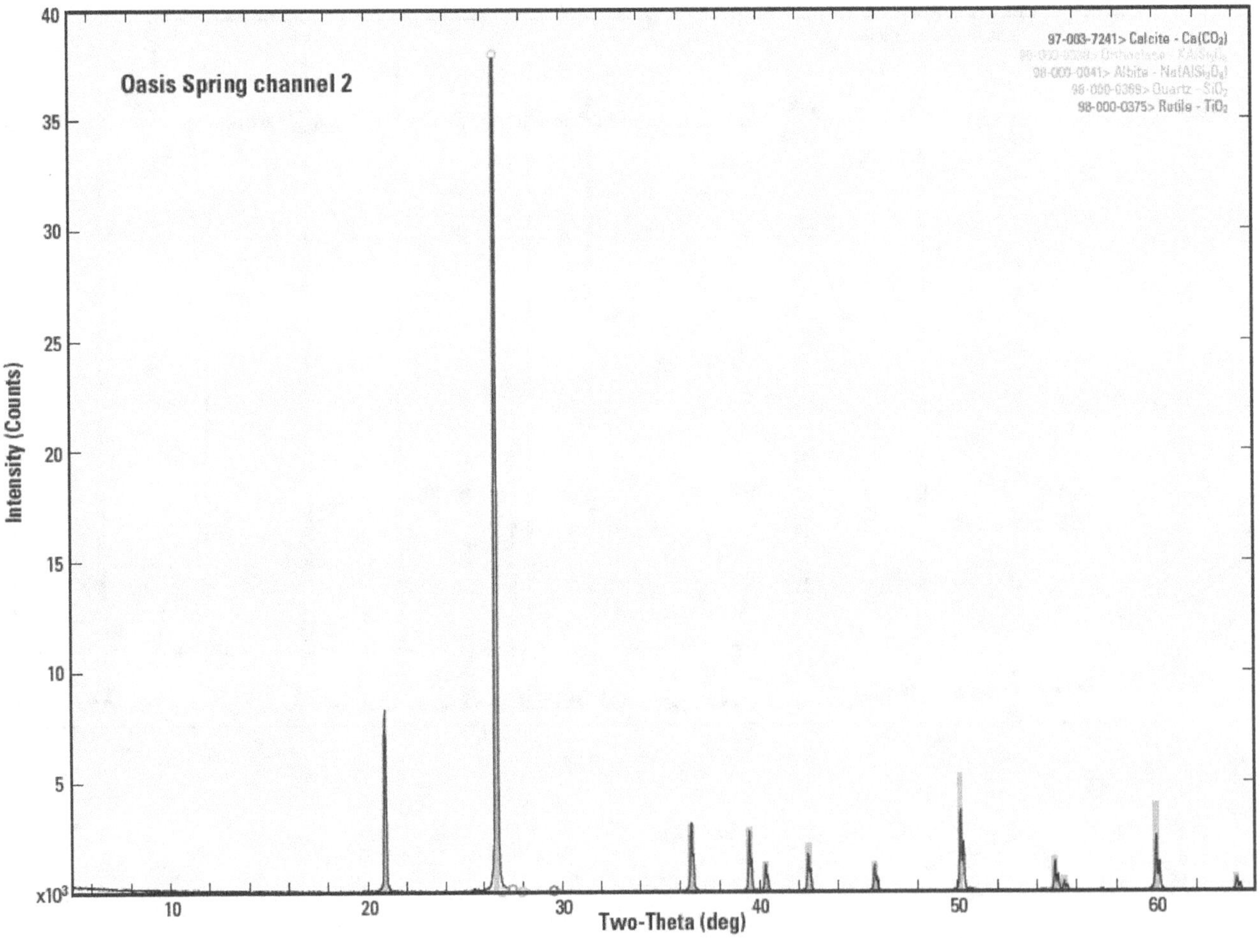

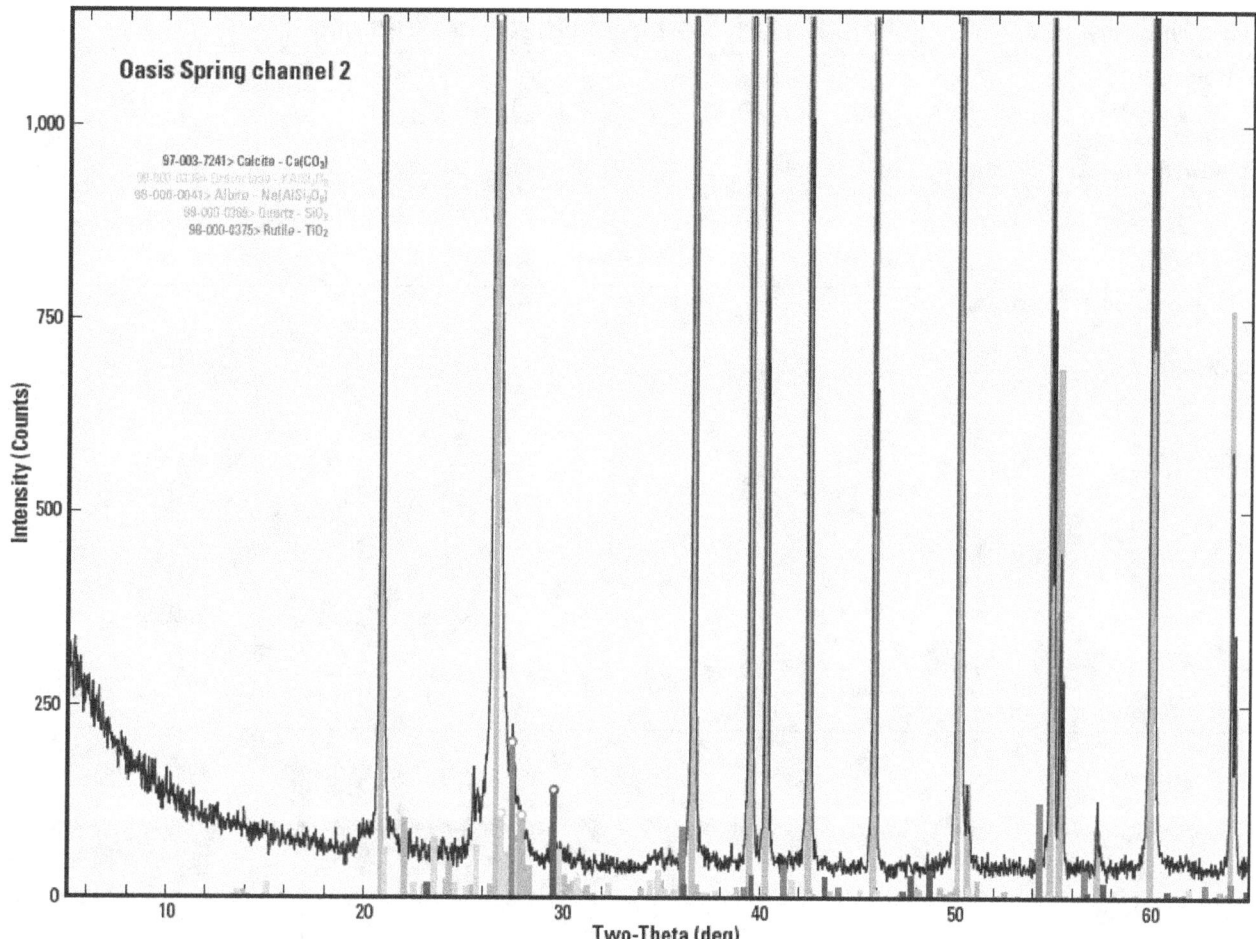

Oasis Spring channel 2

97-003-7241> Calcite - Ca(CO$_3$)
98-000-0036> Orthoclase - KAlSi$_3$O$_8$
98-000-0041> Albite - Na(AlSi$_3$O$_8$)
98-000-0369> Quartz - SiO$_2$
98-000-0375> Rutile - TiO$_2$

Oasis Spring channel 2   ## Whole Pattern Fitting and Rietveld Refinement

Phase ID (6)	Source	I/Ic	Wt%	#L
■ Calcite - $Ca(CO_3)$	FIZ#37241	3.19 (0%)	0.2 (0.0)	17
☐ Orthoclase - $KAlSi_3O_8$	JCS#338	0.69 (0%)	1.4 (0.2)	138
▨ Albite - $Na(AlSi_3O_8)$	JCS#41	0.66 (0%)	1.7 (0.2)	234
▨ Quartz - SiO_2	JCS#369	4.22 (0%)	96.0 (0.5)	40
▨ Anhydrite - $CaSO_4$	JCS#90	1.86 (0%)	0.1 (0.1)	35
■ Rutile - TiO_2	JCS#375	3.41 (0%)	0.6 (0.1)	9

XRF(Wt%): TiO_2 = 0.6%, CaO = 0.2%, K_2O = 0.2%, SO_2 = 0.1%, SiO_2 = 98.0%, Al_2O_3 = 0.6%, Na_2O = 0.2%, CO_2 = 0.1%

NOTE: Fitting Halted at Iteration 22(4): R = 9.84% (E = 6.62%, R/E = 1.49, P = 44, EPS = 0.5)

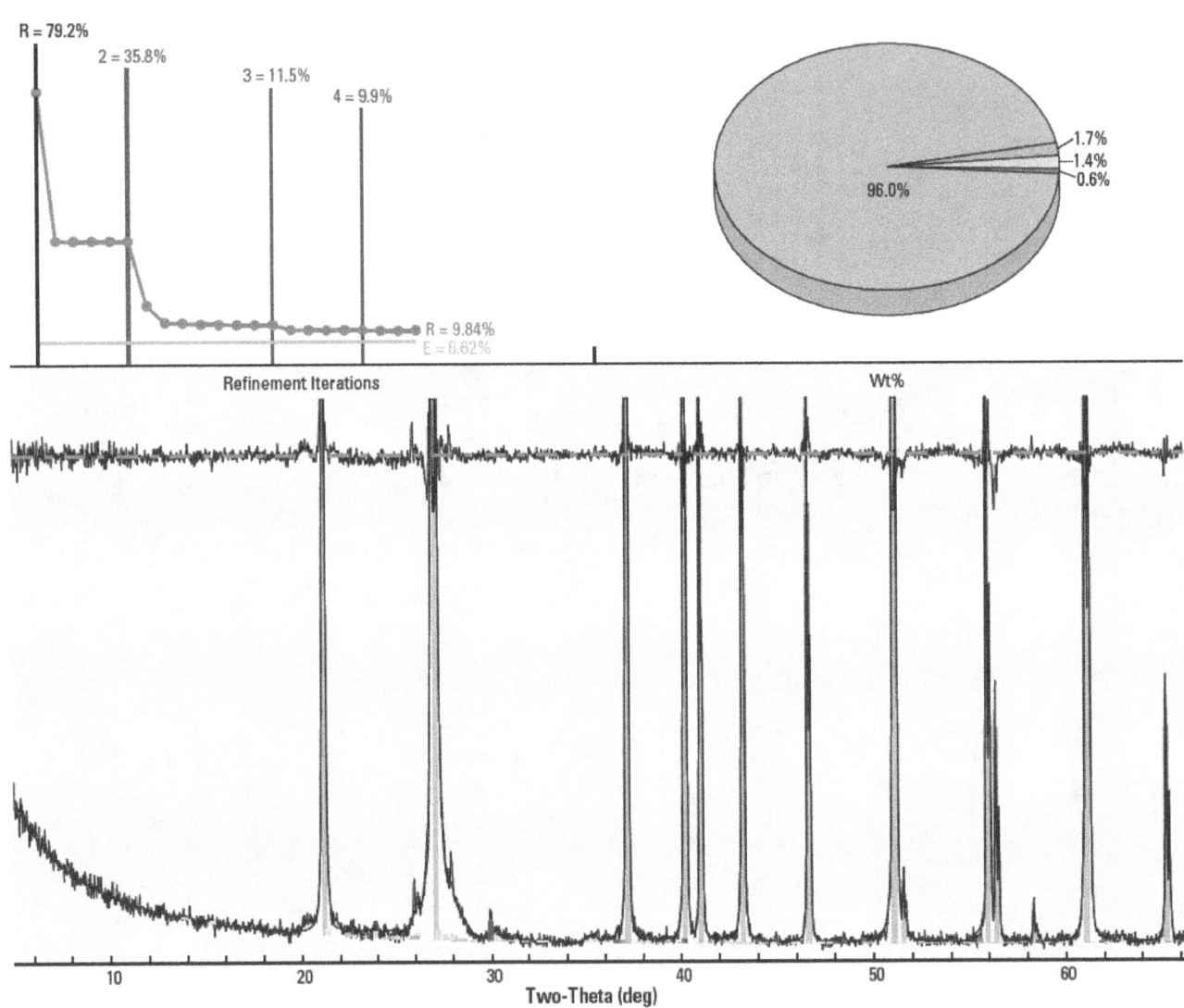

Cow Spring chert

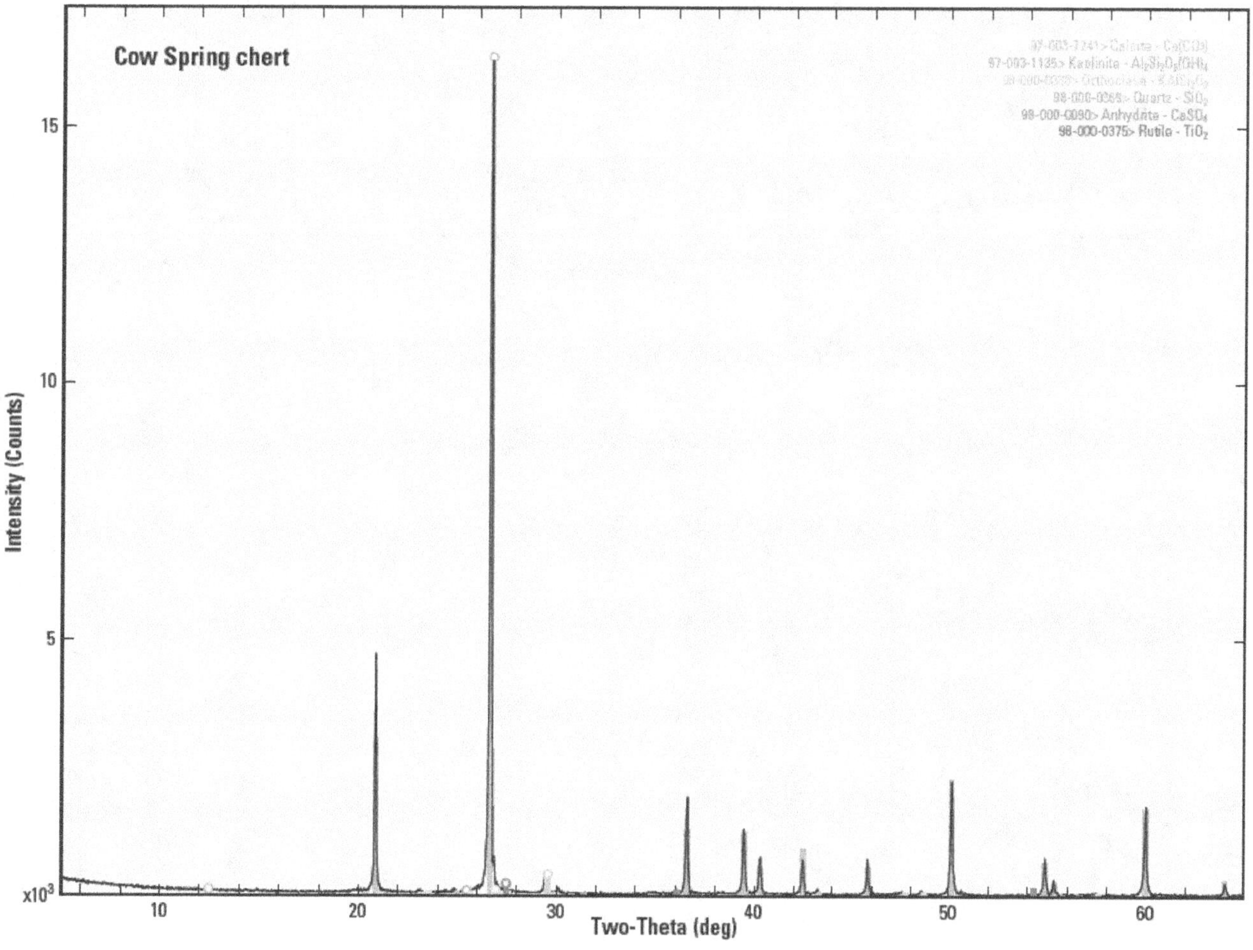

97-003-7241> Calcite - Ca(CO3)
97-003-1135> Kaolinite - Al2Si2O5(OH)4
98-000-0350> Orthoclase - KAlSi3O8
98-000-0369> Quartz - SiO2
98-000-0090> Anhydrite - CaSO4
98-000-0375> Rutile - TiO2

Cow Spring chert

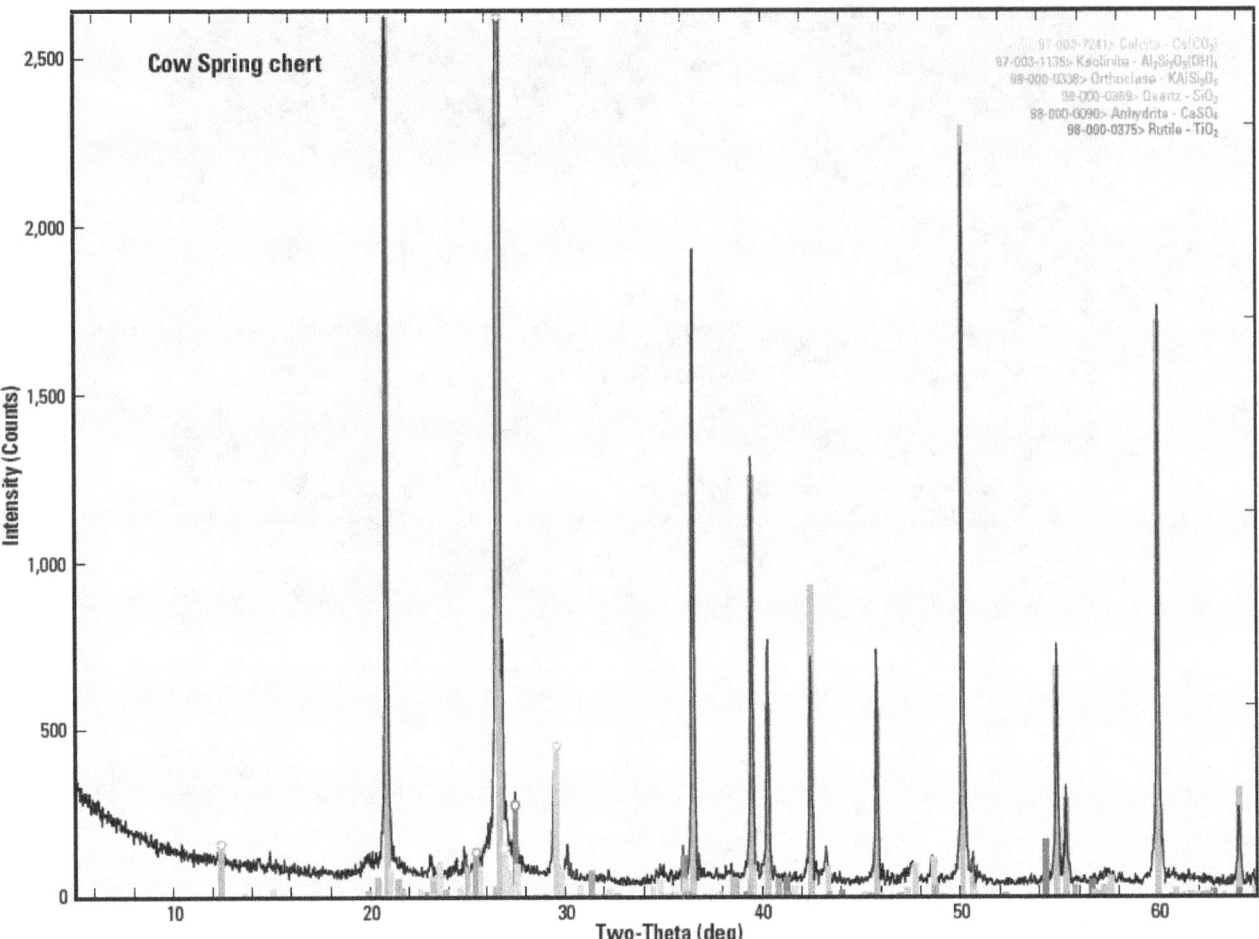

Cow Spring chert

Whole Pattern Fitting and Rietveld Refinement

Phase ID (4)

- Calcite - Ca(CO₃)
- Kaolinite 1A - Al₂Si₂O₅(OH)₄
- Quartz - SiO₂
- Rutile - TiO₂

Source	I/Ic	Wt%	#L
FIZ#37241	3.20 (0%)	3.6 (0.2)	17
FIZ#31135	1.05 (0%)	2.2 (0.3)	242
JCS#369	4.23 (0%)	92.9 (0.8)	40
JCS#375	3.42 (0%)	1.2 (0.2)	9

XRF(Wt%): TiO₂ = 1.2%, CaO = 2.0%, SiO₂ = 94.0%, Al₂O₃ = 0.9%, CO₂ = 1.6%

NOTE: Fitting Halted at Iteration 20(4): R = 14.86% (E = 7.85%, R/E = 1.89, P = 33, EPS = 0.5)

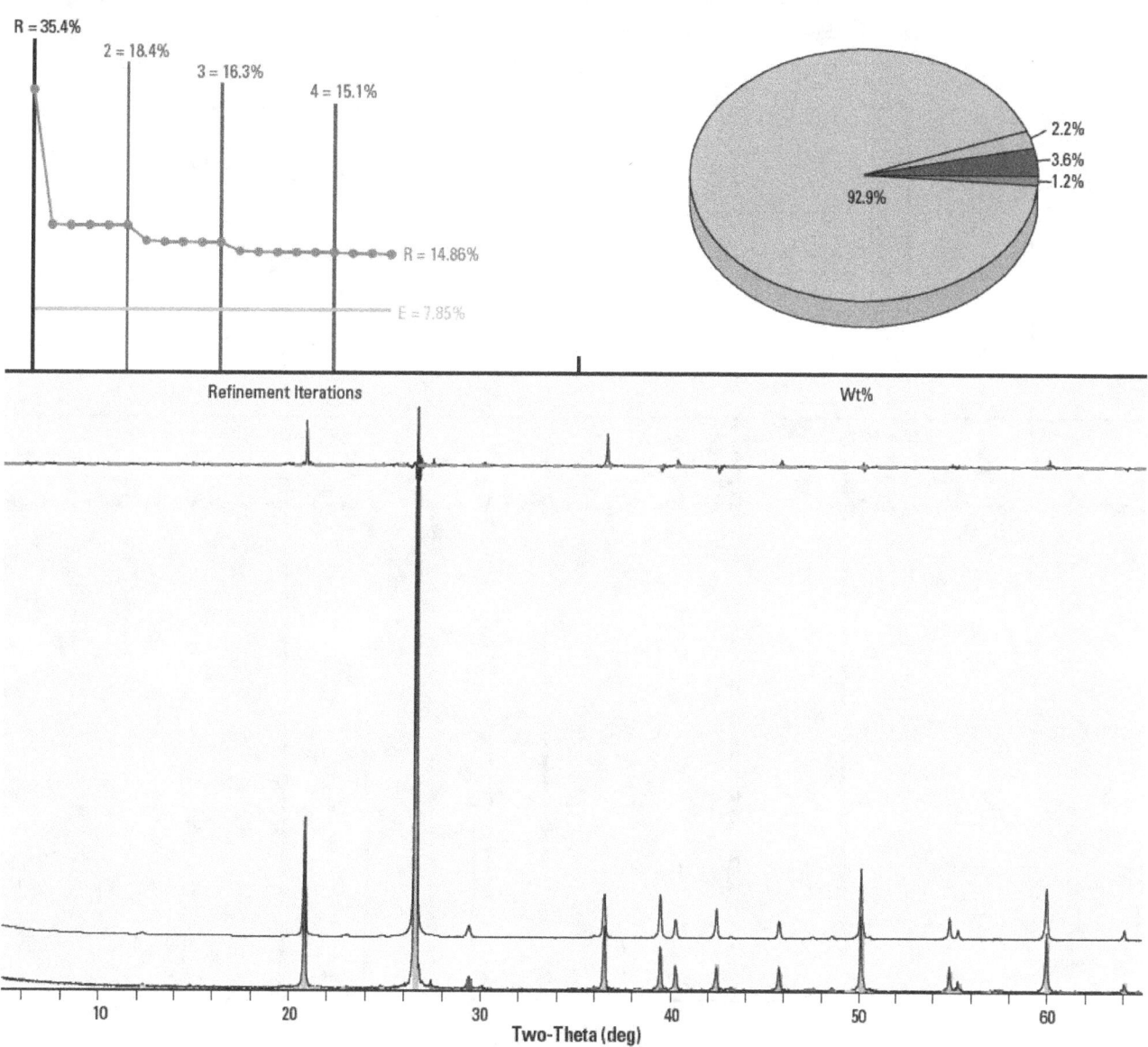

Cow Spring chert

Whole Pattern Fitting and Rietveld Refinement

Phase ID (6)		Source	I/Ic	Wt%	#L
▢ Calcite - $Ca(CO_3)$		FIZ#37241	3.20 (0%)	3.4 (0.1)	17
▣ Kaolinite 1A - $Al_2Si_2O_5(OH)_4$		FIZ#31135	1.05 (0%)	2.0 (0.3)	242
▢ Orthoclase - $KAlSi_3O_8$		JCS#338	0.69 (0%)	5.7 (0.8)	138
▨ Quartz - SiO_2		JCS#369	4.23 (0%)	88.6 (1.0)	40
▨ Anhydrite - $CaSO_4$		JCS#90	1.86 (0%)	0.1 (0.1)	35
▨ Rutile - TiO_2		JCS#375	3.39 (0%)	0.2 (0.0)	9

XRF(Wt%): TiO_2 = 0.2%, CaO = 2.0%, K_2O = 1.0%, SO_2 = 0.1%, SiO_2 = 93.2%, Al_2O_3 = 1.9%, CO_2 = 1.5%

NOTE: Fitting Halted at Iteration 22(4): R = 14.55% (E = 7.83%, R/E = 1.86, P = 44, EPS = 0.5)

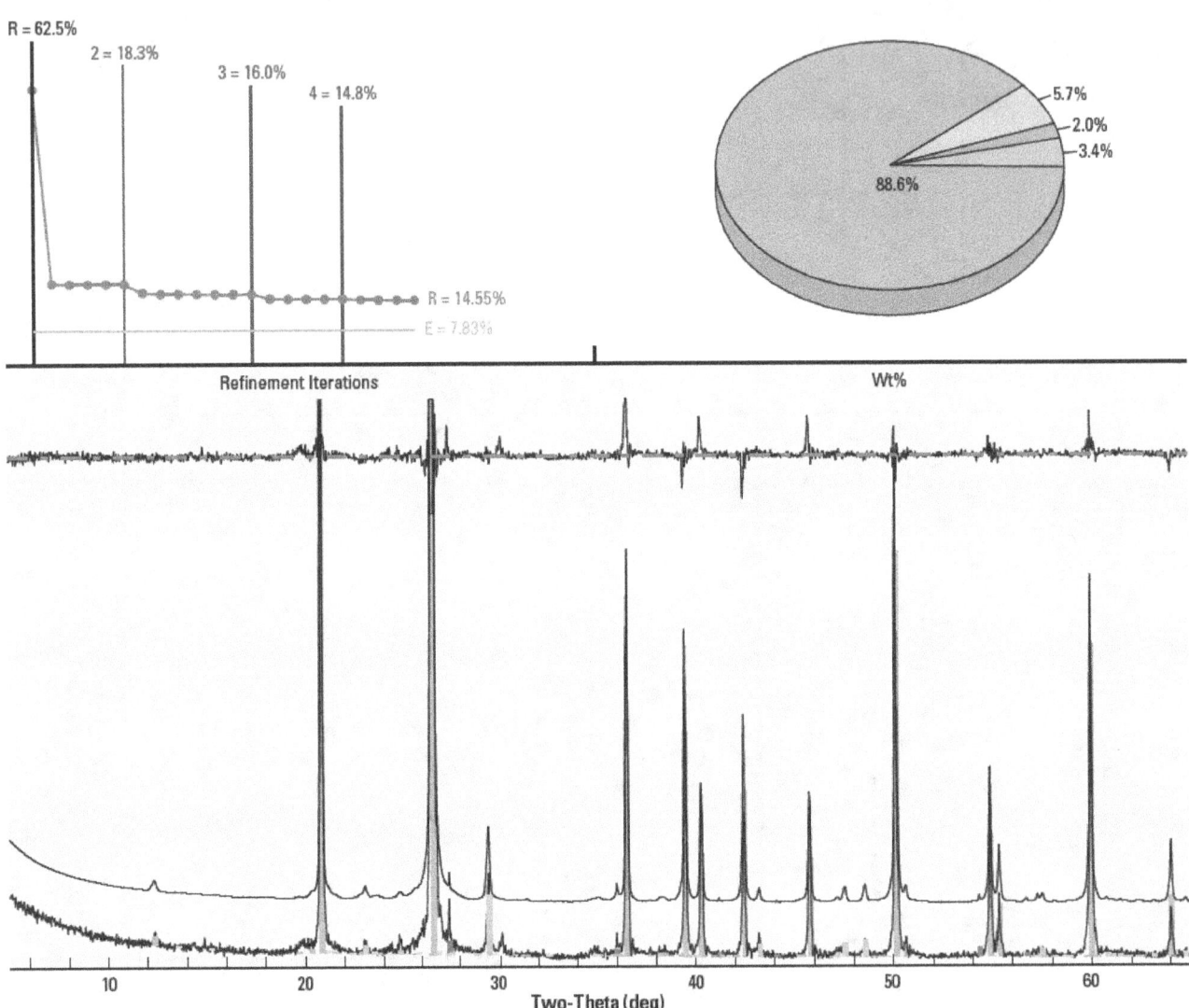

Appendix 4

Appendix 4. Percent ash and chemical composition of new growth from sagebrush plants near the White Mesa uranium mill, San Juan County, Utah, September 2009.

[All analyses of biota tissue in dry weight. **Abbreviations**: ins, insufficient sample amount; mm/dd/yyyy, month/day/year; µg/g, micrograms per gram; >, greater than; <, less than]

Field ID	Station number	Sample date (mm/dd/yyyy)	Ash, (percent)	Aluminum, total digestion, (percent)	Calcium, recoverable, (percent)	Iron, total digestion, (percent)	Potassium, recoverable, (percent)	Magnesium, recoverable, (percent)	Sodium, recoverable, (percent)	Sulfur, total digestion, (percent)	Titanium, total digestion, (percent)
1-0	373233109314301	09/01/2009	4 68	0 97	9	0 47	13 2	2 24	0 17	2 67	0 05
2-0	373231109312101	09/01/2009	4 54	0 72	10 7	0 36	14	2 37	0 15	3 17	0 04
3-0	373233109304901	09/03/2009	4 13	0 68	10 1	0 35	14 7	3 17	0 17	3 87	0 04
4-0	373233109303101	09/03/2009	4 8	0 82	8 64	0 41	13 9	2 95	0 2	3 17	0 04
5-0	373233109301002	09/03/2009	4 8	1 28	8 79	0 6	14	2 72	0 22	2 77	0 05
6-0	373233109294701	09/03/2009	4 97	1 02	9 14	0 49	13 6	2 59	0 18	2 48	0 03
7-0	373233109292201	09/01/2009	4 46	0 68	11	0 34	13 2	2 95	0 13	3 97	0 04
8-0	373217109314401	09/01/2009	4 75	0 68	8 59	0 33	>15	2 22	0 12	3	0 03
9-0	373217109311501	09/02/2009	4 92	0 48	11 1	0 25	13 8	2 42	0 11	2 88	0 03
10-0	373217109294501	09/03/2009	4 81	1 31	9 1	0 62	12 9	3 11	0 22	2 68	0 04
10-1a	373221109295201	09/03/2009	4 36	0 67	10 7	0 33	12 3	4 98	0 18	3 97	0 03
10-1b	373221109295201	09/03/2009	4 29	0 61	10 4	0 3	13 1	4 9	0 17	4	0 03
10-2	373214109295201	09/03/2009	4 17	0 59	9 94	0 28	14 1	3 87	0 15	3 83	0 02
11-0	373218109292101	09/01/2009	4 71	0 91	9 73	0 44	14 7	2 43	0 14	2 65	0 05
12-0	373202109314301	09/01/2009	4 25	0 68	11 3	0 36	13 4	2 89	0 13	2 94	0 04
12-1a	373203109313701	09/02/2009	4 26	0 43	9 94	0 23	>15	2 18	0 11	3 25	0 02
12-1b	373203109313701	09/02/2009	4 27	0 44	9 89	0 23	13 7	2 19	0 1	3 28	0 02
12-2	373158109314801	09/02/2009	4 36	0 32	9 66	0 18	>15	2 58	0 07	2 29	0 02
13-0	373203109312001	09/01/2009	4 35	1 24	10 9	0 6	12 4	2 45	0 19	2 66	0 07
14-0	373202109294401	09/03/2009	4 68	1 34	9 7	0 69	11 8	2 67	0 26	2 75	0 05
14-1	373159109295001	09/03/2009	4 91	1 31	9 25	0 67	12 6	2 73	0 25	2 41	0 04
14-2a	373159109294001	09/03/2009	4 82	1 16	10 4	0 58	12 3	2 97	0 24	3 18	0 06
14-2b	373159109294001	09/03/2009	4 8	1 04	10 6	0 53	12 5	3 02	0 22	3 22	0 05
15-0	373202109292201	09/01/2009	4 35	0 58	9 69	0 31	12 6	2 87	0 11	3 09	0 03
15-1	373159109292801	09/01/2009	4 96	0 78	9 68	0 4	9 84	2 89	0 13	2 56	0 04
15-2	373159109291701	09/01/2009	5 12	0 3	7 43	0 18	>15	3 88	0 67	>5	0 02
16-0	373147109314301	09/02/2009	4 29	0 71	10 8	0 36	10 2	2 97	0 14	2 66	0 04
17-0	373147109311901	09/02/2009	4 73	0 73	8 44	0 38	12 2	2 77	0 17	3 02	0 03
17-1	373151109312601	09/01/2009	4 55	0 62	9 5	0 31	12 6	2 73	0 12	2 31	0 03
17-2	373143109312601	09/01/2009	4 79	0 52	10 3	0 28	11 4	2 83	0 11	3 2	0 02
18-0	373146109294401	09/03/2009	4 73	1 23	9 2	0 61	11 4	2 71	0 24	2 8	0 03
19-0	373148109292201	09/01/2009	4 4	0 67	9 58	0 34	11 7	1 71	0 1	2 61	0 03
20-0	373132109314301	09/02/2009	4 48	0 53	8 85	0 28	12 9	3 29	0 12	2 57	0 03
21-0	373132109312001	09/02/2009	4 54	0 75	9 63	0 38	13 4	2 75	0 19	2 79	0 03
22-0	373131109294501	09/03/2009	4 68	0 98	9	0 57	12 5	2 79	0 24	1 99	0 02
22-1	373122109294801	09/03/2009	4 69	1 03	11 4	0 53	10 8	2 49	0 18	2 77	0 04
22-2	373129109294201	09/03/2009	4 65	1 02	10 1	0 54	11 4	2 9	0 23	2 55	0 03
23-0	373132109292201	09/01/2009	4 36	0 89	9 66	0 45	12 6	2 95	0 15	2 96	0 04
23-1	373135109291701	09/01/2009	4 21	0 65	10 2	0 35	13	2 73	0 11	2 83	0 03
23-2	373128109292801	09/01/2009	4 34	0 53	10 9	0 28	11 5	2 99	0 11	3 14	0 02

Appendix 4.   Percent ash and chemical composition of new growth from sagebrush plants near the White Mesa uranium mill, San Juan County, Utah, September 2009.—Continued

[All analyses of biota tissue in dry weight. **Abbreviations**: ins, insufficient sample amount; mm/dd/yyyy, month/day/year; µg/g, micrograms per gram; >, greater than; <, less than]

Field ID	Station number	Sample date (mm/dd/yyyy)	Ash, (percent)	Aluminum, total digestion, (percent)	Calcium, recoverable, (percent)	Iron, total digestion, (percent)	Potassium, recoverable, (percent)	Magnesium, recoverable, (percent)	Sodium, recoverable, (percent)	Sulfur, total digestion, (percent)	Titanium, total digestion, (percent)
24-0	373116109314201	09/02/2009	4 49	0 61	8 36	0 34	14 6	2 58	0 15	3 08	0 03
25-0	373116109311901	09/02/2009	4 49	1 07	9 31	0 51	12	2 68	0 27	2 97	0 03
26-0	373110109305501	09/02/2009	4 7	0 81	9 89	0 42	11 3	2 84	0 18	3 04	0 03
27-0	373114109303101	09/03/2009	4 06	0 9	9 39	0 46	11 2	3 42	0 22	3 17	0 03
28-0	373115109300901	09/03/2009	4 6	1 16	8 77	0 58	12	2 28	0 21	2 32	0 03
29-0	373116109294501	09/03/2009	4 25	0 73	9 09	0 37	11 3	2 39	0 14	2 14	0 02
30-0	373116109292201	09/02/2009	4 43	0 69	9 95	0 36	13 9	2 41	0 12	3 05	0 02
31-0	373106109314701	09/02/2009	4 39	0 64	10 6	0 33	11 5	2 37	0 13	2 89	0 03
31-1a	373101109314201	09/02/2009	4 55	0 68	9 75	0 35	10 9	3 27	0 16	3 07	0 03
31-1b	373101109314201	09/02/2009	4 46	0 65	9 68	0 34	11	3 27	0 16	3 09	0 03
31-2	373058109314901	09/02/2009	4 69	0 9	8	0 45	12 2	3 01	0 17	2 9	0 03
32-0	373100109311801	09/02/2009	4 55	0 84	9 42	0 43	12 3	3	0 18	2 93	0 03
33-0	373100109305701	09/02/2009	4 39	1 01	9 5	0 51	11 6	3 29	0 19	3	0 04
34-0	373100109303201	09/02/2009	4 49	1 06	9 35	0 53	10 8	2 68	0 2	2 67	0 03
35-0	373101109300901	09/02/2009	3 97	0 96	8 1	0 47	13 3	2 82	0 15	2 94	0 03
36-0	373101109294601	09/02/2009	4 39	0 96	9 33	0 47	11 7	2 61	0 15	2 34	0 02
37-0	373101109292201	09/01/2009	4 34	0 72	8 37	0 37	11 6	2 32	0 13	2 68	0 01
38-0	373045109314301	09/02/2009	4 66	0 75	10	0 37	8 92	2 88	0 13	3 02	0 02
38-1a	373049109313701	09/02/2009	4 55	0 81	9 55	0 41	9 81	3 01	0 15	3 14	0 02
38-1b	373049109313701	09/02/2009	4 56	0 79	9 37	0 4	11 4	3	0 15	3 02	0 02
38-2	373041109314901	09/02/2009	4 52	0 78	9 84	0 39	10 8	3 26	0 14	3	0 03
39-0	373045109312001	09/02/2009	4 76	1 02	10 8	0 51	9 83	3 48	0 17	3 3	0 03
40-0	373045109305601	09/02/2009	4 57	0 56	11 2	0 29	11 9	3	0 09	3 12	0 02
40-1	373049109310201	09/02/2009	4 74	0 55	10	0 29	13 1	3 96	0 12	3 13	0 03
40-2a	373042109305001	09/02/2009	4 62	0 87	7 91	0 45	>15	2 38	0 13	2 3	0 03
40-2b	373042109305001	09/02/2009	4 66	0 82	7 84	0 43	15	2 31	0 12	2 23	0 02
41-0	373045109303201	09/02/2009	4 34	0 68	10 4	0 33	12 1	2 69	0 1	3 12	0 02
42-0	373045109300901	09/02/2009	4 44	1 1	7 84	0 54	>15	2 28	0 16	2 56	0 03
43-0	373045109294501	09/02/2009	4 04	1 24	8 68	0 58	13 7	2 74	0 18	2 78	0 04
44-0	373045109292201	09/01/2009	4 7	1 1	8 13	0 54	13 7	2 32	0 15	2 15	0 03

Appendix 4. Percent ash and chemical composition of new growth from sagebrush plants near the White Mesa uranium mill, San Juan County, Utah, September 2009.—Continued

[All analyses of biota tissue in dry weight. **Abbreviations**: ins, insufficient sample amount; mm/dd/yyyy, month/day/year; µg/g, micrograms per gram; >, greater than; <, less than]

Field ID	Station number	Sample date (mm/dd/yyyy)	Silver, total digestion, (µg/g)	Barium, total digestion, (µg/g)	Beryllium, total digestion, (µg/g)	Bismuth, total digestion, (µg/g)	Cadmium, total digestion, (µg/g)	Cerium, total digestion, (µg/g)	Cobalt, total digestion, (µg/g)	Chromium, total digestion, (µg/g)	Cesium, total digestion, (µg/g)
1-0	373233109314301	09/01/2009	<1	275	0 4	0 16	0 9	10	2 4	7	<5
2-0	373231109312101	09/01/2009	<1	617	0 4	0 14	1 7	8 51	2 8	8	<5
3-0	373233109304901	09/03/2009	<1	263	0 4	0 18	0 9	8 1	3 3	6	<5
4-0	373233109303101	09/03/2009	<1	243	0 4	0 38	1 2	10 9	4 7	9	<5
5-0	373233109301002	09/03/2009	<1	320	0 5	0 64	1 5	18 5	5 3	10	<5
6-0	373233109294701	09/03/2009	<1	287	0 5	0 93	1 7	14 1	4 3	8	<5
7-0	373233109292201	09/01/2009	<1	289	0 3	0 29	1 3	8 38	2 2	6	<5
8-0	373217109314401	09/01/2009	<1	290	0 4	0 1	0 9	7 23	1 8	7	<5
9-0	373217109311501	09/02/2009	<1	224	0 3	0 14	0 9	5 26	2 2	5	<5
10-0	373217109294501	09/03/2009	<1	370	0 6	3	1 9	25	6 7	11	<5
10-1a	373221109295201	09/03/2009	<1	262	0 4	0 66	1 2	11	3 7	7	<5
10-1b	373221109295201	09/03/2009	<1	258	0 3	0 63	1 1	10 5	3 6	8	<5
10-2	373214109295201	09/03/2009	<1	174	0 3	1 46	1	11 7	3 7	17	<5
11-0	373218109292101	09/01/2009	<1	227	0 4	0 3	1 7	10 4	2 4	8	<5
12-0	373202109314301	09/01/2009	<1	573	0 2	0 16	1 4	7 57	2 1	4	<5
12-1a	373203109313701	09/02/2009	<1	370	<0 1	0 12	1 2	5 06	1 5	5	<5
12-1b	373203109313701	09/02/2009	<1	369	<0 1	0 11	1 1	4 99	1 4	4	<5
12-2	373158109314801	09/02/2009	<1	185	<0 1	0 12	1 7	3 91	1 3	4	<5
13-0	373203109312001	09/01/2009	<1	303	0 3	0 19	1 6	14 2	3 3	8	<5
14-0	373202109294401	09/03/2009	<1	356	0 4	0 81	1	16 3	5 6	10	<5
14-1	373159109295001	09/03/2009	<1	329	0 4	1 25	1 7	22 5	6 9	10	<5
14-2a	373159109294001	09/03/2009	<1	388	0 3	0 45	1 6	13 7	4 4	15	<5
14-2b	373159109294001	09/03/2009	<1	377	0 3	0 43	1 6	12 8	4 2	8	<5
15-0	373202109292201	09/01/2009	<1	277	0 1	0 29	5	7 34	6 7	6	<5
15-1	373159109292801	09/01/2009	<1	294	0 2	0 36	1 5	9 23	2 7	7	<5
15-2	373159109291701	09/01/2009	<1	143	<0 1	0 16	1 5	3 23	1 4	4	<5
16-0	373147109314301	09/02/2009	<1	309	0 2	0 18	1 4	8 31	2 2	6	<5
17-0	373147109311901	09/02/2009	<1	270	0 2	0 16	1 2	7 98	3 1	5	<5
17-1	373151109312601	09/01/2009	<1	234	0 1	0 1	0 9	7 36	2	6	<5
17-2	373143109312601	09/01/2009	<1	330	0 1	0 08	2	6 04	2 3	6	<5
18-0	373146109294401	09/03/2009	<1	368	0 3	0 77	2 2	18	4 9	11	<5
19-0	373148109292201	09/01/2009	<1	192	0 1	0 12	2 6	6 71	2 3	6	<5
20-0	373132109314301	09/02/2009	<1	217	0 1	0 08	0 7	5 99	1 9	6	<5
21-0	373132109312001	09/02/2009	<1	286	0 2	0 12	1 4	8 71	4 1	10	<5
22-0	373131109294501	09/03/2009	<1	303	0 3	0 32	2 3	12 5	4	14	<5
22-1	373122109294801	09/03/2009	<1	310	0 2	0 12	1 6	12 7	3 2	10	<5
22-2	373129109294201	09/03/2009	<1	359	0 2	0 36	2 3	13 3	3 9	10	<5
23-0	373132109292201	09/01/2009	<1	287	0 2	0 17	1 1	10 5	2 5	10	<5
23-1	373135109291701	09/01/2009	<1	293	0 1	0 14	2 2	8 25	2 3	6	<5
23-2	373128109292801	09/01/2009	<1	376	<0 1	0 18	1 5	6 77	2	5	<5
24-0	373116109314201	09/02/2009	<1	248	0 1	0 08	1 2	7 06	2 6	6	<5
25-0	373116109311901	09/02/2009	<1	341	0 3	0 13	2 1	11 5	4 6	9	<5
26-0	373110109305501	09/02/2009	<1	319	0 2	0 14	2 3	10 1	4 2	7	<5
27-0	373114109303101	09/03/2009	<1	418	0 2	0 14	1 6	10	4 1	10	<5
28-0	373115109300901	09/03/2009	<1	372	0 5	0 13	2 5	13 7	4 7	11	<5
29-0	373116109294501	09/03/2009	4	310	0 2	0 11	2 4	8 34	2 7	6	<5
30-0	373116109292201	09/02/2009	<1	339	0 2	0 21	2	5 05	2 1	6	<5

Appendix 4.   Percent ash and chemical composition of new growth from sagebrush plants near the White Mesa uranium mill, San Juan County, Utah, September 2009.—Continued

[All analyses of biota tissue in dry weight. **Abbreviations**: ins, insufficient sample amount; mm/dd/yyyy, month/day/year; μg/g, micrograms per gram; >, greater than; <, less than]

Field ID	Station number	Sample date (mm/dd/yyyy)	Silver, total digestion, (μg/g)	Barium, total digestion, (μg/g)	Beryllium, total digestion, (μg/g)	Bismuth, total digestion, (μg/g)	Cadmium, total digestion, (μg/g)	Cerium, total digestion, (μg/g)	Cobalt, total digestion, (μg/g)	Chromium, total digestion, (μg/g)	Cesium, total digestion, (μg/g)
31-0	373106109314701	09/02/2009	<1	259	0 2	0 09	1 4	7 96	2	6	<5
31-1a	373101109314201	09/02/2009	<1	356	0 2	0 12	1 5	7 58	3 3	9	<5
31-1b	373101109314201	09/02/2009	<1	357	0 2	0 11	1 5	7 24	3 2	6	<5
31-2	373058109314901	09/02/2009	<1	266	0 2	0 12	1 3	9 91	2 7	8	<5
32-0	373100109311801	09/02/2009	<1	299	0 2	0 14	1 4	9 14	3 5	8	<5
33-0	373100109305701	09/02/2009	<1	394	0 3	0 14	1 7	11 3	3 7	9	<5
34-0	373100109303201	09/02/2009	<1	389	0 3	0 14	1 8	11 5	3 7	10	<5
35-0	373101109300901	09/02/2009	<1	370	0 2	0 12	1 3	11 1	3 1	9	<5
36-0	373101109294601	09/02/2009	<1	327	0 2	0 11	2	11 3	3 2	8	<5
37-0	373101109292201	09/01/2009	<1	293	0 2	0 22	1 7	8 17	2 6	9	<5
38-0	373045109314301	09/02/2009	<1	325	0 2	0 09	1	8 15	2 5	8	<5
38-1a	373049109313701	09/02/2009	<1	335	0 2	0 09	1 1	7 95	2 6	7	<5
38-1b	373049109313701	09/02/2009	<1	347	0 2	0 1	1 2	8 14	2 7	7	<5
38-2	373041109314901	09/02/2009	<1	377	0 2	0 1	1 3	8 29	2 6	7	<5
39-0	373045109312001	09/02/2009	<1	427	0 2	0 12	1 5	11 3	3 1	10	<5
40-0	373045109305601	09/02/2009	5	324	<0 1	0 07	1 6	6 15	2 2	6	<5
40-1	373049109310201	09/02/2009	<1	332	0 1	0 09	0 6	6 22	2 4	5	<5
40-2a	373042109305001	09/02/2009	<1	313	0 2	0 08	1 7	9 69	2 8	24	<5
40-2b	373042109305001	09/02/2009	<1	304	0 2	0 06	1 6	9 05	2 6	8	<5
41-0	373045109303201	09/02/2009	<1	418	0 2	0 05	2 2	7 67	2	6	<5
42-0	373045109300901	09/02/2009	<1	390	0 5	0 09	1 8	13 3	3 3	10	<5
43-0	373045109294501	09/02/2009	<1	413	0 4	0 11	1 4	14 2	3 2	10	<5
44-0	373045109292201	09/01/2009	<1	339	0 3	0 13	2	13 7	2 7	10	<5

Appendix 4. Percent ash and chemical composition of new growth from sagebrush plants near the White Mesa uranium mill, San Juan County, Utah, September 2009.—Continued

[All analyses of biota tissue in dry weight. **Abbreviations**: ins, insufficient sample amount; mm/dd/yyyy, month/day/year; µg/g, micrograms per gram; >, greater than; <, less than]

Field ID	Station number	Sample date (mm/dd/yyyy)	Copper, total digestion, (µg/g)	Gallium, total digestion, (µg/g)	Indium, total digestion, (µg/g)	Lanthanum, total digestion, (µg/g)	Lithium, total digestion, (µg/g)	Manganese, total digestion, (µg/g)	Molybdenum, total digestion, (µg/g)	Niobium, total digestion, (µg/g)	Nickel, total digestion, (µg/g)
1-0	373233109314301	09/01/2009	186	2 71	<0 02	4 9	15	1,020	10 7	1 8	15 8
2-0	373231109312101	09/01/2009	170	2 32	<0 02	4 5	10	928	10 4	1 7	20 6
3-0	373233109304901	09/03/2009	206	2 21	<0 02	4 8	8	558	36	1 7	22 6
4-0	373233109303101	09/03/2009	190	2 59	0 02	6 3	11	588	40 8	1 9	25 6
5-0	373233109301002	09/03/2009	175	3 63	0 02	11 4	13	825	23 7	2	23 7
6-0	373233109294701	09/03/2009	153	2 9	0 02	8 5	15	804	14 6	1 6	24 6
7-0	373233109292201	09/01/2009	221	2 06	<0 02	4 9	13	1,030	18 7	1 9	15 1
8-0	373217109314401	09/01/2009	204	1 91	<0 02	3 6	11	1,010	5 86	1	22 8
9-0	373217109311501	09/02/2009	207	1 59	<0 02	2 9	8	286	41 5	1	37 1
10-0	373217109294501	09/03/2009	166	3 64	0 83	16 1	25	700	35	19 2	28 9
10-1a	373221109295201	09/03/2009	207	2 05	<0 02	7 4	11	780	26 9	2 9	28 8
10-1b	373221109295201	09/03/2009	202	1 9	<0 02	7	12	761	27 1	2 1	28 6
10-2	373214109295201	09/03/2009	264	1 96	0 03	7 9	27	760	42	2 4	27 6
11-0	373218109292101	09/01/2009	129	2 5	0 03	6 5	19	641	23 3	1 9	13 9
12-0	373202109314301	09/01/2009	192	2 03	<0 02	4	16	941	21 7	1 3	11 9
12-1a	373203109313701	09/02/2009	203	1 42	<0 02	2 6	58	755	21 7	0 9	9 7
12-1b	373203109313701	09/02/2009	206	1 38	<0 02	2 6	60	764	20 9	0 9	9 6
12-2	373158109314801	09/02/2009	171	1 1	<0 02	2 1	12	851	12 7	0 7	17 7
13-0	373203109312001	09/01/2009	131	3 33	0 02	7 7	14	807	23 4	2 4	18 4
14-0	373202109294401	09/03/2009	250	3 49	0 03	12 6	20	798	45 3	1 9	23 3
14-1	373159109295001	09/03/2009	195	3 72	0 03	14 5	17	957	50 2	1 9	24 3
14-2a	373159109294001	09/03/2009	196	3 11	0 03	9 5	17	944	27 6	2 1	18 9
14-2b	373159109294001	09/03/2009	195	3 04	0 09	9 4	18	979	31	2 2	17 7
15-0	373202109292201	09/01/2009	235	1 63	<0 02	4 2	43	678	10 7	1 5	40
15-1	373159109292801	09/01/2009	220	2 19	<0 02	6 6	35	696	28 1	2	13 8
15-2	373159109291701	09/01/2009	246	1 12	<0 02	1 8	134	570	7 5	0 6	23 1
16-0	373147109314301	09/02/2009	170	2 17	<0 02	4 4	12	938	10 9	1 5	17 2
17-0	373147109311901	09/02/2009	185	2 07	<0 02	4 4	12	723	15	1 5	23 6
17-1	373151109312601	09/01/2009	145	1 7	<0 02	4 1	15	429	17	1 4	12 8
17-2	373143109312601	09/01/2009	175	1 56	<0 02	3 4	8	707	12 3	1 2	21 9
18-0	373146109294401	09/03/2009	152	3 47	0 02	11 9	15	982	17 1	1 4	26 5
19-0	373148109292201	09/01/2009	208	1 67	<0 02	3 8	8	629	13 8	1 4	16 7
20-0	373132109314301	09/02/2009	178	1 47	<0 02	3 5	20	524	27 4	1 1	11 5
21-0	373132109312001	09/02/2009	184	2 07	<0 02	5	12	532	11 6	2	27 3
22-0	373131109294501	09/03/2009	199	2 64	<0 02	7 5	15	869	18 7	1 3	49 8
22-1	373122109294801	09/03/2009	182	2 69	<0 02	7 1	12	700	11 8	1 6	25 2
22-2	373129109294201	09/03/2009	176	2 82	<0 02	7 6	14	665	43 6	1 5	44 7
23-0	373132109292201	09/01/2009	157	2 37	<0 02	6 3	19	587	15 8	1 6	20 7
23-1	373135109291701	09/01/2009	183	1 82	<0 02	4 8	9	1,010	11 9	1 4	16 4
23-2	373128109292801	09/01/2009	158	1 46	<0 02	4 2	9	741	10 2	1 4	15 4
24-0	373116109314201	09/02/2009	189	1 72	<0 02	3 9	12	675	9 68	1 1	18 6
25-0	373116109311901	09/02/2009	149	2 76	<0 02	7	11	683	13 7	1 2	22 5
26-0	373110109305501	09/02/2009	175	2 44	<0 02	6 1	9	901	9 62	2	23 3
27-0	373114109303101	09/03/2009	208	2 23	<0 02	6 3	29	771	18 5	1 3	22 3
28-0	373115109300901	09/03/2009	143	2 86	<0 02	8 1	14	881	12 7	1 1	26 9
29-0	373116109294501	09/03/2009	157	1 98	<0 02	5	11	1,170	9 54	1 1	18 6
30-0	373116109292201	09/02/2009	166	1 93	<0 02	5 1	10	720	12 4	1 4	14 2

Appendix 4. Percent ash and chemical composition of new growth from sagebrush plants near the White Mesa uranium mill, San Juan County, Utah, September 2009.—Continued

[All analyses of biota tissue in dry weight. **Abbreviations**: ins, insufficient sample amount; mm/dd/yyyy, month/day/year; µg/g, micrograms per gram; >, greater than; <, less than]

Field ID	Station number	Sample date (mm/dd/yyyy)	Copper, total digestion, (µg/g)	Gallium, total digestion, (µg/g)	Indium, total digestion, (µg/g)	Lanthanum, total digestion, (µg/g)	Lithium, total digestion, (µg/g)	Manganese, total digestion, (µg/g)	Molybdenum, total digestion, (µg/g)	Niobium, total digestion, (µg/g)	Nickel, total digestion, (µg/g)
31-0	373106109314701	09/02/2009	168	1 71	<0 02	4 7	12	599	12 1	1 3	14 3
31-1a	373101109314201	09/02/2009	151	1 83	<0 02	4 7	19	649	17 4	1 5	23 3
31-1b	373101109314201	09/02/2009	145	1 77	<0 02	4 5	16	647	17 1	1 5	23 2
31-2	373058109314901	09/02/2009	163	2 29	<0 02	5 9	13	920	13 1	1 7	13 7
32-0	373100109311801	09/02/2009	167	2 19	<0 02	5 4	11	699	10 8	1 4	22 5
33-0	373100109305701	09/02/2009	167	2 58	<0 02	6 9	20	783	17 1	1 8	17 2
34-0	373100109303201	09/02/2009	164	2 48	<0 02	7 3	12	816	9 36	1 3	23 4
35-0	373101109300901	09/02/2009	172	2 44	<0 02	6 9	21	780	32 3	1 7	14 6
36-0	373101109294601	09/02/2009	153	2 5	<0 02	6 9	17	1,010	12 1	1 3	15 2
37-0	373101109292201	09/01/2009	142	2 1	<0 02	5	12	978	14 1	1 3	17 8
38-0	373045109314301	09/02/2009	151	1 92	<0 02	5	14	679	13 6	1 5	16 6
38-1a	373049109313701	09/02/2009	191	1 96	<0 02	5 2	16	731	13 2	1 2	13 8
38-1b	373049109313701	09/02/2009	187	2 1	<0 02	5 3	17	724	14 8	1 5	13 7
38-2	373041109314901	09/02/2009	155	2	<0 02	5 1	14	837	15 3	1 5	14 8
39-0	373045109312001	09/02/2009	197	2 62	<0 02	6 9	14	847	12 5	1 8	16 7
40-0	373045109305601	09/02/2009	155	1 63	<0 02	3 8	9	834	20 3	1	22
40-1	373049109310201	09/02/2009	120	1 5	<0 02	3 9	13	563	14 3	1 1	33 6
40-2a	373042109305001	09/02/2009	141	2 27	<0 02	6	10	789	11 6	1 2	21 3
40-2b	373042109305001	09/02/2009	136	2 07	<0 02	5 6	9	773	11 2	1 1	20 6
41-0	373045109303201	09/02/2009	139	1 69	<0 02	4 7	10	620	18 9	1 3	18 3
42-0	373045109300901	09/02/2009	153	2 69	<0 02	7 9	16	789	11 7	1 6	20 1
43-0	373045109294501	09/02/2009	166	3 05	<0 02	9 4	31	738	21 9	1 7	18 7
44-0	373045109292201	09/01/2009	125	2 82	<0 02	8 4	11	898	9 76	1 4	17 7

Appendix 4. Percent ash and chemical composition of new growth from sagebrush plants near the White Mesa uranium mill, San Juan County, Utah, September 2009.—Continued

[All analyses of biota tissue in dry weight. **Abbreviations**: ins, insufficient sample amount; mm/dd/yyyy, month/day/year; µg/g, micrograms per gram; >, greater than; <, less than]

Field ID	Station number	Sample date (mm/dd/yyyy)	Phosphorus, total digestion, (µg/g)	Lead, total digestion, (µg/g)	Rubidium, total digestion, (µg/g)	Antimony, total digestion, (µg/g)	Scandium, total digestion, (µg/g)	Tin, total digestion, (µg/g)	Strontium, total digestion, (µg/g)	Tellurium, total digestion, (µg/g)	Thorium, recoverable, (µg/g)
1-0	373233109314301	09/01/2009	>10,000	5 3	48 8	0 31	1 8	0 6	1,880	<0 1	1 5
2-0	373231109312101	09/01/2009	>10,000	4 4	32	0 34	1 5	0 5	1,590	<0 1	1 2
3-0	373233109304901	09/03/2009	>10,000	5 7	26 5	0 35	1 5	0 9	1,160	<0 1	1 3
4-0	373233109303101	09/03/2009	>10,000	9	59 7	0 33	1 8	1 5	679	<0 1	1 8
5-0	373233109301002	09/03/2009	>10,000	16 3	47 7	0 29	2 7	2 2	826	<0 1	3 1
6-0	373233109294701	09/03/2009	>10,000	13 6	42 5	0 22	2 1	2	670	<0 1	2 3
7-0	373233109292201	09/01/2009	>10,000	6 6	22 3	0 52	1 4	4	824	<0 1	1 3
8-0	373217109314401	09/01/2009	>10,000	3 5	40 3	0 17	1 1	0 3	1,150	<0 1	1 1
9-0	373217109311501	09/02/2009	>10,000	3	14 8	0 17	1 1	0 5	735	<0 1	0 8
10-0	373217109294501	09/03/2009	>10,000	33 3	34	1 44	3 7	84	872	<0 1	5 1
10-1a	373221109295201	09/03/2009	>10,000	13 8	26 8	0 37	1 7	8 1	1,220	<0 1	1 8
10-1b	373221109295201	09/03/2009	>10,000	12 4	23 9	0 29	1 5	2 1	1,210	<0 1	1 7
10-2	373214109295201	09/03/2009	>10,000	15 4	41 4	0 31	1 9	4 2	1,790	<0 1	2 4
11-0	373218109292101	09/01/2009	>10,000	6 7	28	0 26	1 8	1 4	971	<0 1	1 5
12-0	373202109314301	09/01/2009	>10,000	3 7	30 6	0 29	1 3	0 4	1,560	<0 1	1 1
12-1a	373203109313701	09/02/2009	>10,000	3 7	41 2	0 25	0 9	0 3	1,360	<0 1	0 8
12-1b	373203109313701	09/02/2009	>10,000	4 4	38	0 28	0 8	0 4	1,380	<0 1	0 7
12-2	373158109314801	09/02/2009	>10,000	2 1	37 7	0 3	0 7	0 2	734	<0 1	0 6
13-0	373203109312001	09/01/2009	>10,000	6 6	20 6	0 43	2 3	0 8	758	<0 1	2 1
14-0	373202109294401	09/03/2009	>10,000	17 7	26 7	0 35	2 7	2 7	1,080	<0 1	2 8
14-1	373159109295001	09/03/2009	>10,000	25	42 4	0 46	3 1	2 8	713	<0 1	3 7
14-2a	373159109294001	09/03/2009	>10,000	12 3	34 8	0 33	2 2	1 7	1,100	<0 1	2 2
14-2b	373159109294001	09/03/2009	>10,000	11 4	34 9	0 31	2 2	1 7	1,110	<0 1	2
15-0	373202109292201	09/01/2009	>10,000	5 9	36 7	0 28	1 2	3	1,290	<0 1	1 2
15-1	373159109292801	09/01/2009	>10,000	8 7	28 3	0 3	1 7	1 6	1,510	<0 1	1 5
15-2	373159109291701	09/01/2009	>10,000	2 7	51 1	0 2	0 7	0 4	1,010	<0 1	0 5
16-0	373147109314301	09/02/2009	>10,000	4 3	21 7	0 44	1 4	0 4	854	<0 1	1 3
17-0	373147109311901	09/02/2009	>10,000	4 5	31 7	0 37	1 4	0 4	1,080	<0 1	1 3
17-1	373151109312601	09/01/2009	>10,000	3 8	18 2	0 22	1 1	0 3	785	<0 1	1 1
17-2	373143109312601	09/01/2009	>10,000	3 2	28 8	0 3	1	0 3	1,280	<0 1	1
18-0	373146109294401	09/03/2009	>10,000	17 4	33 9	0 47	2 5	1 7	726	<0 1	3
19-0	373148109292201	09/01/2009	>10,000	4 4	52	0 37	1 2	0 5	584	<0 1	1 2
20-0	373132109314301	09/02/2009	>10,000	3	34 5	0 33	1	0 9	1,460	<0 1	0 9
21-0	373132109312001	09/02/2009	>10,000	4 4	56 4	0 27	1 3	0 4	1,040	<0 1	1 4
22-0	373131109294501	09/03/2009	>10,000	8 7	43 8	0 29	1 7	2 9	800	<0 1	2
22-1	373122109294801	09/03/2009	>10,000	6 6	24 2	0 31	1 7	0 6	1,070	<0 1	2
22-2	373129109294201	09/03/2009	>10,000	9 1	33 7	0 43	1 9	0 9	983	<0 1	2 1
23-0	373132109292201	09/01/2009	>10,000	6 6	26 2	0 51	1 5	0 7	901	<0 1	1 7
23-1	373135109291701	09/01/2009	>10,000	4 9	37 3	0 31	1 2	1 1	840	<0 1	1 3
23-2	373128109292801	09/01/2009	>10,000	5	36 7	0 53	1	0 7	1,050	<0 1	1
24-0	373116109314201	09/02/2009	>10,000	3 8	46 5	0 25	1 1	0 5	793	<0 1	1 1
25-0	373116109311901	09/02/2009	>10,000	5 7	40 6	0 33	1 7	0 7	1,050	<0 1	1 8
26-0	373110109305501	09/02/2009	>10,000	5 2	35	0 36	1 6	0 6	911	<0 1	1 6
27-0	373114109303101	09/03/2009	>10,000	6 3	30 8	0 26	1 5	0 5	1,840	<0 1	1 7
28-0	373115109300901	09/03/2009	>10,000	7 4	34 1	0 2	1 9	0 6	759	<0 1	2 1
29-0	373116109294501	09/03/2009	>10,000	8 6	30 4	0 28	1 2	0 7	720	<0 1	1 3
30-0	373116109292201	09/02/2009	>10,000	3 2	38	0 29	1	0 8	988	<0 1	0 8

Appendix 4.   Percent ash and chemical composition of new growth from sagebrush plants near the White Mesa uranium mill, San Juan County, Utah, September 2009.—Continued

[All analyses of biota tissue in dry weight. **Abbreviations**: ins, insufficient sample amount; mm/dd/yyyy, month/day/year; μg/g, micrograms per gram; >, greater than; <, less than]

Field ID	Station number	Sample date (mm/dd/yyyy)	Phosphorus, total digestion, (μg/g)	Lead, total digestion, (μg/g)	Rubidium, total digestion, (μg/g)	Antimony, total digestion, (μg/g)	Scandium, total digestion, (μg/g)	Tin, total digestion, (μg/g)	Strontium, total digestion, (μg/g)	Tellurium, total digestion, (μg/g)	Thorium, recoverable, (μg/g)
31-0	373106109314701	09/02/2009	>10,000	3 9	25 3	0 2	1	0 3	865	<0 1	1 3
31-1a	373101109314201	09/02/2009	>10,000	4 6	23	0 36	1 2	0 3	1,030	<0 1	1 3
31-1b	373101109314201	09/02/2009	>10,000	4 2	22 5	0 43	1 2	0 4	1,020	<0 1	1 2
31-2	373058109314901	09/02/2009	>10,000	5 3	22 4	0 33	1 5	0 4	646	<0 1	1 7
32-0	373100109311801	09/02/2009	>10,000	5 3	26 8	0 33	1 5	0 6	915	<0 1	1 6
33-0	373100109305701	09/02/2009	>10,000	6 3	37	0 35	1 8	0 5	1,160	<0 1	1 8
34-0	373100109303201	09/02/2009	>10,000	6 7	24 9	0 36	1 6	0 6	857	<0 1	2
35-0	373101109300901	09/02/2009	>10,000	6 1	41 4	0 46	1 6	0 5	1,050	<0 1	1 9
36-0	373101109294601	09/02/2009	>10,000	7	31 9	0 36	1 6	0 5	837	<0 1	1 8
37-0	373101109292201	09/01/2009	>10,000	5 6	26 2	0 29	1 3	0 5	828	<0 1	1 4
38-0	373045109314301	09/02/2009	>10,000	4 1	28	0 29	1 3	0 5	1,050	<0 1	1 4
38-1a	373049109313701	09/02/2009	>10,000	4 4	25 1	0 29	1 3	0 3	1,090	<0 1	1 3
38-1b	373049109313701	09/02/2009	>10,000	4 7	30 4	0 39	1 5	0 3	1,110	<0 1	1 3
38-2	373041109314901	09/02/2009	>10,000	4 8	26 8	0 26	1 4	0 3	1,190	<0 1	1 4
39-0	373045109312001	09/02/2009	>10,000	5 8	29 5	0 46	1 8	0 5	1,270	<0 1	1 9
40-0	373045109305601	09/02/2009	>10,000	5 2	44 5	0 22	1	0 2	1,040	<0 1	1
40-1	373049109310201	09/02/2009	>10,000	3 9	33 3	0 28	1	0 4	1,870	<0 1	1 1
40-2a	373042109305001	09/02/2009	>10,000	5 3	34 2	0 32	1 4	0 4	604	<0 1	1 6
40-2b	373042109305001	09/02/2009	>10,000	4 8	31 4	0 33	1 3	0 4	603	<0 1	1 5
41-0	373045109303201	09/02/2009	>10,000	4	36 6	0 24	1	0 4	1,400	<0 1	1 3
42-0	373045109300901	09/02/2009	>10,000	6 4	42 6	0 34	1 7	0 4	937	<0 1	2 1
43-0	373045109294501	09/02/2009	>10,000	7 2	33 8	0 48	1 9	0 6	1,440	<0 1	2 3
44-0	373045109292201	09/01/2009	>10,000	7 1	37 4	0 25	1 7	0 6	680	<0 1	2 2

Appendix 4. Percent ash and chemical composition of new growth from sagebrush plants near the White Mesa uranium mill, San Juan County, Utah, September 2009.—Continued

[All analyses of biota tissue in dry weight. **Abbreviations**: ins, insufficient sample amount; mm/dd/yyyy, month/day/year; µg/g, micrograms per gram; >, greater than; <, less than]

Field ID	Station number	Sample date (mm/dd/yyyy)	Thallium, total digestion, (µg/g)	Uranium, total digestion, (µg/g)	Vanadium, total digestion, (µg/g)	Tungsten, total digestion, (µg/g)	Yttrium, total digestion, (µg/g)	Zinc, total digestion, (µg/g)	Arsenic, total digestion, (µg/g)	Selenium, total digestion, (µg/g)
1-0	373233109314301	09/01/2009	<0 1	3 2	25	0 4	3 4	459	ins	ins
2-0	373231109312101	09/01/2009	<0 1	5 3	30	0 4	3	519	1	0 8
3-0	373233109304901	09/03/2009	<0 1	19	78	0 4	2 9	410	0 9	<0 2
4-0	373233109303101	09/03/2009	<0 1	36 3	131	1 3	3 9	365	1 6	0 3
5-0	373233109301002	09/03/2009	0 1	56 8	297	2 9	5 8	422	1 6	0 3
6-0	373233109294701	09/03/2009	<0 1	52 9	259	2 4	4 3	377	0 8	0 6
7-0	373233109292201	09/01/2009	<0 1	18 4	70	1 3	2 8	598	0 8	<0 2
8-0	373217109314401	09/01/2009	<0 1	2 1	17	0 3	2 5	411	2 1	0 4
9-0	373217109311501	09/02/2009	<0 1	7 1	34	0 3	1 9	236	1 2	0 2
10-0	373217109294501	09/03/2009	0 1	171	582	11 5	7	515	1 2	0 6
10-1a	373221109295201	09/03/2009	<0 1	56 8	250	2 7	3 2	447	1 1	0 4
10-1b	373221109295201	09/03/2009	<0 1	49 5	229	2 4	2 9	443	0 9	0 3
10-2	373214109295201	09/03/2009	<0 1	74	220	3 1	3 5	474	0 8	0 5
11-0	373218109292101	09/01/2009	<0 1	16 4	69	0 9	3 6	517	<0 6	0 3
12-0	373202109314301	09/01/2009	<0 1	3	19	0 3	2 7	421	<0 6	3 3
12-1a	373203109313701	09/02/2009	<0 1	2 3	14	0 2	1 7	556	2	0 5
12-1b	373203109313701	09/02/2009	<0 1	2 2	14	0 2	1 7	563	<0 6	0 5
12-2	373158109314801	09/02/2009	<0 1	1 3	9	0 2	1 3	712	<0 6	<0 2
13-0	373203109312001	09/01/2009	<0 1	7	44	0 5	4 8	502	<0 6	<0 2
14-0	373202109294401	09/03/2009	0 1	72 8	278	3 9	5 9	352	1 5	1
14-1	373159109295001	09/03/2009	0 2	100	319	4 9	6 6	392	0 8	0 6
14-2a	373159109294001	09/03/2009	<0 1	44 9	165	2 1	4 8	340	0 8	0 7
14-2b	373159109294001	09/03/2009	<0 1	40 6	150	2	4 6	329	0 9	0 7
15-0	373202109292201	09/01/2009	<0 1	15 7	55	1	2 4	615	1 7	0 2
15-1	373159109292801	09/01/2009	0 1	25	110	1 6	3 4	646	<0 6	0 7
15-2	373159109291701	09/01/2009	<0 1	5	15	0 3	1 1	679	0 7	<0 2
16-0	373147109314301	09/02/2009	<0 1	4 3	19	0 4	2 9	372	0 6	<0 2
17-0	373147109311901	09/02/2009	<0 1	17 8	54	0 4	2 8	317	<0 6	0 2
17-1	373151109312601	09/01/2009	<0 1	6 3	27	0 3	2 5	459	<0 6	<0 2
17-2	373143109312601	09/01/2009	<0 1	9 4	31	0 3	2 2	285	1	0 4
18-0	373146109294401	09/03/2009	0 1	72 5	201	3 3	5 7	354	0 8	0 4
19-0	373148109292201	09/01/2009	<0 1	8 6	31	0 5	2 3	606	<0 6	<0 2
20-0	373132109314301	09/02/2009	0 1	6 1	28	0 3	2 1	472	<0 6	0 3
21-0	373132109312001	09/02/2009	<0 1	19 8	76	0 3	3 4	360	1 1	0 4
22-0	373131109294501	09/03/2009	<0 1	41 9	91	1 4	3 9	286	0 9	0 3
22-1	373122109294801	09/03/2009	<0 1	32 7	57	0 5	4 2	495	0 7	0 4
22-2	373129109294201	09/03/2009	<0 1	40 5	80	1 5	4	237	1 6	0 4
23-0	373132109292201	09/01/2009	<0 1	15 3	45	0 7	3 5	294	0 7	<0 2
23-1	373135109291701	09/01/2009	<0 1	10 8	31	0 6	2 7	298	<0 6	<0 2
23-2	373128109292801	09/01/2009	<0 1	13 4	41	0 8	2 1	240	<0 6	<0 2
24-0	373116109314201	09/02/2009	<0 1	11 1	39	0 3	2 5	306	2 8	0 3
25-0	373116109311901	09/02/2009	<0 1	21 9	92	0 4	4 4	290	1 6	0 3
26-0	373110109305501	09/02/2009	<0 1	18 3	74	0 4	3 8	306	0 7	0 2
27-0	373114109303101	09/03/2009	<0 1	24 5	90	0 4	3 7	395	0 8	0 3
28-0	373115109300901	09/03/2009	0 3	40	84	0 5	4 7	240	0 9	0 3
29-0	373116109294501	09/03/2009	<0 1	19	33	0 5	2 9	283	0 9	<0 2
30-0	373116109292201	09/02/2009	<0 1	7 5	35	0 5	2 7	223	0 7	0 6

Appendix 4.   Percent ash and chemical composition of new growth from sagebrush plants near the White Mesa uranium mill, San Juan County, Utah, September 2009.—Continued

[All analyses of biota tissue in dry weight. **Abbreviations**: ins, insufficient sample amount; mm/dd/yyyy, month/day/year; µg/g, micrograms per gram; >, greater than; <, less than]

Field ID	Station number	Sample date (mm/dd/yyyy)	Thallium, total digestion, (µg/g)	Uranium, total digestion, (µg/g)	Vanadium, total digestion, (µg/g)	Tungsten, total digestion, (µg/g)	Yttrium, total digestion, (µg/g)	Zinc, total digestion, (µg/g)	Arsenic, total digestion, (µg/g)	Selenium, total digestion, (µg/g)
31-0	373106109314701	09/02/2009	<0 1	6 6	31	0 3	2 7	390	<0 6	<0 2
31-1a	373101109314201	09/02/2009	<0 1	15 3	61	0 3	2 9	271	<0 6	0 2
31-1b	373101109314201	09/02/2009	<0 1	14 9	59	0 3	2 7	268	1 5	0 2
31-2	373058109314901	09/02/2009	<0 1	9 9	44	0 4	3 5	329	<0 6	<0 2
32-0	373100109311801	09/02/2009	<0 1	18 7	69	0 4	3 2	285	0 7	0 3
33-0	373100109305701	09/02/2009	<0 1	17 7	73	0 5	4 1	306	<0 6	<0 2
34-0	373100109303201	09/02/2009	<0 1	28	82	0 4	4	325	0 6	0 2
35-0	373101109300901	09/02/2009	<0 1	23 4	49	0 4	3 9	355	<0 6	<0 2
36-0	373101109294601	09/02/2009	<0 1	21 1	39	0 6	3 8	311	0 7	<0 2
37-0	373101109292201	09/01/2009	<0 1	14 5	36	0 7	2 8	268	<0 6	1 1
38-0	373045109314301	09/02/2009	<0 1	7 3	31	0 3	2 8	262	<0 6	0 4
38-1a	373049109313701	09/02/2009	<0 1	8 1	40	0 4	2 9	329	0 7	0 2
38-1b	373049109313701	09/02/2009	<0 1	8 4	39	0 4	3 2	329	0 7	<0 2
38-2	373041109314901	09/02/2009	0 3	7 1	32	0 3	2 9	281	<0 6	0 2
39-0	373045109312001	09/02/2009	<0 1	10 5	47	0 5	3 8	296	<0 6	0 3
40-0	373045109305601	09/02/2009	<0 1	7	22	0 3	2 2	229	<0 6	0 4
40-1	373049109310201	09/02/2009	<0 1	10 9	36	0 3	2 2	173	<0 6	0 7
40-2a	373042109305001	09/02/2009	<0 1	7 6	31	0 4	3 4	261	<0 6	0 4
40-2b	373042109305001	09/02/2009	<0 1	6 7	29	0 4	3 1	253	<0 6	0 4
41-0	373045109303201	09/02/2009	<0 1	5	20	0 3	2 6	237	1 5	0 3
42-0	373045109300901	09/02/2009	<0 1	23 1	46	0 4	4 5	247	<0 6	0 4
43-0	373045109294501	09/02/2009	0 2	27	42	0 5	5 3	256	0 6	<0 2
44-0	373045109292201	09/01/2009	<0 1	10 8	32	0 6	4 5	297	1 6	<0 2

www.ingramcontent.com/pod-product-compliance
Lightning Source LLC
Chambersburg PA
CBHW081452170526
45166CB00008B/2400